ECOLOGY OF
FRESH WATERS

ECOLOGY OF
FRESH WATERS

BRIAN MOSS

Reader in Environmental Sciences
University of East Anglia
Norwich

A HALSTED PRESS BOOK

JOHN WILEY & SONS

NEW YORK – TORONTO

First published 1980

Published in the U.S.A. and Canada
by Halsted Press,
a Division of John Wiley & Sons, Inc.,
New York

Library of Congress
Cataloging in Publication Data

Moss, Brian.
  Ecology of fresh waters.
  "A Halsted Press book."
  Bibliography: p.
  Includes index.
  1. Freshwater ecology.   I. Title.
QH541.5.F7M67   1980   574.5′2632   80–11259

ISBN 0–470–26942–1

Printed in Great Britain

TO MY WIFE, JOYCE,

for her love,

TO GABRIEL FAURÉ

for the gift of his Requiem

AND TO H. E. BATES

who knew what it was all about

without the help of this

'Water attracts me as women attract men, as cherries attract blackbirds. I fall for it every time.

And, as I hope these pages will show, I fall for it still. Water has some kind of powerful mystery about it. Still waters, moving waters, dark waters: the words themselves have a mysterious, almost dying fall. Roads, meadows, towns, gardens, woods are man-made; a river is a primeval piece of work. It is ageless but, at the same time, perpetually young. It travels, but remains. It is a paradox of eternal age and eternal youth, of change and changelessness, of permanence and transience. And if there is, perhaps, a dogmatic flavour about these remarks it is comforting to reflect that they will be true, roughly, barring astronomical accidents, in ten thousand years.'

H. E. BATES: *Down the River*, Gollancz, 1937

CONTENTS

vii

PREFACE

I have written this book mainly for undergraduates and also for post-graduate students who are perhaps just starting a research project or career in freshwater ecology. Its contents have been subjected to a sort of educational natural selection for it represents the distillation of what several groups of students and I have found mutually interesting. My own predilections naturally are the more prominent—'Thou canst not speak of that thou dost not feel'.

My interest in freshwater ecology, built on what might be a genetic predisposition to the wet and muddy, was fired by two people in particular and this seems the right opportunity to thank them. The earlier was Charles Sinker of the Field Studies Council, whose course on 'Meres and Mosses' at Preston Montford Field Centre in 1960 was my first real contact with aquatic ecology. The other was Dr Frank Round of the University of Bristol who introduced to me in my second undergraduate year, what I recall as the unbelievably interesting, almost romantic notion of lake stratification and the great changes in algal populations throughout the year, based largely on his own work and that of Dr J. W. G. Lund.

Since these first stimuli a very large number of people have helped me and interested me and thus each has made his or her contribution to this book. I hope, at least, that some student reading it might, as a result, start to feel for the subject the same sort of passion that I and many of my fellow freshwater ecologists do.

Norwich, February 1979 BRIAN MOSS

xiii

ACKNOWLEDGMENTS

Many people have helped with the preparation of this book and I should like to thank them all very much. In particular Mrs Sue Winston and Mrs Pauline Blanch have, between them, undertaken all of the typing of the manuscript with cheerful acceptance of alteration upon alteration. Dr A. J. McLachlan, Dr R. T. Leah and Diana Forrest (Mrs Thomas) have made valuable suggestions and improvements, and Mr R. Campbell and Mrs Anne Brown have given much encouragement and help. All of my graduate students and research associates have helped in widening my interests and my wife, Joyce has helped with the tedious tasks of proofing and indexing. Errors and sins of omission nonetheless remain mine.

Dr W. Pennington Tutin very kindly supplied a draft of Fig. 8.2 and a large number of authors, acknowledged in the appropriate places, more than willingly allowed me to use figures from their original work. I am also grateful to the following for granting me permission to use copyrighted material: American Association for the Advancement of Science, Figs. 1.4, 3.15; Akademie-Verlag, Fig. 2.4; Blackwell Scientific Publications, Figs. 2.5, 2.9, 2.6 (part), 3.13, 5.1 (part), 6.3, 6.9, 7.8; Fisheries Research Board of Canada, Figs. 2.7, 2.11, 5.7, 12.2; Elsevier Ltd, Figs. 2.9 (part), 8.6; J. Wiley & Sons, 2.9 (part), 7.1; Duke University Press, Fig. 3.6; The Royal Society, Figs. 3.12 (part), 4.4 (d); Springer-Verlag, Fig. 3.12 (part); American Society for Limnology & Oceanography, Figs. 3.14, 8.1, 12.1; University of Notre Dame, Indiana, Fig. 3.17; The New York Botanical Garden, Fig. 4.2; Liverpool University Press, Fig. 4.4 (b, c); Freshwater Biological Association, Figs. 4.4 (d), 7.4; Academic Press, Figs. 4.6, 5.2; Field Studies Council, Figs. 8.7, 8.8; Sierra Club, Figs. 12.4, 12.5; W. Collins and Dr E. A. Ellis, Fig. 12.6; Dr P. L. Osborne, Fig. 5.1 (part); Oikos, Fig. 5.4; E. Scheweitzerbart'sche Verlagsbuchhandlung, Fig. 5.5; Methuen & Co, Fig. 6.10; Macmillan Ltd, Fig. 6.11; George Allen and Unwin, Fig. 6.12; Oliver and Boyd, Figs. 7.3, 7.5, 9.2, 9.5; Zoological Society of London, Fig. 7.6; *Nature* (Macmillan Journals), Figs. 8.2, 8.3; *Canadian Journal of Botany*, Fig. 8.4; Institute of Biology, Fig. 11.4; Dr W. Junk b.v., Figs. 11.2, 11.3; Cambridge University Press, Fig. 11.5.

CHAPTER 1
LAKES, RIVERS AND
CATCHMENT AREAS

1.1. Introduction

In a classic paper of 1887[140], Stephen Forbes described lakes as microcosms. He laid down some of the ideas of interaction between species and the integration of the ecosystem which are now very familiar to ecologists. He also treated the lake as an ecosystem with a boundary, the water's edge. This was very natural—on one side you got your feet wet, on the other you did not! It is not an idea we hold any longer, however, because quite apart from any pedantic considerations about the transition between open water, marshy fringes and, eventually, dry soil, a lake cannot be understood in isolation, nor even the lake and its inflowing streams. The real unit of study is the catchment area, or drainage basin from which, via its feeder streams, the lake takes its water—water which owes much of its chemical composition to the geology, geography and cultural development of the catchment. Even then the catchment area unit, for which the lake acts as a sink, a rubbish bin for what is washed out of the catchment, does not have firm boundaries. The atmosphere has a role to play and the composition of waters entering lakes may be changed by industrial gases drifting in from kilometres away and dissolving in rainwater falling on the catchment. In lakes which lie in the paths of bird migrations, roosting waterfowl may bring salts in their excreta from feeding areas outside the drainage basin.

1.2 Some basic terms and ideas (Fig. 1.1)

Rain, and any other form of precipitation, is not pure water. It contains significant quantities of dissolved gases (O_2, N_2, CO_2), cations (e.g. H^+, Na^+, K^+, Ca^+, Mg^{2+}), anions (e.g. SO_4^{2-}, Cl^-, NO_3^-, PO_4^{3-}), cation trace elements (e.g. Cu^{2+}, Mn^{3+}, Fe^{3+}, Fe^{3+}), organic compounds, and both organic and inorganic particles. Some of these the rain picks up from tiny droplets of sea spray carried upwards into the atmosphere by wind, some from dust blown from the land, others from the atmosphere itself and products formed from the atmospheric gases in it by lightning. Water dissolves almost everything and limnologists—those who study fresh waters—despite sophisticated techniques, commonly analyse for only a few dozen substances. They could analyse for a few hundred more, and only suspect the existence of many more, particularly organic substances, in rain and natural waters.

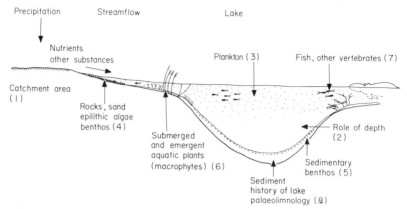

Fig. 1.1. Some limnological terms referred to in Chapter I. Numbers refer to chapters in which these subjects are discussed in detail.

The concentrations of commonly analysed substances are generally low in rainwater, but as the water percolates through or runs off a catchment it changes chemically through leaching of substances from soil and rocks. The harder the rock, the less the chemical modification, but on softer rocks with deep soils and consequent cultivation there are greater changes (Table 1.1).

Particles are added to the water as it drains through the catchment. These may be inorganic—clay, silt or even sand particles may be carried down, dependent on the violence of the flow—or they may be organic. These will include fine, even colloidal, remains of the decomposition of organic matter in soil which the soil microorganisms are unable further to degrade, and also large and small pieces of barely degraded plant litter washed into the streams. Abscissed leaves are common in forested catchments in autumn, but there is a continual loss of branches, leaves, senescent flowers and pollen to some streams.

In the stream itself there will be changes in the water composition and in the particles carried. Streams high in the catchment will flow turbulently and prevent

Table 1.1. The effect of catchment area on some aspects of drainage water chemistry. All values in mg l^{-1} except pH.

	Undisturbed forest on igneous rock, New Hampshire, U.S.A. (Likens et al.[295])		Lowland chalk and glacial drift, agricultural Norfolk, U.K. (Edwards[122])	
	Rainfall	Stream	Rainfall	Stream
Na$^+$	0.12	0.87	1.2	32.5
K$^+$	0.07	0.23	0.74	3.1
Mg^{2+}	0.04	0.38	0.21	6.9
Ca^{2+}	0.16	1.65	3.7	100.0
Cl$^-$	0.47	0.55	< 1.0	47.0
HCO$_3^-$	0.006	0.92	0	288.0
pH	4.14	4.92	3.5	7.7

much silt deposition, but benthic (i.e. bottom-living) invertebrates including some caddis fly (Trichoptera) and blackfly (*Simulium* spp., Diptera) larvae filter fine organic particles from the water and some of the dissolved substances are used for growth by aquatic mosses and liverworts (Bryophyta), or by layers of photosynthetic algae attached to the rocks. These are termed epilithic algae and they are grazed by other benthic invertebrates, e.g. mayfly (Ephemeroptera) and stonefly (Plecoptera) nymphs, freshwater limpets and snails (Mollusca). Other invertebrates, like the freshwater shrimps (*Gammarus* spp., Crustacea) may feed on the microorganisms, particularly fungi, which colonize pieces of litter lodging in crevices between the rocks on the stream-bed. Carnivorous invertebrates (leeches (Annelida), flatworms (Platyhelminthes)) and some insect larvae prey on the herbivores and detritivores, and fish on any invertebrates or plant material they can use.

As well as taking materials from the water (inorganic nutrients by algae, dissolved inorganic and organic substances by fungi and associated bacteria), this community also adds to it dissolved excretory products, particulate faeces, and the dislodged waste from the break-up of large organic particles during feeding. It also adds members of its own species. Thousands of invertebrates, accidentally dislodged by the current, despite their many adaptations to avoid this, form the 'drift', a downstream flood of floundering benthic animals which form easy prey for the fish.

Organisms adapted to live suspended in water (plankton) do not have time to develop significant populations in fast currents but as the moving water acquires greater volume or discharge and areas of reduced current, there is time for build-up of distinctive plankton populations. Phytoplankton, the microscopic, or sometimes just visible to the unaided eye, photosynthetic plankton comprises an array of algae and photosynthetic bacteria, which is physiologically extremely diverse and encompasses more than a dozen phyla. Phytoplankters are generally adapted to prolonged suspension, not for floating in the precise sense of the word. They may be much denser than water and kept in suspension by convection or wind-induced currents. The zooplankton includes mostly small Crustacea and Rotifera in fresh waters. Traditionally these have been thought to feed on phytoplankton, either by filtration of the smaller ones or by grasping and chewing the larger ones, or, if carnivorous, on each other. The suspended organic detritus brought in continually from upstream is also a component of the plankton community available to the zooplankton, as are the bacteria and fungi which colonize the detritus or live free on dissolved organic substances in the water. Fish may feed on phytoplankton, particularly in the Tropics, and on zooplankton, and in turn fish-eating fish, reptiles, birds and mammals may join the ecosystem.

As lower water flows develop at the edges of a growing river and plankton develops, so also may sediment be deposited on the bottom, providing a rooting medium for larger aquatic plants (macrophytes) and a habitat for mud-living benthic invertebrates. These burrow through the rich deposit or feed on the new supply at the surface of it and include oligochaete worms (Annelida), chironomid larvae (Diptera) and bivalve molluscs. The emergent (from the water surface) reed beds and submerged macrophyte beds also provide a physical habitat for a great diversity of invertebrates and fish and a food supply in the form of epiphytic (attached to plants) algae and bacteria.

A larger and larger river begins to bear fewer and fewer differences from a shallow lake, with its plankton, submerged weedbeds and sediment deposits. It is only when a relatively deep lake basin has been created by natural or human action that a new stage in the river-lake continuum can be distinguished. Such lakes are really rivers in which the flow is so reduced (sometimes becoming zero for lengthy periods) that the water body can acquire a physical structure, either horizontally through the entry of water supplies of different origins, or more usually vertically with a complex layering or stratification. This vertical stratification is created since water rapidly absorbs light and heat radiation, which leads to a layering of illuminated, warmer, less dense water on deeper, darker, colder water. In turn this creates a range of conditions for different chemical processes and living organisms. Because of this, and because of the necessarily longer time that a parcel of water is retained in the water course, the chemical changes imposed on the water in a lake are much greater than in the riverine stretches.

This generalized sequence will not, of course, be found in every river and lake system. In a rocky, mountain catchment, fast flowing streams may discharge directly into a rocky lake basin, carved out by previous glacial action. Such a lake may have few rooted aquatic macrophytes because little silt may have collected from the small upland drainage area, but there may nevertheless be a distinct plankton community. At another extreme, certain lakes in arid regions lose water not by an overflow, but only by evaporation; they may lack vertical stratification because they are shallow and wind can easily disturb any layering temporarily set up; macrophytes may be absent too because such endorheic (internal flow) lakes are usually very saline.

Very deep lakes such as Tanganyika (max. 1470 m) and Malawi (max. 706 m) in East Africa were formed by earth plate separation and are so deep that vertical stratification, established many thousands of years ago, is now disturbed only at the surface and is permanent below 100 m or so. The sea itself can be regarded as a very large endorheic lake, fed by the world's rivers. It is so large relative to the latter that its chemical composition is very constant over much of its volume, thus belying the generalization that chemical changes are greater in lakes than in rivers.

All of these examples emphasize that an understanding of freshwater ecosystems is best obtained in a framework of physico-chemical and biological processes going on in a continuum with as many dimensions as there are components. A traditional classification into streams, rivers and lakes, and into different sorts of lakes defined by certain combinations of features is no longer useful. I am not sure what the 'aim' of limnological research should be; I have always thought that it should be defined by that which interests the individual limnologist, but many of us are interested in what controls organic production in fresh waters and this is a convenient starting point.

1.3 Production in fresh waters

In the 1960's an international effort was made to collect data on productivity from a wide variety of lakes under the auspices of the International Biological Programme. The methods used varied but a synthesis of results made by Brylinsky & Mann[52] seems to include generalizations which most limnologists would accept. Most data were on phytoplankton production from natural lakes, reservoirs and slow-flowing rivers, and Brylinsky & Mann calculated the extent to which productivity, measured as the rate of photosynthesis in the surface waters was correlated with features of the catchment area, water chemistry and geographical location. The sites studied were in North America, Europe, Africa and Asia, with relatively fewer from the tropics than from the temperate zone. This largely reflects the distribution of limnological laboratories.

In the analysis carried out, results were expressed as the degree to which gross (i.e. no correction for respiration has been applied) phytoplankton photosynthesis in a given lake can be predicted from a knowledge of certain factors

Table 1.2. Results of multiple regression analyses comparing the importance of variables for lakes on a global basis compared with those in temperate (39°N–55°N) latitudes (adapted from Brylinsky & Mann[52]).

Variable	Variance explained by each variable (%)	
	All lakes	Temperate lakes
1 Latitude	56	2
Altitude	1	10
Conductivity	8	25
2 Latitude	56	2
Altitude	2	13
Total phosphorus	7	17
3 Latitude	32	6
Altitude	1	4
Chlorophyll *a*	47	27

such as latitude or the total amount of dissolved substances in the water. This degree is expressed as the percentage of the variance in the data explained by each of these environmental factors alone or in combination. The greater the percentage the more closely is phytoplankton photosynthesis correlated with that factor and the more likely is there to be a close causative link.

Table 1.2 gives these data for all of the water bodies (up to 54 of them) studied and for lakes in northern temperate latitudes (39°N–55°N). Conductivity is a measure of the total amount of dissolved electrolytes in the water and phosphorus is an element essential for algal growth but also one of the scarcest. Chlorophyll content is a measure of the biomass of the phytoplankton present.

Globally, latitude is important in determining phytoplankton photosynthesis (and in turn, in a general way, the productivity of higher trophic levels). Latitude is a summary measure of several energy-related variables, being itself a measure of the angle the sun subtends with the vertical at a point on the earth's surface when it is overhead at the equator. Lower latitudes (Equator 0°) have greater solar radiation and higher temperatures than higher latitudes. It is the availability of light energy to power photosynthesis and perhaps the greater rate at which the scarcer nutrients might be recycled at higher water temperatures which might underly the high correlation with latitude. Clearly also the amount of phyto-plankton biomass present also helps determine photosynthetic rate and a puzzling feature of the data is that biomass is normally closely related to chemical factors like conductivity and phosphorus availability. Yet these factors seem to be less valuable in predicting photosynthesis on a global basis.

Within a narrow range of latitudes, however, the potential photosynthesis which might be attained with the light energy available may be seriously limited by the supply of nutrients for subsequent growth. In temperate regions (Table 1.2), chemical factors assume a much greater importance, and so does altitude, reflecting perhaps the decrease in mean water temperature in upland lakes.

The data available to Brylinsky & Mann were not perfect, indeed they were probably very varied in quality, but such analysis had not previously been attempted. A subsequent analysis has been carried out by Schindler[451] using part of the original data used by Brylinsky & Mann and many new data, all of which have been screened for their reliability. This analysis seems to have eliminated many of the inconsistencies of the earlier one and points to a much greater role of nutrient availability, particularly of phosphorus. It also finds a possible correlation between nutrient supply and decreasing latitude related perhaps not only to rates of recycling but also to a larger concentration of phosphorus compounds in tropical rainfall. In the new analysis light seems much less likely than nutrient supply to limit production in fresh waters. Nonetheless it has great importance in determining events within lakes and, with nutrients, must be given further consideration.

1.4 Light

The electromagnetic radiation received from the sun at the top of the atmosphere amounts to rather more than is received at the surface of a water body. Some is reflected and scattered and certain frequencies (particularly the very high (ultraviolet) and very low (infra red)) are selectively absorbed by the atmos-phere. What remains is rapidly absorbed by water itself and except in the surface few centimetres the light climate underwater is confined approximately to the visible frequencies (13350 cm^{-1} – 28600 cm^{-1}). Expressed as wavelengths these are 750 nm and 350 nm.

Visible light is absorbed by the water itself, dissolved substances and par-ticles suspended in it. In general, the highest and lowest wavelengths (reds and blues) are absorbed most rapidly, most strongly by dissolved organic substances and particles and the water itself. The middle wavebands (yellows and greens) penetrate deepest until they themselves are absorbed. In waters deeply stained with yellow and brown organic compounds which drain from peaty catchments, red, orange and yellow wavelengths may penetrate relatively deeper than any others.

The amounts of dissolved and suspended substances present in a water will determine the total depth to which light can penetrate. Absorption in a uniform water column is exponential. This means that in equal successive increments of water depth, light of a given wavelength is reduced by a fixed proportion for each

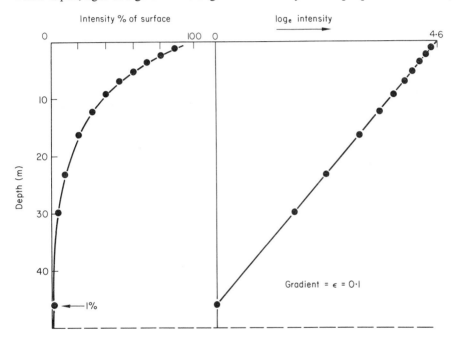

Fig. 1.2. Absorption of light in a uniform water column. The left-hand panel shows light intensity as percentages of that found at the surface (after reflection losses have been allowed for). The right-hand panel shows the same data with light intensities expressed as natural logarithms. The gradient of the line is the extinction coefficient.

increment. Thus light at the surface, after losses by reflection are allowed for, might be reduced by a factor of 10% in the uppermost metre, leaving 90% of the original intensity unabsorbed. In the second metre a further 10% of this 90% will be absorbed leaving only 81% of the surface intensity at 2 m. At 3 m, 81–8·1 or 76·9% will remain, and so on.

If the percentages remaining of the surface light at successive depths are plotted against depth (Fig. 1.2), an exponential or logarithmic curve can be constructed, described by the equation:

$$I = I_0 e^{-\epsilon z} \dots\dots\dots\dots\dots\dots\dots\dots\dots(a)$$

where I_0 is the light intensity at any given depth, I the intensity at a depth z metres below and e the base of natural logarithms. ϵ is the exponential coefficient, or the fraction absorbed per metre, 0.1 in the above example. ϵ can easily be calculated from an expansion of equation (a) which allows the data to be plotted as a straight line graph (Fig. 1.2):

$$\epsilon = \frac{1.}{z} \, 2.303 \, (\log_{10} I_0 - \log_{10} I) \dots\dots\dots\dots\dots\dots(b)$$

Theoretically light, is never totally extinguished. In practice, before it reaches even the undetectable levels at the limits of available instruments, it reaches a level at which only 1% of the surface intensity remains. The 1% level of the most penetrative wavelength (or waveband usually since underwater light intensities are normally measured with a light meter fitted with coloured filters each of which passes a range of wavelengths) has a conventional significance in limnology. *Approximately* it describes the level where algal photosynthesis is reduced to the point where energy fixation is matched by the organisms' own respiratory requirements; this is called the compensation point. Below this point algal growth cannot occur, but above it, in what is called the euphotic zone net growth can take place, provided, of course that the nutrient materials are available. The depth of the euphotic zone can be calculated from equation (b), using $I_0 = 100$ and $I = 1$, and is given by:

$$z_{eu} = \frac{4.6}{\epsilon'} \dots\dots\dots\dots\dots\dots\dots\dots\dots\dots\dots\dots(c)$$

where ϵ' is the extinction coefficient of light detectable with an unselective filter. It has been found empirically[487] that:

$$z_{eu} = \frac{3.7}{\epsilon''}$$

where ϵ'' is the extinction coefficient of the most penetrative wavelength or band.

If a light meter is not available, an estimate of z_{eu} may be obtained by dangling a weighted white-painted disc about 25 cm in diameter in the water. Named after Professor Secchi, who developed the idea from an Italian admiral who dangled a dinner plate in the Mediterranean Sea, the Secchi disc is lowered until it just disappears from view and then raised until it reappears. The mean of these two depths is 0.3–0.5 of z_{eu}. The disc is useful for relative measures over time in the same lake or river or in expedition work where heavy equipment cannot be carried. The depth measured is to some extent a function of the person using the disc, the ambient light intensity, and the nature of the water surface (calm or roiled) at the time, however.

z_{eu} may be less than a metre in lowland rivers charged with silt or in lakes supporting large populations of phytoplankton, or in excess of 50 m in upland lakes with very clear waters. A theoretical maximum of around 200 m is placed on it by the absorptive properties of pure water itself. Fig. 1.3 shows the ideal relationship of z_{eu} and ϵ' with the relative position of some particular water bodies indicated. In most of the world's water bodies, photosynthetic production is confined to less than half of the mass of water, and in some only to a thin surface skin, below which lies a huge dark world.

Any role of latitude in determining productivity through light availability is thus not straightforward. Quite apart from the effects of cloud in intercepting light and upsetting the theoretical smooth decrease in available energy from

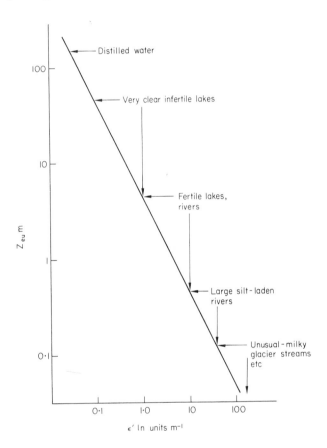

Fig. 1.3. Relationship between the euphotic depth, Z_{eu} in metres, taken as the level at which 1% of photosynthetically usable radiation still remains unabsorbed, and the extinction coefficient, ϵ^1 for light detected by an instrument with no selective filter.

Equator to Poles, many factors may conspire to deprive the algae and plants by competing with them for the absorption of light. Mostly these factors are related to the nature of the catchment area and its effect on water chemistry. This, certainly within a narrow range of latitude and probably on a wider scale will determine the extent to which the potential set by the supply of light energy will be attained. The role of the catchment area will now be considered.

1.5 Catchment areas and water chemistry

Some of the components of rain are shown in Table 1.1. Compare them with the composition of stream water which has percolated through an undisturbed forest in a hilly area and with that from a lowland river. Dust from the drier lowland increases the levels of all the ions measured in the rain falling on it compared with those in the wetter upland area, but the overall concentrations in both samples of rain are low. The contrast between the drainage waters is very marked indeed. Percolation through the soils on igneous granites and hard metamorphic rocks has added usually less than 1 mg l^{-1} of any ion to the water. Weathering of such rocks is slow. Tens or even hundreds of mg l^{-1} of most ions boost concentrations in the lowland area. The rain percolates through deep soils deposited by glacial action and containing material from soft, easily weathered formations like chalk. Carbon dioxide dissolved from the atmosphere in the rain easily dissolves calcium carbonate to release soluble bicarbonate and calcium ions. In turn these form part of a chemical system which removes H$^+$ ions and increases pH (see Chapter 3).

The ions listed in Table 1.1 are part of a set, which also includes SO$_4^{2-}$, called major ions, because proportionately they are the most abundant in the majority of, if not all, fresh waters. Analysis of them is easy and they are frequently measured in routine limnological investigations. This fact, and their collective name, gives them an apparent importance which they do not really have for even at their lowest concentrations there is generally such a large supply of them in relation to the requirements of the organisms present that they have little influence on the productivity and processes going on in freshwater lakes. If the concentration of major ions is very high however, that is when they reach levels that would give the water a brackish (salty) taste with concentrations of g l^{-1} rather than mg l^{-1}, they become very important in determining particularly the species of plants and animals that can live in the water through their role in osmo-regulation. In the extreme case of endorheic lakes where levels may be so high that chlorides and carbonates crystallize in white crusts at the edges, only a handful of species may be able to survive. At low levels also, major ions may determine the nature of the fauna. A correlation has been found[40] between calcium concentrations and the diversity and abundance of freshwater molluscs. Snails, limpets and bivalves need much calcium for their shell formation and this correlation is undoubtedly causative. Other correlations between major ion levels, particular organisms and events in fresh waters may easily be found. These are usually not causative but reflect the general relationship that the supply of major ions has to the supply of important (since they are relatively scarce) plant nutrients such as compounds of nitrogen and phosphorus. These have a much more important influence on production in fresh waters than the major ions.

1.6 Key nutrients

Take a sample of water from almost any water body, dispense it into clean glass flasks and add, in some convenient replicated experimental design, a range of ions alone and in combination at concentrations of a few mg l^{-1}. Leave the flasks in good light for a week or two and the chances are very high that you will notice much greater growth of green or yellowish green algae in flasks to which phosphate has been added than in those to which it has not. For some waters it may have been necessary to add both phosphate and nitrate or ammonium ions, and in others, particularly some tropical ones, nitrogen compounds alone may suffice.

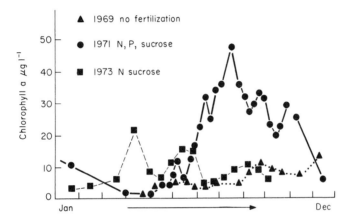

Fig. 1.4. Effects on biomass of phytoplankton (measured as its chlorophyll a content) of fertilizing Lake 304, Canadian Experimental Lakes area. In 1969 there was no fertilization; in 1971, totals of 0.4 g $PO_4 - P$, 5.2 g $NH_4 - N$ and 5.5 g sucrose were added per m^2 in 20 equal weekly increments. In 1973 $NH_4 - N$ and sucrose were added at the same rates as in 1971, but phosphate was not added. (Based on Schindler[449], Fig. 2. Copyright 1974, American Association for the Advancement of Sciences.)

The experiment is even more convincing if done on whole lakes! In Canada a lake shaped roughly like a violin, though with a much narrower waist at the middle, was divided into two with a vinyl reinforced nylon curtain sealed into the mud and to the banks in the region of the waist[449]. Phosphate, nitrate and sucrose were added to one half, and only nitrate and sucrose to the other. (The point of adding sucrose was to test a hypothesis that carbon might have been in short supply for algal growth.) Only in the side to which phosphate was added was there a dramatic increase in phytoplankton growth. The other side did not change from its previous state. Fig. 1.4 shows the results of a similar experiment in which a lake was fertilized with phosphate, nitrate and sucrose in some years, and only nitrate and sucrose or not at all in others. Significantly, phytoplankton (measured as its chlorophyll a content) rose in abundance only where phosphate was added.

Phosphorus is, on average, the scarcest element in the earth's crust of those required absolutely for algal and higher plant growth. It is also relatively insoluble, being readily precipitated as complexes of iron, aluminium, calcium and other relatively abundant metals in rocks and soils, and there is no reservoir of gaseous phosphorus compounds available in the atmosphere. Table 1.3 shows the ratios of amounts of various other necessary elements to the amounts of phosphorus required for algal and plant growth. It shows also the same ratio, weight for weight, in the soils and rocks which make up the lithospheric supply

Table 1.3. The relative supply and demand of elements required by plants and algae and derived from soils and rocks (lithosphere) of the catchment area (modified from Hutchinson [238]).

Element	(1) Ratio of amount of element to that of phosphorus in the lithosphere	(2) Ratio of amount required of element to amount required of phosphorus in plants and algae	Ratio of (1) to (2)
Na	32.5	0.52	43
Mg	22.2	1.39	16
Si	268.1	0.65	410
P	1.0	1.0	1.0
K	19.9	6.1	3.3
Ca	39.5	7.8	5.1
Mn	0.90	0.27	3.3
Fe	53.6	0.06	880
Co	0.02	0.0002	110
Cu	0.05	0.006	8.5
Zn	0.07	0.04	1.5
Mo	0.0014	0.0004	3.6

of these elements. The third column gives the ratio of the first two columns (lithosphere: plants), and indicates the relative supply to the relative need. Ratios greater than 1 in the third column indicate that the need is more likely to be met by the supply than it is in the case of phosphorus, and all of the elements supplied from the lithosphere that plants require are in this category. Certain elements are scarce in the earth's crust though necessary for plant growth and are not included in the table. Sulphur (as sulphate), chlorine (as chloride), and boron (as borate) have vast reserves in the sea, which, through upward drift of droplets into the atmosphere and transport by air currents, subsidize the crustal supply enough for the ratio of supply to need always to be higher than that for phosphorus. Similarly, nitrogen and carbon have atmospheric reserves though they are relatively scarce in the crust. The atmospheric reserve of carbon dioxide is directly accessible to water bodies through diffusion, and, dissolved in rain, helps mobilize bicarbonate ions, which are a source of carbon in fresh waters, from the lithosphere. Plants and algae in fresh waters are never more than temporarily short of this element. The supply to need ratio is about 30.

The case of nitrogen is more complex since despite the huge atmospheric supply (nitrogen constitutes nearly 80% of the atmosphere) elemental nitrogen is relatively unreactive and available as such only to a few organisms (nitrogen fixers), all of them prokaryotic and including certain bacteria and blue-green algae (Cyanophyta). Nitrogen is converted to available compounds by atmospheric lighting sparks and perhaps ultraviolet radiation and by nitrogen fixers. These compounds (most importantly nitrate and ammonium) are very soluble and hence are readily transported into waterways. Nevertheless the supply to need ratio for nitrogen is not much greater than that for phosphorus, and in some areas, particularly in East and Central Africa where rocks particularly rich in phosphate occur, and in areas where man enriches water with phosphate from sewage effluent, the *local* supply to need ratio of nitrogen may be less than the *local* ratio for phosphorus.

As a generalization of wide validity, however, the extent to which the potential productivity of a water body is realized is set by the supply of available phosphorus compounds. The supply of phosphorus bears a general relationship to the supply of less scarce elements such as calcium and potassium, since catchments of easily weatherable rocks release more of everything than catchments of hard rocks. In the past, productive lakes have been characterized by limnologists as those having a relatively high calcium content, for example. Such a correlation does not imply causation and may be very misleading. In some lakes, called marl lakes, where calcium concentrations are very high owing to their situation among soils rich in fine limestone particles, phosphate may be so readily precipitated by inorganic chemical reactions that extremely low productivity is characteristic.

1.7 How the total phosphorus content of water is established

Phosphorus enters water bodies not only as inorganic phosphate ions (PO_4^{3-}, $H_2PO_4^-$, HPO_4^{2-}) but also in inorganic polymers, organic phosphorus compounds, in living microorganisms and dead detritus. Only some of these forms are immediately available for plant and algal growth, but others may become so through microbial activity in the water. The processes are complex and discussed in Chapter 3. The sum of all the forms of phosphorus content, P_{tot}, is a reasonable measure of the fertility of a water. A rather infertile lake may have only about $1\ \mu g P_{tot} l^{-1}$, while the most fertile may have $1000\ \mu g P_{tot} l^{-1}$ or more, with an unbroken continuum in between. P_{tot} is determined by a number of factors which can be quantitatively related in a general model, or equation, derivation of which, following the work of Vollenweider[513], is given below. The theory can be applied to the behaviour of any element in a water body, and so is couched in general terms.

The amount of a substance entering a water body per unit time, t, and per unit area of a lake is called the areal loading, L. There are usually several sources of loadings, including subdivisions of the catchment area if it is not uniform— its land use or geology may vary considerably—the atmosphere, including dry fallout and rainfall directly onto the lake, the excretion and defaecation of visiting birds and other vertebrates like hippopotami, and, at certain times, release of the substance from the underlying sediment into the water. The latter is called an internal loading, the others are external. Substances cannot, of course, be indefinitely released without replacement from the water mass or external sources. With the possible exception of shallow endorheic lakes (see Chapter 5), loss to the sediments far outweighs release from them over periods as long as a year. Over shorter periods, however, there may be a net gain from the sediment.

Consider a lake receiving an element M from various catchment sources. Let M_w be the total amount of element in the water body, M_{in} the total amount entering, M_{out} the total amount lost through the outflows, and σ the fraction of M_w lost to the sediments during time t. The rate of change in M_w is then given by:

$$\frac{d(M_w)}{dt} = M_{in} - M_{out} - \sigma M_w \dots\dots\dots\dots\dots\dots(d)$$

If v_{in} and v_{out} are the volumes of the inflows and outflows and $[M_{in}]$ and $[m_{out}]$ the respective *concentrations* of M:

$$\frac{d(M_w)}{dt} = v_{in}[m_{in}] - v_{out}[m_{out}] - \sigma M_w \dots\dots\dots\dots\dots(e)$$

Let V be the total volume of the water body and $[m_w]$ the concentration of M in the water of the water body. Therefore, dividing (e) by V:

$$\frac{d[m_w]}{dt} = \frac{v_{in}[m_{in}]}{V} - \frac{v_{out}[m_{out}]}{V} - \sigma[m_w] \dots\dots\dots\dots\dots(f)$$

Since v_{out}/V equals ρ, the flushing coefficient or number of times the volume of the water body is replaced in time t, $(v_{in}[m_{in}])/V$ is the loading per unit volume of water body, designated Q. If it is assumed (reasonably) that $[m_{out}] = [m_w]$, then:

$$\frac{d[m_w]}{dt} = Q - \rho[m_w] - \sigma[m_w] \dots\dots\dots\dots\dots\dots(g)$$

Over periods of several weeks the total concentration of the element M (though not necessarily of each of the sub-components that make up the total pool of M) often remains reasonably steady and in terms of order of magnitude it can be regarded as approximately constant over much longer periods. During each of the periods of relative stability:

$$\frac{d[m_w]}{dt} = 0$$

$$\text{and } Q - \rho[m_w] - \sigma[m_w] = 0 \quad \ldots\ldots\ldots\ldots\ldots\ldots(h)$$

re-arranging this:

$$[m_w] = \frac{Q}{(\rho + \sigma)} \quad \ldots\ldots\ldots\ldots\ldots\ldots\ldots\ldots\ldots(i)$$

Since water bodies have very varied volumes it is more useful to express $[m_w]$ in terms of loading per unit area of surface of the water body instead of per unit volume. If the loading per unit area is L, and since $V = Az$, where A is the area of water body and z its mean depth:

$$Q = \frac{[m_{in}]v_{in}}{V} = \frac{[m_{in}]v_{in}}{Az} = \frac{L}{z} \quad \ldots\ldots\ldots\ldots\ldots\ldots(j)$$

Thus (i) becomes:

$$[m_w] = \frac{L}{z(\sigma + \rho)} \quad \ldots\ldots\ldots\ldots\ldots\ldots\ldots\ldots(k)$$

This is the basic model and L can be expanded to equal the sum of the separate loadings from different sources; each can usually be measured by straightforward techniques. z and ρ can be readily obtained from survey and standard hydrological techniques. To validate the model fully, however, σ must be measured and this is not always easy. It is perhaps better to replace it in the model with R, the fraction of the loading that is retained in the lake, not sedimented but ultimately destined to be washed from the lake through the overflows.

$$R = \frac{[m_w]V\rho}{L.A.} = \frac{[m_w]z\rho}{L} \quad \ldots\ldots\ldots\ldots\ldots\ldots \quad (l)$$

Since, from (k):

$$\frac{z[m_w]}{L} = \frac{1}{(\rho + \sigma)}$$

$$R = \frac{\rho}{(\sigma + \rho)} \text{ and } \frac{R}{\rho} = \frac{1}{(\rho + \sigma)}$$

Therefore:

$$[m_w] = \frac{LR}{z\rho} \quad \ldots\ldots\ldots\ldots\ldots\ldots\ldots\ldots(m)$$

R can be obtained as the ratio of total efflux of M via the overflows ($v_{out}[m_{out}]$) divided by the total influx, $L.A.$

This general model is important in that it allows prediction of $[m_w]$ under varying conditions of water flow and changes in L. This not only helps to explain why water bodies have the characteristics that they do but allows the effects of man-made changes to be determined in advance (see Chapters 10 & 12). It is useful to see how the parameters of the Vollenweider model combine to determine the absolute values of $[m_w]$ for various elements.

1.8 The effects of L

The loading of a substance carried in water from the land draining into a water body depends on two things—the size of the catchment in relation to the area of the lake, and the rate of removal of the substance per unit area of catchment. If the lake is large in relation to its catchment area the loading will be relatively low from this source. Crater Lake, Oregon, a lake set in the crater of an extinct

Table 1.4. Rates of removal of substances in mg m^{-2} yr^{-1} (kg km^{-2} yr^{-1}) from different sorts of catchment area (variously compiled from Likens et al.[295], and Omernik[389] and other sources).

	Phosphorus	Nitrogen	Sodium	Calcium
Forest (U.S.A.)	8.3	440		
Mostly agricultural (U.S.A.)	22.7	630		
Entirely agricultural (U.S.A.)	30.8	982		
Tundra, Sweden	4.1	97		
Temperate deciduous forest	1.9–26.0	100–560	750–6200	690–18200
Temperate coniferous forest	2.0–42.0	40–230	600–3840	600–5720
Upland temperate moorland	40.0	30	4520	5380
Tropical forest	10.0	470	6450	470–4310
Lowland agriculture, Norfolk, U.K.	13.0			
Forest on igneous rock (range and mean)	4.7 (0.7–8.8)			
Forest on sedimentary rock (range and mean)	11.7 (6.7–18.3)			
Forest plus pasture on igneous rock	10.2 (5.9–16.0)			
Forest plus pasture on sedimentary rock	23.2 (11.1–37.0)			

volcano, with only the thin rim of the crater as its drainage area, is an example. Conversely, for stretches of a river the loading will be relatively high, but the effects of the loading on $[m_w]$ will be mitigated, in regions that are not endorheic, by a relatively high flushing rate.

The effects of L are most readily seen where markedly different catchments have correspondingly varied removal rates. Table 1.4 provides some examples. Two points are clear. Sedimentary rocks have greater removal rates than igneous rocks, and agricultural areas have greater rates than undisturbed ecosystems. In the latter the amount exported to streams is often balanced by the amounts received in rainfall for nitrogen and phosphorus, though not for major ions which are supplied by continued rock weathering. Ecosystems have evolved in such a way as to conserve scarce nutrients like phosphorus and nitrogen which are rapidly taken up and 'stored' in the plant biomass. Agricultural disturbance of the soil and fertilization at times when the biomass is insufficient to store these substances leads to increased removal rates.

Various factors complicate interpretation of these figures in terms of the relevance of the Vollenweider model to prediction of fertility and productivity of a water body. The amounts of substances removed, in the cases of relatively insoluble elements like phosphorus, are not all available for plant or algal growth in the lake to which they are delivered. Some phosphorus is adsorbed onto eroded soil particles. Surveys in the U.S.A. show that perhaps only 40% of the total phosphorus carried by streams is in an available form, the rest rapidly becoming locked into the sediments of the rivers and lakes. Climate will also influence removal rates. Rainfall distributed evenly throughout the year will usually lead to lower removal rates than a similar amount of rain concentrated into a short season of violent and erosive storms.

Despite these complications, it is clear that the higher concentrations of major ions like Na^+ and Ca^{2+} in lake waters reflect higher removal rates of these readily soluble ions than those of the much more insoluble phosphorus and the only slightly less scarce nitrogen. In the latter case, its scarcity reflects not the insolubility of its compounds but the gaseous nature of the element at normal temperatures and pressures. It was lost from the earth's crust to its atmosphere early in geological time and does not readily react in its elemental state.

The loading rates of phosphorus and nitrogen on a water body may be independent of the catchment area as such under certain circumstances. These are when concentrated groups of men, domestic or wild animals discharge excreta, directly or as effluent via a treatment works, to a watercourse. Human activities are generally the cause, directly or indirectly, in all of these cases. The maximum potential sizes of populations of wild animals and pre-technological man are determined ultimately by the resources locally available to them though other factors may serve to regulate their populations at levels much lower than the potential. They thus recycle available substances within the catchment area, having evolved as part of its ecosystem, and their activities are included in the natural removal rates discussed above.

If, for some reason, man, or animals, or both are concentrated near a waterway—in a town, an intensive stock-rearing unit, or a game park or wildlife reserve, the population no longer reflects the nature and size of the catchment, for substances like phosphorus and nitrogen are imported as food from elsewhere. They are excreted in almost the same quantity as they are imported in a growing population (in the sense of net growth of body tissue) and at the same rate where the population is stable. The use of domestic detergents containing phosphorus by human populations also increases the loading. In all these cases the loading rates of phosphorus (particularly) and nitrogen (sometimes), on a local waterway may be several times higher than even those derived from agricultural land.

Man may cause similar high loading rates by encouraging the concentration of migratory birds on lakes designated as sanctuaries or nature reserves. Migratory geese are attracted to Wintergreen Lake, Michigan by a resident wing-clipped flock and fertilize the lake heavily[330], and black-headed gulls (*Larus ridibundus* L.) have congregated on Hickling Broad, in eastern England in increasing numbers in recent winters, substantially boosting the phosphorus loading[281]. Gull populations have increased markedly in Europe as a whole in the 20th century but the reasons are obscure. They have moved inland from their original coastal habitat to feed in ploughed fields and waste tips. Burgeoning human populations creating more and more edible waste are thus probably one of the causes of the increase in gull numbers.

1.9 The effects of σ and ρ

Just as the ratio of catchment area to lake area affects the loading rate, the catchment area to lake volume ratio helps determine the flushing rate, and its reciprocal, the mean residence time of water in the lake. The amount of rainfall, or rather the fraction of it that is discharged from the catchment after evapo-transpiration of the land vegetation, is also crucial. Sections of rivers (e.g. a few km in length) have very high flushing rates, perhaps more than once per day at times, but only once per week or more at times of very low flow. For lakes the mean residence time may be weeks or years. Barton Broad, a shallow $z \approx 1$m), small (63·4 ha) lake situated astride an English river, the Ant, is flushed out between 10 and 30 times per year. Lake Tanganyika, a vast body of water nearly 2 km deep in parts, would take over a thousand years to refill with present catchment rainfall, if all its waters were drained out. In practice, since most of its water is in permanent deep layers, only the surface few hundred metres are flushed but even so the residence time is many years.

The effects of ρ in determining substance concentrations are straightforward. The higher the flushing rate, the smaller the effective concentration of a substance for a given loading. ρ also has a regulating role. If rainfall is high in a particular year, and the removal rate of a substance from a catchment is thus increased, so that the loading also increases, the effective concentration $[m_w]$ in the water will not increase as the higher rainfall will lead also to a greater flushing rate (equation (k)). In practice $[m_w]$ may decrease since the rate of removal may not increase proportionately with rainfall. It is likely to fall since when rainfall increases, more of it runs off the land surface and does not pass through the soil where most substances are picked up.

The effects of flushing also mean that the concentration of a substance in a lake (in the absence of loading sources other than land drainage) cannot exceed the concentration of that substance in the water percolating to the catchment streams and in the streams themselves. Such concentrations vary, as shown for nitrogen and phosphorus in Table 1.5. For most of those temperate areas where concentrations of total phosphorus approach about 100 μg l^{-1} or perhaps a little less in the water reaching a lake, it is likely that a source of phosphorus from some human or animal source, unrelated areally to the catchment, is present. Such effluent or excreta have the important property that they cannot be regulated by flushing rate, though their effects on lake fertility may be reduced if the flushing rate is high.

Relatively little is known of the role of σ, the sedimentation coefficient, in determining concentrations in a water body. For soluble ions like the major ions Na$^+$, K$^+$, Ca^{2+}, Mg^{2+}, Cl^{-1}, SO$_4{}^{2-}$, HCO$_3{}^-$, it is likely to be very small, but for readily precipitable trace elements like iron, and scarce nutrients like phosphorus, which is rapidly taken up into the plankton and thence stands a chance of sedimentation, it will be higher. It seems logical that the longer water remains

Table 1.5. Mean concentration of phosphorus and nitrogen in stream waters analysed in the U.S.A. Values in μg l^{-1} (after Omernik[389]).

	Total P	Soluble Inorganic P	Total N	Soluble Inorganic N
Forest	14	6	850	232
Mostly forest	35	14	885	347
Mostly agricultural	66	27	1812	1049

in a lake the more likely it is that a higher proportion of a substance will be sedimented out. Thus the higher the flushing rate, the lower the sedimentation rate will be. It is also probably true that σ will have relatively little effect where ρ is high (say 10 yr^{-1}). At low values of ρ ($<$ 1.0 yr^{-1}) the two coefficients will have a buffering effect on each other in determining [m_w], but as ρ increases in riverine lakes and rivers, σ may be neglected.

The last component of the Vollenweider model (equation (k)) for determining substance concentrations in water bodies is z, the mean depth. z influences ρ through its effects on lake volume, and σ through possible recycling of substances from the sediment in shallow lakes. It also determines the physical structure of a water mass through stratification and hence complicates prediction of substance concentrations because many substances are not uniformly distributed when the water mass is stratified. z is dealt with in the next chapter.

1.10 Further reading

Further reading appropriate to this chapter falls into two categories—basic limnology textbooks giving a more detailed treatment of water chemistry, and works amplifying the use of models in predicting the all-important substance concentrations in lakes. In the first category the standard work has long been Hutchinson[234], and Wetzel[530] and Golterman[176] provide more recent, chemically orientated texts. Stumm & Morgan[485] give a very detailed account of water chemistry. General introductions to limnology which are more region-ally based are, for African waters, Beadle[24] and for Australasia, Bayly & Williams[23]. Admirable texts on river ecology are Hynes[242] and Whitton[536]. Books dealing with the general relationships of water chemistry and catchment area are Hasler[202], H.M.S.O.[210], Likens & Borman[294], and Likens *et al.*[295]. Useful studies relating catchment area geology to water chemistry are Douglas[108] and Mackereth[316] (see also Macan[310]). Experimental studies on the relationship between productivity and water chemistry may be ap-proached through Larsen & Mercier[279], Schindler *et al.*[452], and Schindler & Fee[454]. Validation of the Vollenweider model is dealt with by Dillon & Rigler[101] and Vollenweider & Dillon[514]. Source books for methods of chemical analysis of water are the American Public Health Association[9], Golterman *et al.*[178], Mackereth *et al.*[32] and Strickland & Parsons[484].

Some of the prominent limnological research journals are *Limnology and Oceanography*; *Freshwater Biology*; *Archiv für Hydrobiologie*; *Internationale Revue der Gesamten Hydrobiologie*; *Hydrobiologia*; *Internationale Verhandlungen der Vereinigung theoretische und angewandte Limnologie*, and *Journal of the Fisheries Research Board of Canada*.

CHAPTER 2
THE ROLE OF DEPTH

2.1 Introduction—the origin of water depth

Water bodies may be only a few cm deep, or even zero when certain arid zone lakes seasonally dry up, or they may have many hundreds of metres of water. Depth is determined partly by the means by which the lake basin was formed, partly by the extent to which accumulating sediment has reduced this, and partly by climatic fluctuations which have caused major changes in the water balance of many areas from time to time.

There is no rigid correlation between mean depth of a basin and the way it was formed, though, in general, the larger a water body, the deeper it is. Earth movements in which plates of the earth's crust moved sideways apart produced the immensely deep African rift lakes Tanganyika and Malawi (Fig. 2.1), but the very shallow lakes Nakuru and Hannington also have basins in the rift valley. L. Victoria, moderately deep, and L. Chilwa, very shallow, were both produced by a gentle local sinking of the land, perhaps as it resettled after the plate movements. Movement of the continental glaciers during the last glaciation carved out the deep L. Superior and also tens of thousands of small shallow lakes in Canada, New England and mid-western America (Fig. 2.2). Damming by man has created the large, deep lakes Kariba and Nasser, and innumerable local 'fish ponds' a metre or two deep throughout Africa and Asia.

The rate of sediment accumulation, also, is governed by many factors. Lakes fed by run-off streams from glaciers may accumulate centimetres per year of fine inorganic rock flour carried in the milky streams, whilst other lakes may lay down only a fraction of a mm per year. Large catchment areas result in high sedimentation rates if the basins they feed are small, or if their rainfall is concentrated into a period of time and is highly erosive. Disturbance of a catchment by clearing of the natural vegetation leads to increased sedimentation and the greater the phytoplankton production the greater is the build-up of organic sediment. A median sedimentation rate for the lakes of the world is probably in the range 1–10 mm yr^{-1}, and hence makes a substantial difference to the mean depth of most lakes only on a long-term basis.

Streams and rivers are, predictably, relatively shallow but the depth of a lake is almost an accident of birth. Nevertheless its magnitude determines much of how a lake ecosystem functions. In the completely mixed lake for which the Vollenweider model (equation (k) Chapter 1) is directly applicable, mean depth helps determine the effect of a given loading on substance concentration by determining the extent to which that loading will be diluted in a greater or lesser water volume. Through volume also it helps determine the flushing coefficient, ρ. The product of mean depth (z) and ρ is to some extent self-buffering—the greater the depth, the lower ρ will tend to be, other things being equal. Depth also controls the structure of the water mass, and the balance of planktonic and bottom-living communities. These aspects will now be considered.

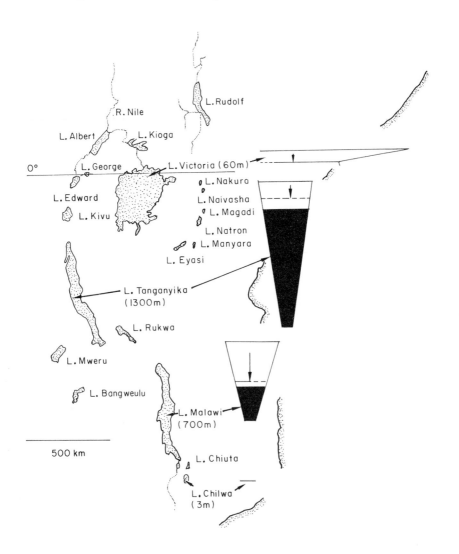

Fig. 2.1. Some major East African Lakes. For Lakes Victoria, Tanganyika, Malawi and Chilwa, scaled profiles are given, with arrows indicating the maximum depth to which the water column is mixed during the year. The darkened parts of the profiles indicate permanently unmixed and deoxygenated water (based partly on Beauchamp[25]).

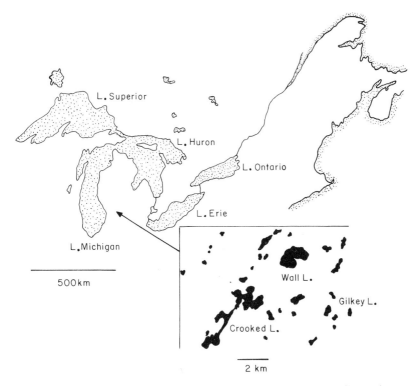

Fig. 2.2. The St Lawrence Great Lakes of Canada. The inset shows a small moraine area in Michigan, U.S.A., where glaciation has formed very many smalls shallow lakes.

2.2 Vertical stratification and the structure of water masses

Turbulently flowing streams are isothermal; the water movement is sufficient to break down any temperature gradients. However, in very slow-flowing backwaters of rivers and in lakes the surface water tends to heat up more rapidly than that below it, and at some times of the year in deep enough water, wind and water movements are insufficient completely to destroy this gradient, which progressively intensifies.

Like all electromagnetic radiation, the longer wavelengths (above 700 nm) which impart most heat are absorbed exponentially by the water, its dissolved substances, and suspended particles (see Chapter 1). The extinction coefficients for such wavelengths are high (10–100 times greater than those for wavelengths in the middle of the visible and photosynthetically active ranges) and heat radiation is effectively completely absorbed in the top metre or two of water. Ideally this would create an exponential fall in temperature from the surface in a still water mass, but there is generally some wind disturbance. The wind mixes the heated surface layers downwards but may be insufficient to mix them to the bottom of the basin. This results in an upper, warmer, isothermal layer, a few metres or tens of metres deep, which, because water decreases in density with temperature above about 3·94°C, floats on cooler, denser, usually also isothermal water. Traces of the idealized exponential temperature curve may be detectable in the lower layer, however. Between the upper layer, the epilimnion, and the lower layer, the hypolimnion, is a transitional zone called the metalimnion. In this zone a temperature gradient, the thermocline, of as much, or more than $1°C$ m^{-1} may be detected. This structure is called direct stratification and is often recorded in lakes in warmer climates, and in temperate lakes during the spring to autumn period. It may be very short-lived (a day or so) on calm days in lakes of only a few metres depth so that a slight increase in wind action will mix the water completely again. On the other hand in many lakes it persists for all of the warmer part of the year, or almost all of the year in some tropical lakes. Even in lakes as shallow as 3–4 m it may be semi-permanent if the lake is very well sheltered by dense forest. Conversely in somewhat deeper lakes in open landscapes swept by the wind, no persistent stratification may ever form.

Temperature ranges between epilimnion and hypolimnion may be from 20°C to 4°C in temperate continental lakes with severe winters and hot summers, for example in mid-western North America, and from 18°C to 10°C in maritime temperate lakes where the water mass does not cool down so much in winter, nor heat up so much in summer. Such conditions are typical of lakes in the English Lake District and the western Scottish lochs. In the tropics the range may be only a few degrees, from around 29°C to 25°C, but the stratification may be almost as stable as one supported by a much greater temperature range in temperate regions. This is because the change in water density per degree change in temperature is so much greater at high water temperatures than it is at lower ones (Fig. 2.3).

The annual course of stratification is illustrated for three lakes, L. Victoria (Uganda), L. Windermere (U.K.) and Gull Lake (Michigan, U.S.A.), in Figs. 2.4–2.7, by means of depth-time diagrams. To construct such diagrams, temperatures, obtained from a succession of depths by suitable remote recording thermometers or thermistors, are plotted against the dates on which they were obtained throughout the year. Isotherms are then drawn connecting points of equal temperature, much as contours are drawn along equal heights in making a map. The greater the vertical slope of the lines, the less is the temperature gradient, so that vertical lines indicate completely mixed (isothermal) water masses. The more the lines tend to the horizontal, the greater the vertical temperature gradient in the water mass, and the stronger the stratification.

A period of direct stratification ends when the water column is mixed from top to bottom. In temperate lakes this is usually caused by a combination of cooling from the surface in autumn, as the solar angle decreases, and windier weather. Eventually the density gradient becomes too small to be stable under the prevailing wind conditions. In L. Victoria (Fig. 2.4) which lies astride the equator, the surface fall in temperature is provided by the evaporative cooling of the Trade Winds which begin to blow just after the middle of the year. The two shorter periods of mixing in L. Victoria are probably not features of the main

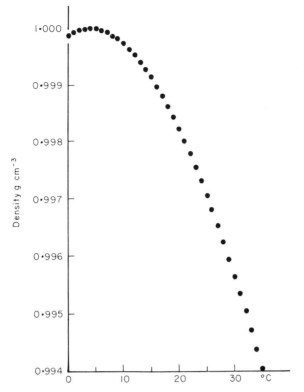

Fig. 2.3. Graph of density of pure water, under one standard atmosphere pressure in relation to temperature. The peak density is reached at 3.94°C.

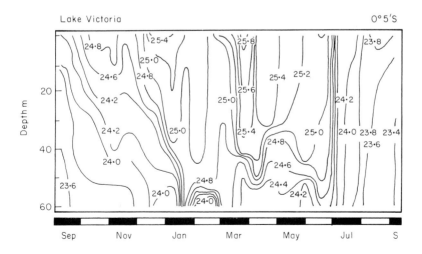

Fig. 2.4. Depth-time diagram of thermal stratification in a tropical lake, L. Victoria, Uganda. (Based on Talling[490].) Isotherms in Figs. 2.4–2.7 are in °C.

body of the lake. Regular measurements in this huge lake (25 600 km²) could only be made a few km offshore, where temporary movements of permanently isothermal shallow water from the margins sometimes displaced the 'usual' water mass. This emphasizes that there may be horizontal as well as vertical differences in the structure of water masses.

The period of top to bottom mixing in temperate lakes may last from one period of direct summer stratification to the next such period (Fig. 2.5). This is also true of many tropical lakes, but not of many temperate continental lakes or of polar lakes. The peculiar properties of water lead to its having a maximum density at around 3.94°C (or in practical terms 4°C). This leads, in these lakes, to a period of inverse stratification during winter with the warmest and densest water usually at 4°C at the bottom (though it may be cooler) and with an upward gradient of colder and less dense water to the surface where ice usually forms at 0°C.

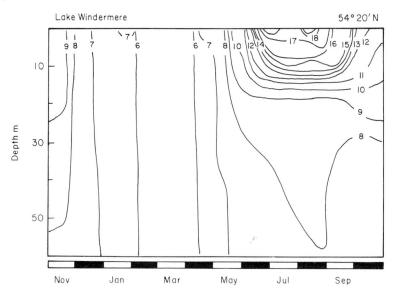

Fig. 2.5. Depth-time diagram of thermal stratification in a temperate lake experiencing a maritime climate, L. Windermere, English Lake District. (Based on Jenkin[256].)

The density–temperature relationship (Fig. 2.3) of water comes from its not being an amorphous liquid. Its molecular structure is such as to allow electrostatic attraction and association between the separate molecules, giving a loose structure to the fluid. This structure is most pronounced in the various crystalline forms of ice, which is much less dense than liquid water, and in which the molecules are held well apart. When ice melts some of the crystalline organization is retained, but progressively breaks down as the temperature rises. The molecules thus collapse into the spaces which had separated them in the ice crystal structure. More are packed into a given volume, and the density should

rise because of this process. At the same time increasing temperature increases
the kinetic energies of the molecules and causes them to move further apart.
This will tend to decrease the density. As ice melts two processes are operating,
one to increase, the other to decrease the density. It seems that the first process is
predominant up to 3.94°C, when the maximum density of water is reached, and
the second above this temperature, when density progressively decreases. The
association of molecules in liquid water underlies also the relatively high vis-
cosity and high specific heat of water. Much more energy is needed to separate
the molecules than is needed for the hydrides of elements related to oxygen, like
sulphur and selenium. Lakes are consequently well buffered to temperature
changes in the overlying atmosphere and aquatic organisms experience much
narrower temperature ranges than terrestrial ones. The molecular association
also explains how water (the substance) remains water (the liquid) at normal
earth surface temperatures. H_2S and H_2Se are gases.

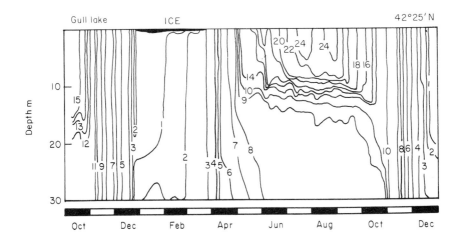

Fig. 2.6. Depth-time diagram of thermal stratification in a temperate lake experiencing a
continental climate, Gull Lake, Michigan. (Based on Moss[362].)

The consequences of water's properties are immense for lakes in very cold
regions. Ice forms at the surface as the water cools below 4°C and since ice and
the snow which may accumulate on it act to some extent as insulators to further
heat loss, it is rare for a lake to freeze to the bottom. Fish and other organisms
have therefore not had to evolve means of overwintering frozen solid.

The period of inverse stratification, where it occurs (Fig. 2.6) is broken by
the progressive melting of the ice and warming of the surface water as the sun's
angle increases in spring. A period of spring mixing, at first at 4°C, afterwards at
higher temperatures, precedes sufficient warming to create direct stratification
in most sufficiently temperate lakes. In polar regions, the summer is too short

for this and after a period of mixing at a maximum temperature perhaps around 10°C, inverse stratification sets in again for the autumn and winter (Fig. 2.7).

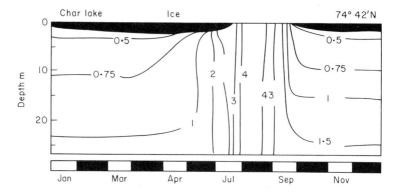

Fig. 2.7. Depth-time diagram of thermal stratification in a polar lake, Char Lake, Canadian Arctic. (Based on Schindler *et al.*[455].)

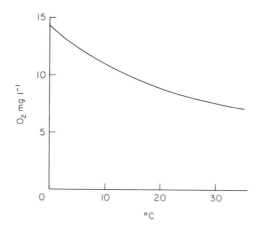

Fig. 2.8. Solubility of oxygen in pure water under one standard atmosphere pressure.

2.3 The consequences of thermal stratification for water chemistry.

Gases in the atmosphere maintain dynamic equilibria with the same gases, N_2, O_2, CO_2 and the inert gases, dissolved in waters, such that the concentrations of all but CO_2 in water freely mixing and exposed to the atmosphere are predictable almost entirely from the water temperature. Fig. 2.8 shows the relationship between saturation levels of dissolved oxygen and temperature. Carbon dioxide is part of a series of equilibria involving carbonate, bicarbonate, hydroxyl and hydrogen ions and the levels of these (see Chapter 3) as well as the atmospheric concentration of carbon dioxide also significantly determine its equilibrium concentration in waters. Respiration and photosynthesis may lead to temporary departures from equilibrium of oxygen and carbon dioxide over a daily period even in well mixed, open water, or for longer in calm periods when gas exchange is limited to simple diffusion through a flat, almost static water surface. In fertile lakes supersaturation of oxygen in the epilimnion is common for long periods. There will, however, always be a tendency to return to equilibrium conditions.

This is not so in water isolated from the atmosphere in hypolimnia or under ice. Inverse stratification takes place during cold periods when light intensity and day length are also low and short respectively, and hypolimnia include the least well-illuminated parts of a lake. Photosynthetic oxygen production is small or zero yet respiration of bacteria associated with falling detritus or sediments and benthic animals continues. Oxygen concentrations decrease and carbon dioxide concentrations increase under these conditions. Levels of molecular nitrogen are largely unaffected since, apart from the small quantities removed by nitrogen fixers (see Chapter 3) or produced by denitrifying bacteria (see Chapter 5), nitrogen is biologically rather inert.

Since oxygen is not particularly soluble in water (Fig. 2.8) it may be completely exhausted in water isolated from the air. As its concentration approaches and falls below 1 mg l^{-1}, chemical changes at the sediment surface (see Chapter 5) may result in release of inorganic ions (FeII, MnII, PO$_4^{3-}$) previously locked into insoluble oxidized complexes in the sediment, and populations of invertebrates and fish may be greatly affected (see Chapters 5 & 7).

2.4 Lake fertility and hypolimnial oxygen depletion

The rate of deoxygenation of the hypolimnion in terms of the oxygen consumed per unit area of lake per unit time is closely related to the rate of production of organic matter in the epilimnion, subject to two provisos. These are that large quantities of readily degradable, dissolved or particulate organic matter are not being washed in from the catchment area, and that flushing is not removing a large proportion of the phytoplankton production through the overflow during the summer. These two conditions are most likely to be met in large lakes which are not situated in peaty moorland or blanket bog catchments.

A continual rain of phytoplankton cells, detritus, zooplankton and fish faeces and corpses and associated bacteria falls through the metalimnion, where the abrupt change in density may cause a pause in the rate of fall and some accumulation of organic matter, to the hypolimnion. The greater the epilimnial production, the greater the supply of organic matter to the hypolimnion and the greater the demand on the irreplaceable (until overturn) oxygen reserves there. The effect on the state of the oxygen reserve, that is, the concentration of dissolved oxygen in the water, is not so directly connected with the epilimnial production, since the total amount of oxygen available depends partly on the hypolimnion temperature and partly on the hypolimnion volume. A given epi-limnial production may cause complete anaerobiosis of a hypolimnion in a lake only 10 m deep with half its volume below the thermocline and at a temperature of 10°C, but have a negligible effect on oxygen concentrations in a 100m deep lake with 90% of its water at 4°C in the hypolimnion. In both cases, however, the rates of hypolimnial oxygen depletion per m² of water should be comparable.

Hypolimnial oxygen concentration therefore is not a reliable guide to the productivity of a lake. In practice, however, many less productive lakes are situated in deep basins in rocky, upland catchments and for all these reasons have high hypolimnial oxygen levels, whilst many productive lakes have shallow basins in the subdued relief of fertile lowland catchments and greatly deoxygen-ated hypolimnia. On the other hand, shallow unproductive lakes with variably aerobic hypolimnia may be found amid sandy and gravelly glacial outwash over vast areas of Canada, and much of the deep, relatively unproductive L. Tangan-yika is permanently deoxygenated. Wind has not been able to mix the deeper

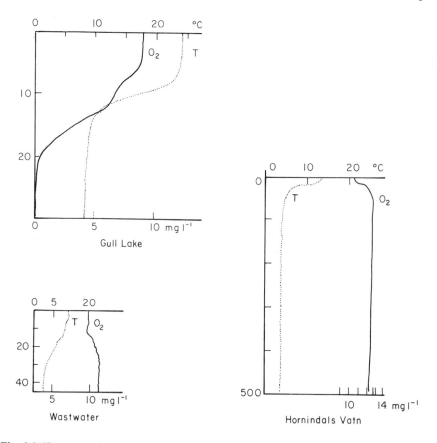

Fig. 2.9. Oxygen and temperature profiles in mid-summer from three lakes. Gull Lake is relatively fertile, but the other two are infertile lakes. (Based on Moss[362], Macan[310], and Hutchinson[234].) Vertical depth scales are in metres.

layers of this lake for many thousands of years, if ever, and despite low epilimnial production, and a once immense reserve of oxygen in the deep layers, the supply has slowly been used up. Such permanent deep layers are not strictly hypolimnia in the usual sense, but monimolimnia (s. monimolimnion); they have usually accumulated sufficient dissolved substances from sediment release and decay of sedimenting particles that their density has been permanently increased. The epilimnion/hypolimnion structure is usually superposed on the monimolimnion of such lakes. Profiles of oxygen concentrations for three lakes in summer are shown in Fig. 2.9.

In inversely stratified waters, trapped under ice, the rate of deoxygenation per unit area again reflects the productivity of the lake, but again the absolute concentration of oxygen may not. Deoxygenation may be barely less rapid than in the same lake in summer, if its hypolimnion is very cold, though often the water mass has heated to well above $4°C$ before stratification is stabilized in summer. The residual organic matter left in the lake as production decreases in autumn is the material available for bacterial respiration in the winter months.

2.5 Effects of stratification on the distribution of organisms

Epilimnia are turbulent and phytoplankters are easily maintained in suspension in them. Hypolimnia, although there may be some water movement in them, have stiller waters and most phytoplankters falling into them cannot remain in suspension and fall to the sediments. There is no return through the metalimnion, except in the case of some buoyant blue-green algae (see Chapters 3 & 10). Plankton concentrations are thus much greater in the epilimnion than in the hypolimnion. If the depth of the epilimnion, or mixed depth, z_m, is much less than the euphotic depth, z_{eu} (see Chapter 1) then a part of the potential production will not be realized, for organisms will not be able to use the light still available in the lower part of the euphotic zone.

In the metalimnion the gradient of dissolved oxygen concentrations, which occurs when the hypolimnion becomes severely deoxygenated, creates a chemically diverse layer where substances produced by decomposition or which are unstable except at low oxygen concentrations can persist. These include FeII and MnII ions which may encourage localized growth of some algae, e.g. *Trachelomonas* which require them for the production of thecae which surround the cells. Detritus temporarily accumulating in this zone may also provide substrates for heterotrophic bacteria. In turn these may produce necessary vitamins for certain phytoplankton flagellates, like *Cryptomonas* spp. which are unable to synthesize their own supplies, but grow and congregate where an external supply is available. If the upper part of the hypolimnion is within the euphotic zone, and if it becomes sufficiently deoxygenated for bacteria, such as *Desulphovibrio*, to convert SO_4^{2-} to HS^- in meeting their own respiratory requirements then photosynthetic bacteria such as *Thiopedia* may form large populations. They use H_2S as a hydrogen donor in photosynthesis, and persist in the hypolimnion because they have intracellular vesicles of gas which make them buoyant. The chemical gradients around the metalimnion thus may create a marked vertical patterning in the occurrence of different species; an example is shown in Fig. 2.10.

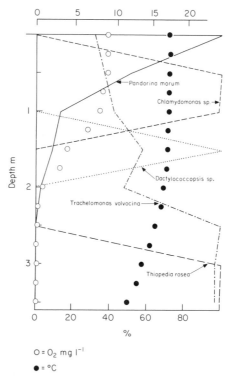

Fig. 2.10. Vertical distribution of some phytoplankters in a small pond, sheltered from wind by woodland (Abbot's pond, Somerset, U.K.). Populations are expressed as percentages of their numerical maxima in the profile (lower scale), temperature as solid circles and oxygen concentrations as open circles (upper scale). *Pandorina* & *Chlamydomonas* are green algal flagellates. *Dactylococcopsis* is a coccoid blue green alga, *Trachelomonas* a Euglenoid flagellate, and *Thiopedia* a purple sulphur (photosynthetic) bacterium. (See Chapter 3 for further details on algae.)

Fish distribution may also be influenced by stratification. Two groups of freshwater fish, the Salmonidae (salmon and trout) and the Coregonidae (whitefish) cannot tolerate such high temperatures nor such low oxygen levels as many other fish (see Chapter 7). In summer they may migrate into the deeper waters of stratified lakes as the surface waters warm and their saturation oxygen levels decrease. If the hypolimnion becomes even moderately deoxygenated this refuge is not open to them, and such fish are excluded from fertile lakes for this reason at least.

2.6 Deep and shallow lakes and the role of the sediment

Substances enter lakes in streams, in percolating water or from the atmosphere. They are lost from the system to the atmosphere as gases, through the overflows in dissolved or particulate form, or, as particles, by incorporation into the sediment. It may be surprising to read of the latter as a loss to the system, but in many senses it is. Although substances may be released from sediments, on balance more of all substances is sedimented than released. Because of this, and losses to the overflows, substances cannot accumulate in the water from year to year so long as the rates of supply from the catchment area and other sources remain steady. The only possible exception is that of endorheic lakes which have no overflow and this is discussed in Chapter 5.

The reasons why substances become trapped in sediments are largely physical and chemical. The dense sediment lies at the bottom of the lake (or large river) where turbulent mixing, induced at the surface by wind, is least. Even if soluble ions are released into the interstitial water of the sediment they have little opportunity of being mixed with the overlying water. Simple outward diffusion must occur but this is a very slow process providing insignificant internal loading as compared with external loadings. Substances may also be locked in the sediments in insoluble forms. For example, though free phosphate ions can exist in the interstitial water made anaerobic by bacterial respiration deep in the sediment, they are precipitated in the surface layers as compounds of FeIII and other metals. The surface sediment layers are usually not anaerobic because oxygen can diffuse or be mixed into them from the overlying water. Under such conditions insoluble ferric and other metal complexes form a crust which is generally coloured brown compared with the darker, often black anaerobic sediment underneath.

Release of substances of which the solubility is, in part, determined by the oxidation state of the sediment (Fe, Mn, PO_4 and others) occurs when the overlying water becomes depleted of oxygen, though not necessarily completely anaerobic. These substances therefore accumulate in deoxygenated hypolimnia, during the summer, but because the hypolimnion does not mix with the epilimnion, they do not become available for phytoplankton growth in the upper, illuminated layers of the lake. The hypolimnial nutrients do mix throughout the water column during the autumn mixing period, but seem then to have little effect on phytoplankton growth for two reasons. Firstly, they are exposed to aerobic conditions again and some of the insoluble complexes reform and are re-precipitated, and secondly, those that escape precipitation are diluted by the flushing through of increasing amounts of run-off water from the catchment.

In lakes too shallow to stratify, the surface of the sediment may also become anaerobic in summer if sufficient sedimenting organic matter is provided for rapid decomposition at high summer temperatures. In certain circumstances (discussed in detail in Chapter 5) phosphate and other nutrients may then be released and become available to the phytoplankton in the overlying short water column.

In all lakes any release of oxidation-sensitive substances from the sediments is dependent on a ready supply of easily decomposable organic matter to the sediment surface where bacteria can decompose it. This supply comes from production in the upper water layers, which in turn depends on the maintenance of pools of the necessary nutrients by external loading. The sediment surface then may act as a temporary store for nutrients ultimately derived externally. If this supply is reduced, production in the phytoplankton falls, the supply of labile organic matter to the sediment surface is diminished, the oxidized surface crust on the sediment is not destroyed, and no marked release of substances sensitive to oxidation can occur. Even if vast reserves of nutrients have accumulated in the sediments of a lake, they remain effectively inert, lost from the ecosystem once, sooner or later, they have been buried beneath the surface of the mud. These principles are extremely important in predicting whether or not a lake much changed by man-made fertilization, for example with sewage effluent, can be restored to a former state (see Chapter 10).

2.7 Depth and the distribution of communities in a lake or river

Submerged aquatic macrophytes may grow on the bed of a water body so long as sufficient light is available, and a shallow water body will, other things being equal, have a greater proportion of its basin floored with plant communities than a deep one. The submerged weed-beds have a rich invertebrate fauna (see Chapter 6) and provide cover from predators and spawning sites for some species of fish. The weed-beds also reduce water movement so that accumulations of metabolic by-products of their activity may give horizontal differences in the water chemistry of the basin. Below the euphotic depth, apparently bare sediment usually harbours a community of invertebrates, less diverse than that of the weed-beds, but nonetheless abundant. This benthic community, often with chironomid (Diptera) larvae and oligochaete worms prominent in it (see Chapter 5), derives its food supply from sedimenting organic matter, and is thus very much dependent on production in the euphotic zone. It may be present, though sparse and quiescent, even in deoxygenated sediments overlain by an anaerobic hypolimnion.

The balance between weed-bed and sedimentary benthic communities obviously depends on the distribution of depth in the basin, on the depth of the euphotic zone, and hence on the amount of phytoplankton production. In turn this is controlled by the levels of key nutrients and hence the parameters of the Vollenweider model. A further dimension is added by the proportion of water less than about 2 m deep in which emergent macrophytes, rooted in sediment but with much of their photosynthetic tissue above the water, can grow. Production in reed swamps is high (see Chapter 6) and they support large populations of invertebrates and fish. In lowland rivers, particularly in the Tropics, seasonal flooding may also extend the aquatic ecosystem over many square kilometres of swamps in the flood plain (see Chapter 9).

Depth helps determine productivity in several ways, some of which are quantifiable, as in the Vollenweider model, other of which yet can only be qualitatively expressed, as in the relative importance of macrophyte beds and phytoplankton. One simple model which probably expresses the importance of the latter for fish production was developed by Ryder (Ryder *et al.*[440]). This relates the fish yield (the harvest obtained from fisheries, which is probably related to absolute production over a large sample of lakes) to the ratio of total dissolved solids concentration in the water to mean depth (Fig. 2.11). The direct proportionality of total dissolved solids crudely expresses the ultimate importance of some substances, like phosphorus, in controlling production, whilst the inverse relationship with depth perhaps reflects the smaller proportion of macrophyte communities providing shelter and spawning sites in deep lakes.

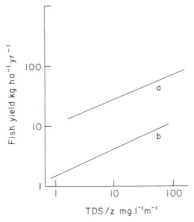

Fig. 2.11. The relationship between fish yield and the ratio of total dissolved solids (TDS) to mean depth (z) in series of tropical (African) lakes (a) and Canadian lakes (b). (After Ryder *et al.*[440].)

2.8 Summary of introductory chapters 1 and 2—the trophic classification of water bodies

Chapters 1 and 2 have drawn attention to the main features which determine the productivity of those water bodies which are large enough to have accumulated plankton populations. The productivity of small streams is discussed in Chapter 4. Subsequent chapters will elaborate on the general principles sketched here so it is now convenient to have a general overview. Potential productivity in lakes and rivers may be described in terms of the amount of carbon fixed by photosynthesis per unit area, or in terms of the total concentration [m_w] of the scarcest substance, usually phosphorus, necessary for plant and algal growth in the lake. My preference is for the latter because it is readily and accurately measurable, which photosynthetic carbon fixation is not (see Chapter 3), and because understanding of the parameters which determine it if not perfect, is well advanced, as indicated by the emphasis placed on the Vollenweider model in these chapters. Table 2.1 shows the range of the total phosphorus continuum. Refinements, even of this, which take into account vertical stratification, are now available[513] and quantitative treatment of the important macrophyte communities in shallow lakes will eventually follow.

Whatever measure is used, however, the emphasis is on a continuum of productivity from the very low to the maximum potential for a given latitude. Such an emphasis differs greatly from that of classic limnology which has attempted to classify water bodies into types. Many attempts have been regionally based and the systems developed for temperate Europe, for example, break

Table 2.1

The range of total phosphorus concentrations found in fresh waters	
1 μg Pl^{-1}	Undisturbed lakes in rocky, remote or upland catchments, e.g. Canadian Shield Lakes, Scottish lochs, L. Superior.
10 μg Pl^{-1}	Relatively undisturbed lakes in fertile lowland catchments, e.g. many mid-western American lakes, some prairie lakes and lakes in New England.
20 μg Pl^{-1}	Similar lakes with cultivated or deforested catchments and often some loading from sewage effluent, e.g. other lakes in the American Mid-West, including L. Erie, Cheshire & Shropshire meres, Norfolk Broads in the 19th century.
100 μg Pl^{-1}	Lakes in developed areas receiving substantial quantities of sewage or stock effluent, which dominates any land-derived loading, e.g. many Norfolk Broads. Possibly some shallow equatorial lakes, like L. George, Uganda, where production and flushing are not seasonally different.
1000 μg Pl^{-1}	Sewage oxidation ponds, heavily fertilized fish ponds in Asia, some endorheic lakes, e.g. Natron, Chilwa.
$\sim$ 10000 μg Pl^{-1} or more	

down in arid Australia or Equatorial Africa. There persists, even so, a touching belief in two lake types, the oligotrophic and the eutrophic, named from Greek words meaning 'providing little nourishment' (oligothophos) and 'well nourished' (euthophos), infertile and fertile. The classic oligotrophic lake is taken to be a deep, rocky basin, low in dissolved substances, with low production, a hypolimnion saturated with oxygen, few or no aquatic macrophytes and salmonid fish. In contrast the typical eutrophic lake is said to be shallow and silted, with high concentrations of dissolved solids, high production, a deoxygenated hypolimnion, extensive weed beds and non-Salmonid fish.

This concept originated in early studies of deep mountain lakes of Norway, Scotland and the English Lake District, and of shallow weedy lakes in the glacial outwash plains around the shores of the Baltic Sea. The problem now is that the exceptions outnumber the lakes that conform to type, as must be expected in view of the several factors, not necessarily mutually correlated, that determine fertility, and the difficulty of putting universally agreed limits on what constitutes the minimum productivity of a lake that can be described as eutrophic.

The recognition of a continuum subconsciously came early with the use of the term 'mesotrophic' to describe the indeterminate middle range.

The words oligotrophic and eutrophic can no longer be unambiguous and are used in a confusing variety of ways. Their best use now is to mean merely infertile and fertile, without any other connotations, and it seems that the latter terms with the recognition of a continuum between their extremes are likely to be less misleading. 'Oligotrophic' and 'eutrophic' will hence not be used in this book.

The nutrient loadings of many lakes have been significantly increased by human activities in recent years (see Chapter 10) and this process has been labelled 'eutrophication'. This term too has its problems—it may mean an increase in total phosphorus from 2 μg Pl^{-1} to 8 μg Pl^{-1} to a Canadian limnologist studying lakes on the hard rock Laurentian shield, or an increase from 20 μg Pl^{-1} to 250 μg Pl^{-1} to someone concerned with lowland lakes in Eastern Britain. It is a term confused also with the inevitable process of filling in of a lake with sediment, since, in the original usage (see above) typical eutrophic lakes were silted. A better term would have been 'fertilization' but whether I like it or not 'eutrophication' is etymologically here to stay!

2.9 Further reading

Further details on the physics of water and on stratification in temperate lakes in particular are comprehensively given in Hutchinson[234] and works by Talling[489, 490, 491] add similar information for tropical waters. The role of sediments in nutrient relationships was first described by Mortimer[358], and a recent symposium edited by Golterman[177] confirms, but extends the classic picture.

Though early views on lake classification are now dated, the history of their development is the story of how thinking in limnology has developed. Hutchinson[237] reviews the use of the terms oligotrophic and eutrophic. The foundations of limnology were laid variously in America, Europe and Britain by men of great creativity and imagination. Works by one of these, Pearsall[396] are readily accessible and exemplary for the degree of understanding which came from samples obtained at no small effort, by bicycle and rowboat. Not all his interpretations would now be accepted, but without the foundations laid by him and others, we should not have even our present understanding. Macan[310] gives an interesting account of the circumstances of this early work, as does Mortimer[359] for that of the equally important American pioneer E. A. Birge.

CHAPTER 3
THE PLANKTON

3.1 The structure of the plankton community

Lake water sparkling in afternoon sunlight hides a miniscule waterscape which is closer to a slum than a paradise. It contains millions of organisms in every cubic centimetre, some of which are photosynthetic, others of which feed on live and dead, dissolved and particulate organic materials. The water contains their excretions and secretions, faeces and corpses, intermixed with debris washed into suspension from the surrounding land. In this melange chemical and biological changes, both cyclic and irreversible, are taking place very rapidly.

Scaling one part of the plankton to human size and considering the rest relative to this will help indicate the structure of this community, the plankton. The rotifer *Keratella quadrata* (Fig. 3.1), a common small animal of ponds, has a body about 125 μm long with spines half as long held out behind it, and is a convenient organism with which to scale up the rest of the plankton. In reality it is about half the size of a full stop on this page. If the body of *Keratella* is the size of a tall man then the rest of the plankton ranges in size from children's marbles to large houses, or, if the fish which move through it be considered, to the size of 'whales' thirty miles long!

3.1.1 Planktonic bacteria and viruses
About the size of marbles in our model, though with varying shapes, are plank-tonic bacteria (Fig. 3.1). They are suspended freely in the water as single cells or small colonies and commonly are studded onto a nucleus of dead organic detritus or other organisms. It is difficult to know at what population densities they occur because methods of study are in their infancy. The most promising current methods of estimating their numbers or biomass use counting, after the staining of centrifugates or filtrates of water with 'vital' stains, or the deter-mination of specific chemical components such as the muramic acid of the cell wall[354].

Many different kinds of bacteria are present, even when shape is the only criterion of diversity, and overtly similar rods or cocci may divide a multitude of different functions among themselves. Some may metabolize mucose sugar polymers or one or more of the many carbohydrate polymers found in algal cell walls; others may break down the chitins of Crustacean exoskeletons and yet

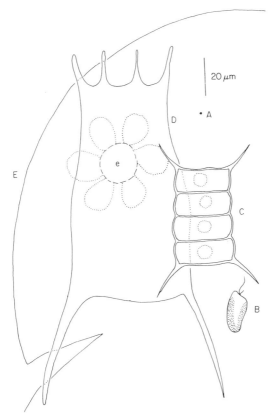

Fig. 3.1. Relative sizes of some major components of the plankton. (A) a bacterium; (B) *Cryptomonas*, a relatively small phytoplankter; (C) *Scenedesmus*, a moderately large phyto-plankter; (D) *Keratella*, a small zooplankter; (E) outline of the head with eye (e), of *Daphnia*, a large zooplankter. The head constitutes about a quarter to a fifth of the total body size.

others the organic phosphate compounds secreted by other components of the plankton.

Viruses have never been intensively investigated in the plankton. Undoubt-edly they are present and infect specifically many other organisms. Known cyanophages[479] lyse some blue-green algae, but the natural occurrence of viruses of other plankton groups remains yet to be recorded.

3.1.2 Phytoplankton
The larger bacteria overlap the size range of the smallest phytoplankters, the photosynthetic component of the plankton. Indeed some bacteria themselves (in addition to the blue-green algae whose cells have a prokaryotic structure like those of bacteria) are photosynthetic but are abundant only in conditions where dissolved oxygen levels are very low. These include the hypolimnia of fertile lakes (see Chapter 5) and some small ponds where much organic matter, e.g. autumn leaves, is washed in. Unlike the blue-green and other algae, such

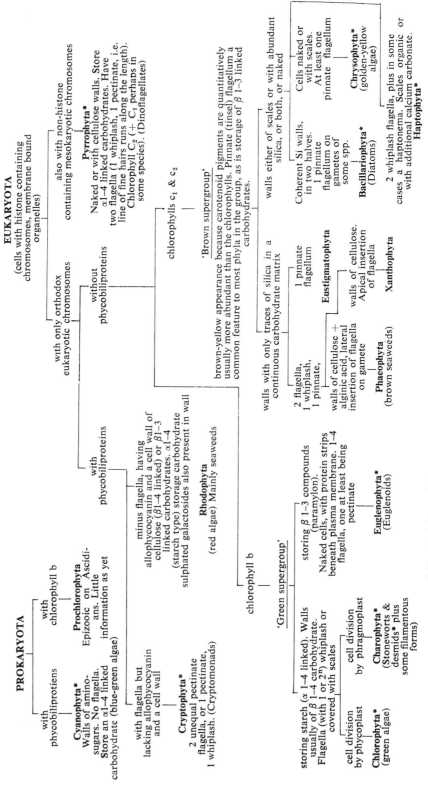

Table 3.1. Differentiation among the phyla of algae. Phyla which are important in the phytoplankton are asterisked.

bacteria use not water but compounds like hydrogen sulphide and thiosulphate as hydrogen donors in photosynthesis. They may be very important for the operation of nutrient cycles within the whole lake, particularly those of sulphur and carbon (see Chapter 5).

Taxonomically, and in size, the phytoplankton is a very diverse group (Fig. 3.2, Table 3.1). In the scaled-up model their sizes range from those of marbles and golf balls (1–5 μm) to those of footballs and water melons (up to 50 μm). When they occur as colonies they may be visible to the naked eye (several hundred μm) and would be scaled as heavy horses or elephants!

A common misconception is that phytoplankters float, with similar densities to that of water. This is generally not so. The blue-green algae, which are some-times very abundant in highly fertile lakes but occur in all other water bodies, may have an organelle called a gas vesicle. It comprises a mass of protein-bound prisms with conical ends, contains air and gives positive buoyancy. The total volume of vesicles per cell is influenced by light intensity and cell division rate, such that under some circumstances blue-green algae may truly float at particu-lar depths in the water column which favour their growth and in other circum-stances may form a paint-like scum or water-bloom at the lake surface.

Excepting *Botryococcus braunii*, an alga which may remain positively buoy-ant by storing large quantities of oil, all other phytoplankters are more dense than water. In the case of diatoms, which have cell walls of silica, they may be considerably more so, having a specific gravity of as much as 0·06 in excess of that of water. These phytoplankters are kept suspended in the water largely by wind-generated eddy currents. Some species have flagella and movement of these may help counteract the tendency to sink whilst the non-flagellated species have evolved cell or colony shapes which decrease the rate of sinking, or delay alignment in the water in a position which would aid sinking. Flat plates, needle shapes with curved ends, and the possession of spines and projections all seem to be advantageous. However, too easy an acceptance of shape as adaptive

---

'The Algae' is not a taxonomically distinct group, but is a collection, like 'the Invertebrates' of not necessarily closely related phyla. The only common characteristics of these phyla are possession of chlorophyll *a*, an ability to use water as a hydrogen donor in photosynthesis, gametes, if produced, not sheathed by non-reproductive cells, and an aquatic habitat for growth. Two groups, the Cyanophyta (blue-green algae) and Prochlorophyta, are prokaryotic, having a typical bacteria-like cell structure. Most members of two phyla, the Rhodophyta (red algae) and Phaeophyta (brown algae) may be typically plants, requiring no organic substances for growth that they cannot themselves synthesize from inorganic substances. The boundary between these and the rest, all grouped in the kingdom Protista (see Whittaker[535]) is indis-tinct. The Protista, by definition, have a mixed feeding mode, variously combining photo-synthesis with an absolute requirement for at least one organic substance (generally a vitamin), facultative heterotrophy of dissolved organic substances, or, in a few cases, phagotrophy of organic particles. From the Protista the green higher plants have developed probably from the Charophyta by loss of obligate heterotrophy, the animals have evolved by loss of the photo-synthetic apparatus from a group unspecified, and the Fungi were perhaps derived from Protista which lost both photosynthetic apparatus and any phagotrophic ability. Further details, from which Table 3.1 is constructed may be obtained in Antia[11], Bold & Wynne[38], Dodge[104], Fogg et al.[138], Fritsch[143], Leedale[287], Lewin[291], Morris[357], Pickett-Heaps[407], Stewart[478], Stewart & Mattox[477], and Werner[521].

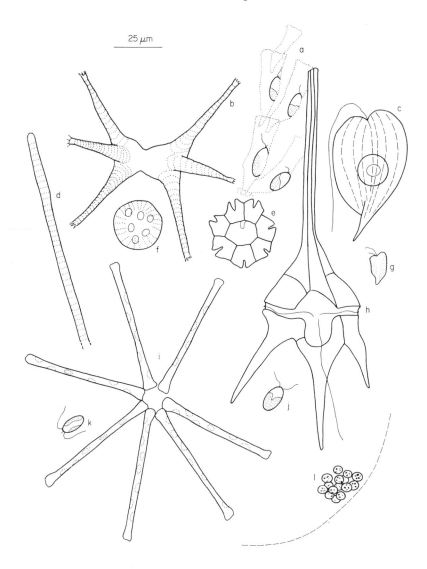

25 μm

Fig. 3.2. Some typical phytoplankton algae, drawn to the same scale.
Cyanophyta (blue-green algae)—(d) *Oscillatoria*, (l) *Microcystis;*
Chrysophyta (yellow-green or golden algae)—(a) *Dinobryon*;
Chlorophyta (green algae)—(e) *Pediastrum*, (b) *Staurastrum* (a member of a group called the desmids strictly now placed in the Charophyta), (j) *Chlamydomonas;*
Bacillariophyta (diatoms)—(f) *Cyclotella*, (i) *Asterionella;*
Euglenophyta (euglenoids)—(c) *Phacus;*
Cryptophyta (cryptomonads)—(g) *Rhodomonas;*
Pyrrophyta (dinoflagellates)—(h) *Ceratium;*
Haptophyta—(k) *Prymnesium.*
Microcystis (l) is a very large alga, of which a diagram of the entire colony could occupy as much as this page. Only a few cells are shown.

should be avoided. Envelopes of mucilage, invisible unless the cells are mounted in Indian ink, may be thick enough to give a spiny cell an effectively spherical shape.

Why have phytoplankton cells not all evolved positive buoyancy when sinking, with its potential for loss from the euphotic zone, so clearly has disadvantages? The answer is that it also has advantages, which must, on balance, outweigh the disadvantages. Phytoplankters need a supply of inorganic nutrients from the water. These they absorb from the water layer, a few micrometres thick, immediately in contact with the cell wall or membrane. Molecular forces tend to preserve this layer intact and it soon becomes depleted of nutrients which are not rapidly replaced by diffusion alone. Continuous movement of the cell through the water, as it sinks and is retrieved by upwardly directed eddy currents, sloughs away the depleted nutrient shell and maintains a continual supply of undepleted water at the cell surface[376, 236].

The phytoplankton was long assumed to comprise organisms which satisfied their energy needs through photosynthesis and absorbed only inorganic nutrients. This is far from the truth. The planktonic algae are not a homogeneous group but belong to at least ten presently recognized phyla (Table 3.1), many members of which require at least one preformed organic compound for their growth. 70% of algae tested until now require one or more of the three vitamins cyanocobalamin (B12), thiamine and biotin. B12 is most commonly needed, perhaps sometimes as a readily soluble source of cobalt which it contains, but often as a complete organic moiety, the co-factor of a necessary enzyme. The test of independence of organic compounds is the ability of the alga to grow in a pure culture through a series of many sub-cultures in entirely inorganic media in the laboratory. This has been shown for a relatively few blue-green algae, diatoms and desmids. Many algae are able to take up simple organic compounds in the dark or light, but the concentrations of such compounds are normally low in natural waters and bacteria successfully compete for them. It is probable that dependence on organic compounds is less marked in the open water phytoplankton than in benthic algal communities.

The smaller phytoplankters (up to 10 μm) may occur in very large numbers: 10^6 per ml is a characteristic upper figure (though subject to wide variation) compared with about 10^4 for the larger phytoplankters. On the scaled up model the smaller species would appear as a population of objects the size of tennis balls spaced at distances of about eight feet in three dimensions. A population of the larger algae can be imagined as a similar constellation of water melons thirty yards apart from each other. Even considering that several species may simultaneously be forming large populations in the water, there is clearly a lot of space between the individuals. This might explain why, despite their ubiquity, parasitic fungi (frequently chytrids) and parasitic protozoa only cause epidemics when algal population densities are very large. Successful infestation of a host cell requires an encounter between the parasite (sized about 5 μm—golf ball size) and the host in an environment where hosts are well spaced out and both host and parasite are continually moved by eddy currents.

The phytoplankters have sometimes been divided into ultraplankton ($< 5 \mu$m or thereabouts), nannoplankton (5–60μm) and net plankton ($> 60 \mu$m). The boundaries between these groups vary between authors perhaps largely as a function of available net mesh sizes. Phytoplankters come in a continuum of sizes, however, and the only real use of this size categorization is to point out the fact that much early work, which used samples taken by nets towed through the water, undoubtedly failed to sample most of the phytoplankton. Nets are nowadays only used ill-advisedly in quantitative phytoplankton sampling.

The very wide size range of phytoplankters (from tennis balls to elephants in the model) is remarkable, particularly since in a nutrient-scarce medium, small bodies with high surface to volume ratios should be able to compete more effectively for nutrients. Large cells also sink faster and hence are more vulnerable to loss from the epilimnion in stratified lakes. There *are* large phytoplankters, however, and they have persisted in the plankton for a very long time. There must be an advantage to large size and this seems to be that big cells are less readily eaten by filter feeding zooplankters. They are mechanically unable to manipulate large cells or colonies into their mouths.

Phytoplankters are also remarkable for their almost complete abandonment of sexual reproduction. They live vulnerable lives, most of them are readily grazed, all may be washed out of their habitat by incoming floods, many may be lost to the sediments by sinking. Such conditions have favoured selection of the most rapid means of reproduction of many individuals to replace those lost from the population. Hence simple cell division with generation times of only hours or a few days is how the phytoplankton reproduces. This means that their potentialities for genetic change are limited, but the environment in which they live is a relatively predictable one, because of the buffering properties of water to climatic changes, for example. Preservation of a fixed, but highly adapted genotype to a relatively specialized habitat—suspension in a very dilute fluid—has therefore been a sound stratagem. And in any case, given the spacing of the cells, and their vulnerability to movement by random eddy currents, the chances of contact for fertilization by cells or their gametes must be quite small.

3.1.3 Zooplankton

The Protozoa, rotifers and crustaceans are the major groups of freshwater zooplankton (Fig. 3.3). Planktonic protozoons (amoeboid forms, Heliozoa, ciliates, colourless flagellates) may be far more important than the limited studies yet carried out upon them would suggest but other animal groups are generally rather infrequent. Freshwater jellyfish, carnivorous on other zooplankters, some flatworms, gastrotrichs and mites do occur, but attention has justifiably been directed towards the rotifers and Crustacea. The rotifers are man-sized to horse-sized on our scaled-up model and are mostly suspension feeders. Their name comes from the rhythmically beating, apparently rotating 'wheels' of cilia close to the mouth. These direct water with its suspended fine particles into the gut. Some rotifers may have more complicated food gathering mechanisms. One grasps the flagellum of phytoplankton cells like *Cryptomonas*,

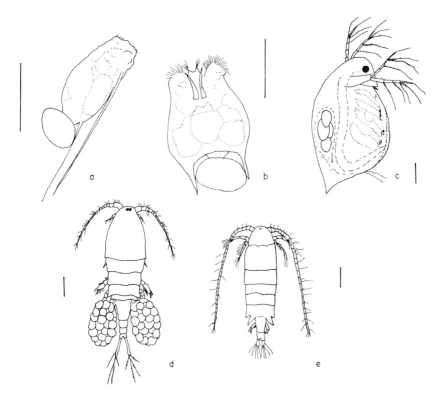

Fig. 3.3. Some representative zooplankters. In all cases the length of the scale lines represents 100 μm.
Rotifera—(a) *Filina*, (b) *Brachionus*. The corona of cilia can be seen in each case, and at the rear, a single egg.
Crustacea (Cladocera)—(c) *Daphnia*. The filtering limbs are enclosed by the carapace which also contains the egg pouch containing a few eggs.
Crustacea (Copepoda)—(d) *Cyclops*. The egg sacs are paired and the antennae, the lower pair of appendages on the head, are not branched, (e) *Diaptomus*, a Calanoid copepod, in which the antennae are branched. When the animal is carrying eggs these are contained in a single egg sac, in contrast to the paired sacs of the Cyclopoid copepods.

and tears the cell open to release its ingestible contents. Rotifers feed on particles from about 1–20 μm in size (Table 3.2), a range shared by the filter feeding Cladocera (Crustacea) or water fleas, which can also take food a little larger, to 50 μm in size.

The Cladocera, which include the well known genus *Daphnia* include some carnivores on smaller zooplankton, *Leptodora* and *Polyphemus* which are raptorial, meaning that they actively grasp their prey. The filter feeders in this group beat the water with thoracic limbs provided with closely spaced filtering setae and eventually convey a bolus of food to their mouths. They move through the water more actively than the smaller rotifers, using a rowing action of the large, branched second antennae. On the scaled model, cladocerans, up to a few mm in length in reality, would be as tall as church steeples in some cases.

The third important group of zooplankters, also Crustacean, is the Copepoda, whose members are usually a little larger than Cladocera, may be filter feeding (mostly the Calanoid copepods, like *Diaptomus*) or raptorial, the Cyclopoid copepods, which include *Cyclops*. The prey of these may be smaller zooplankton or larger colonies, or masses of phytoplankton. Overall the copepods can tackle a wider range of bigger food particles (5–100 μm) than the Cladocera and rotifers. The size of particles filtered or grasped by zooplankters depends on the size of the organism—this has elegantly been shown by feeding plastic beads of known size to them—but there is clearly some separation among the groups in terms of the range eaten, and also in terms of the feeding rate. This can be measured as the volume of water swept clear of particles per animal per

Table 3.2. Comparative features of the three main zooplankton groups. (After Allan[4].)

	Rotifera	Cladocera	Copepoda
Generation time (days)	1.25–7	5.5–24	7–32
Adult body length (mm)	0–2–0.6	0.3–3.0	0.5–5.0
Food size (μm)	1–20	1–50	5–100
Feeding method	Suspension feeding	Filter feeding (raptorial in carnivores)	Filter or raptorial feeding
Filtering rate	Very low	High	Low
Susceptibility to predators:			
Invertebrate	High	Moderate	Moderate (adults) high (juveniles)
Vertebrate	Very low	High	Low

day. Again it depends on a number of factors but is generally low for rotifers, ten to a hundred times higher in Copepods and often even higher than that in Cladocerans.

Life histories differ among the zooplankton also. Although the cytological mechanisms differ, both rotifers and Cladocera are parthenogenetic. Females produce broods of eggs asexually which hatch into more females. This allows rapid replacement of populations vulnerable especially to predation (see later). The eggs are born in sacs by rotifers and in pouches deep in the carapace which characterizes the Cladoceran exoskeleton and young are released which resemble the adults and soon grow large enough to reproduce. The rate of egg production is high, for a new generation of rotifers is produced in only a few days and each female produces up to 25 young in her lifetime of 1–3 weeks. Cladocerans take longer per generation (1–4 weeks), but each have a longer life expectancy (up to 12 weeks or so) and produce up to 700 young per lifetime.

The genetic advantages of sexual reproduction are not lost in these parthenogenetic animals because most produce males during times of food shortage or other inclement conditions. Eggs are then fertilized and do not hatch out for some time, when they form a new season's parthenogenetic females.

The Copepods are quite different. Each generation is sexual, and before the new mature adults are formed, there are eleven successive moults in the life history. The first six after the egg hatches are of juveniles called nauplii, which look quite different from the adults. The next five, the copepodites do look like the sixth copepodite stage which is the reproductive adult. In terms of their longevity and fecundity the Copepods are comparable with the Cladocerans.

The zooplankters thus include a diversity of form and activity in their communities and much of this is ultimately related to the major effects that predation, among themselves, or by vertebrates, has upon their numbers (Table 3.2). They are also much more heterogeneously distributed than the phytoplankters. In a mixed water body the latter are usually randomly distributed and the small rotifers may be also. The Crustacean zooplankton move actively, may shoal both vertically and horizontally and often go through diurnal vertical migrations, reaching the water surface by night and moving down by day. They are tricky to sample quantitatively. Nets are used, for the concentration of zooplankters (up to 1-2 per ml for the larger ones) is not great enough for a sample dipped from the water to contain sufficient for counting.

3.1.4 Detritus

Organic detritus, always present in the plankton, comes in the entire range of sizes occupied by living organisms from the smallest bacteria to the largest zooplankters. Colloidal organic matter is probably also present. Detritus originates both internally (autochthonously) from death of organisms and faeces production by zooplankton and fish, and externally (allochthonously) from litter decomposition in the catchment area. The relative proportions of each kind vary with such factors as the ratio of catchment area to lake area, and the fertility of the lake.

Inorganic particles are also present in varying quantity. Large rivers may carry a great amount and clear lakes very little. Apart from its effect in reducing light penetration, suspended inorganic matter seems to influence the plankton very little. A major exception is when it is formed autochthonously as colloidal particles of calcium and magnesium carbonates. This happens in unusually calcareous lakes following depletion of carbon dioxide through photosynthesis and consequent increase in pH. Trace elements, phosphorus and other nutrients may then be co-precipitated and made unavailable. There is some evidence, also, that inorganic particles may act as attachment sites for bacterial growth, thus enhancing the zooplankton food supply.

3.1.5 Fish

Freshwater fish are not strictly planktonic, even when newly hatched, for they move among other communities as well as in the open water. Their biology is considered in Chapter 7, but their influence on the plankton is considerable and is considered later in this chapter.

3.2 Functioning of the plankton community

The plankton community is a very dynamic one. Not only are the relative positions of all of its particles, live or dead, changing from second to second, but also dozens of chemical changes are going on simultaneously. We can divide the plankton into a series of 'compartments' for the purposes of examining its activities. These are phytoplankton, microorganisms (bacteria and fungi), detritus, zooplankton, dissolved substances and fish. A particular phosphate ion or carbon atom may, in summer, find itself shuttled through several such compartments in a few minutes. Fig. 3.4 shows some of the main pathways between compartments.

Many of the processes illustrated in Fig. 3.4 are very imperfectly understood. They happen very rapidly and steady states, amenable to analysis by the relatively imperfect methods available, are short-lived. Function in the plankton can first be considered on a short-term (hourly, daily) basis, with the better understood processes considered in turn, and then on the more long-term, seasonal basis.

3.2.1 Photosynthetic production of phytoplankton

Photosynthesis in phytoplankton is measured as gross photosynthesis, the total intake of energy or carbon, and as an approximation to net photosynthesis, the effective production of new cell material after respiratory requirements have been met. Photosynthesis and also the uptake of dissolved nutrients by phytoplankton are among the more readily measurable processes in the plankton.

Photosynthesis is generally measured by enclosing a sample of the population in a clear glass bottle (light bottle) and measuring either the release of oxygen or the uptake of carbon dioxide. Measurements are made relative to that of a replicate sample shielded from light, in a dark bottle, by opaque black paint or tape. Lake water is normally withdrawn from specific depths by specially designed samplers and, after filling, the light and dark bottles are resuspended normally at depths from which the water was originally taken.

Oxygen release and carbon dioxide uptake, respectively, measure different aspects of photosynthesis. Theoretically, oxygen evolution gives gross photosynthesis since it includes a correction for respiration. There are no methods which unequivocally measure net production though determination of carbon dioxide uptake using radioactive ^{14}C is believed by some investigators to give a good approximation. Simple assessment of the change in phytoplankton biomass over a period is obviated by the fact that the rates of loss of biomass (by sedimentation or grazing) may be as high as the rate of production.

The oxygen method

This was developed by two Norwegians, Gaarder & Gran in 1927[151] for use in coastal seas. The light and dark bottles are completely filled with water (no air space) and are usually incubated for several hours. After incubation the dissolved oxygen levels are usually measured by chemical titration.

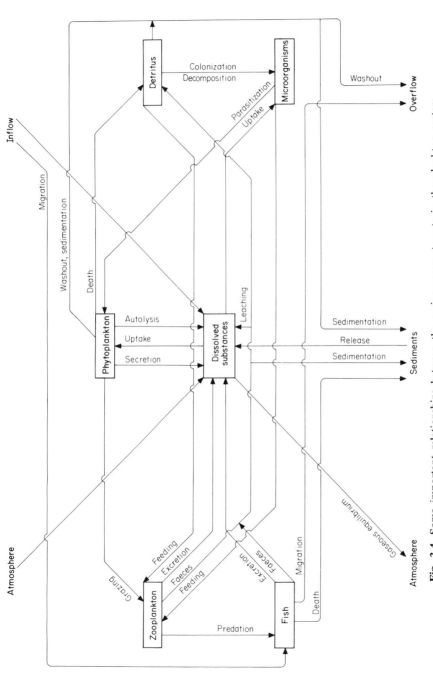

Fig. 3.4. Some important relationships between the major compartments in the plankton system.

In the light bottle, oxygen has been produced in photosynthesis. Not all of the total oxygen production appears in solution, however, since some has simultaneously been used in respiration by the phytoplankton and by the microorganisms and zooplankton also present. The change in oxygen concentration is

$$\Delta O_2{}^{LB} = O_2{}^{LB} - O_2{}^{I},$$

where $O_2{}^{LB}$ and $O_2{}^{I}$ represent the concentrations of oxygen in the bottle at the end and at the start of incubation. In the dark bottle no oxygen is produced but oxygen is absorbed by respiration of phytoplankton, zooplankton and microorganisms. The decrease in oxygen concentration in the dark bottle is

$$\Delta O_2{}^{DB} = O_2{}^{I} - O_2{}^{DB},$$

where $O_2{}^{DB}$ is the concentration of oxygen in the dark bottle at the end of the incubation. The respiration of the plankton community is supposed to have been similar in both dark and light bottles. Total oxygen production, equivalent to gross photosynthesis, is then the amount by which the oxygen concentration in the light bottle has increased plus that amount of oxygen which has not appeared in solution because it has been used in respiration. That is, the amount by which the oxygen concentration has decreased in the dark bottle.

$$P_{gross} = O_2{}^{LB} - O_2{}^{I} + (O_2{}^{I} - O_2{}^{DB}) \dots\dots\dots\dots\dots (a)$$

Thus the difference in oxygen concentrations between the bottles at the end of the incubation gives the gross photosynthesis during the incubation. Rate of oxygen release can be converted to rate of carbon uptake by calculation from the summary equation of photosynthesis ($6CO_2 + 6H_2O \rightarrow 6O_2 + (COH_2)_6$), in which production of 1 mg O_2 is equivalent to uptake of 0.37 mg C. It could also be related to energy transfer from knowledge either of the amount of light energy needed to release a known amount of oxygen or from the calorific value of the photosynthetic products.

Short incubation periods are necessary since the longer the plankton is unnaturally enclosed in a bottle the more likely it is to behave aberrantly. For example, the changes in oxygen concentrations themselves, which would be partly buffered by exchange with the atmosphere in the open lake, may lead to changes in both photosynthetic and respiratory rates. Also the quality of light in the light bottle differs from the natural light climate by the absorptive properties of the glass. Progressively developing differential effects in the bottles include greater bacterial growth on the inner surfaces of the dark bottle. Antibiotic substances may be released by photosynthesizing phytoplankton.

Necessity for short incubations means that the oxygen method is unsuitable for infertile waters where very long incubations would be necessary for measurable oxygen release. In fertile waters, however, it is an excellent method giving a measurement which can be pinpointed to a definite estimate of gross photosynthesis. The decrease in oxygen concentration in the dark bottle does not, of course, give a measure of phytoplankton respiration, except perhaps in such dense blooms of blue-green algae as occur in some highly fertilized or tropical lakes[53] since respiration by microorganisms may be very significant. It is therefore generally not possible to measure net photosynthesis by the oxygen method.

The ^{14}C uptake method

Carbon dioxide enters into complex equilibria with bicarbonate, carbonic acid, carbonate and hydrogen ions in water (see page 71). Levels of CO_2 in the water are thus buffered by these equilibria so that addition and depletion of CO_2 through plankton activity are not fully detectable. If the carbon atoms are radioactively labelled their movements can be detected and a convenient isotope (^{14}C) is readily available and is often used in the form of sodium bicarbonate[476].

Since CO_2 may be taken up by various chemosynthetic bacteria independently of light intensity, both light and dark bottles are prepared containing replicate samples. Into each bottle is injected a small volume of radioactive bicarbonate solution. The bicarbonate rapidly equilibrates with the unlabelled pool of inorganic carbon compounds. The phytoplankton absorb both radioactive and unlabelled CO_2 and sometimes labelled and unlabelled HCO_3^- as well during an incubation of a few hours.

After incubation, water from the bottles is filtered through cellulose ester filters (pore size $<$ I μm). Suction pumps must be used to do this and it is important that the pressure differential between the sides of the filter is kept small or rupture of delicate cells will occur and lead to loss of incorporated radioactive carbon. Withdrawal of CO_2 from bicarbonate at the cell surfaces may have caused carbonate precipitation on the cell walls. The filter is exposed to fumes of concentrated HCl to remove this. Radioactivity present in the cells on the filter may then be measured by placing the filter on a suitable tray against the detector window of a geiger counter, by digesting the material into liquid form for scintillation counting, or by burning it to gas for gas-flow counting. The latter procedure leads to least error, the former to the most but is also cheapest and quickest.

Uptake of carbon is calculated per unit volume of water and time as follows:

$$\frac{\text{Total uptake of inorganic carbon}}{\text{Total inorganic carbon available}} = \frac{\text{Amount of } ^{14}C \text{ taken up}}{\text{Total } ^{14}C \text{ available}}$$

The total inorganic carbon available is calculated from simple chemical determinations on the water (alkalinity and pH) and includes the inorganic carbon added as and with the isotope. Radioactive ^{14}C available is that which was added, and can be assayed by standard techniques. From the uptake of ^{14}C in the light bottle is subtracted that in the dark bottle to correct for chemosynthetic uptake by bacteria and the difference is used in the above equation.

Were the phytoplankton not respiring, the total uptake of inorganic carbon would give a measure of gross photosynthesis (after small corrections for the differential rates of uptake of ^{14}C and the lighter ^{12}C had been made). However phytoplankton respire simultaneously, producing CO_2 as they do so. Were all the CO_2 produced in respiration to be released from the cells directly, the ^{14}C method would measure net photosynthesis. However some of the respiratory CO_2 is internally recycled for use in photosynthesis. The method therefore measures a quantity somewhere between net and gross photosynthesis, and the problem is that this point is not fixed. The only real advantage of the method is its sensitivity—it can be used where phytoplankton populations are very sparse —but it has been widely used even where it need not be, possibly because of the lure of a method which uses sophisticated apparatus!

3.2.2 Factors affecting phytoplankton photosynthesis

Photosynthesis in mixed water columns falls into patterns among which those shown in Fig. 3.5 commonly occur. Maximum photosynthesis may be at the surface, but particularly on bright days it may be some distance below. This is not always due to smaller populations of phytoplankton near the surface since surface inhibition has been noted when uniform suspensions of phytoplankton or of algal cultures have been suspended in the water column. Ultraviolet light, although it is rapidly absorbed by water and dissolved organic compounds, is present in perhaps sufficient quantity towards the surface to cause inhibition.

Integration of the photosynthesis–depth curve (Fig. 3.5) gives a value for primary production in the water column. Its size depends on several factors including the method of measurement chosen. Gross photosynthesis and the ultimate production of new cell material are separate processes, linked together, but by varying factors. High light intensities may stimulate rapid formation of the products of photosynthesis proper—ATP and $NADPH_2$—with concomitant high oxygen production, but there may be a shortage of essential cell components (e.g. PO_4, NO_3) such that new biomass cannot be formed. Cells may then respire rapidly but cell or population growth may be negligible. The ^{14}C method in such cases may indicate much lower levels of photosynthesis than the oxygen method. If light intensities are high and there are no chemical limiting factors, both methods should indicate high rates of photosynthesis.

Two sorts of factors limiting to phytoplankton activity can be distinguished. Rate-limiting factors affect rates of gross photosynthesis and yield-limiting factors set upper limits to the potential production of biomass. Hence they mainly influence net production.

Light is the most common rate-limiting factor. In depth profiles of photosynthesis (Fig. 3.5) it is usual to find an exponential decrease with depth below the level of maximum photosynthesis. This parallels the exponential absorption of light in the water column (Chapter 1). Light may become yield- as well as rate-limiting very near the surface in highly fertile lakes where so much algal crop has been formed that the cells shade one another. In lakes and lowland rivers turbid with allochthonous matter, light may also become yield-limiting.

The most common yield-limiting factors are chemical ones. A wide range of substances is required for phytoplankton growth but most are present in more than adequate quantities. Two main ways of detecting potential limiting factors have been used. The first measures the rate of uptake of ^{14}C in samples of plankton to which various potentially limiting substances have been added, together with some $NaH^{14}CO_3$. Increased ^{14}C uptake over controls, measured over a few hours, is believed to indicate that the substance stimulating the increase was limiting. Many substances have been shown to increase the ^{14}C uptake rate, some of them present in the lake in question in large, though not toxic, amounts. Interpretation of these assays is therefore difficult[164]. Longer experiments in which the increase, if any, of biomass of the phytoplankton is

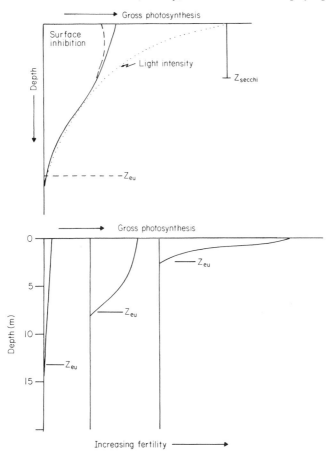

Fig. 3.5. (Upper graph)—typical relation of photosynthesis to depth in a lake. Surface inhibition (hatched line) perhaps due to ultraviolet light, which is readily absorbed by the water, may be found on sunny days, but usually the maximum photosynthesis is at the surface (solid line). The exponential absorption of light (see Chapter 1) is shown by the light intensity curve (dotted line).Zsecchi. is the depth at which a white disc disappears from the view of an observer at the surface, and Z_{eu} is that of the euphotic zone. Z_{eu} is the point at which gross photo-synthesis of the phytoplankton is equal to its respiration (also known as the compensation point). Below this point net photosynthesis is zero. (Lower graph)—characteristic photo-synthesis/depth curves for lakes of increasing fertility. Increasing fertility produces higher phytoplankton biomass which itself progressively makes the euphotic zone shallower as it absorbs the available light.

measured seem more reliable. The range of limiting substances which has been detected by these is much smaller. Phosphate, nitrate or ammonium, and iron and manganese, in that order, have been most frequently identified in temperate lakes, and these results have been confirmed by experiments in which whole lakes or partitioned parts of them have been fertilized (see Chapter 1). In natural, tropical African lakes nitrogen compounds seem to be limiting more frequently than phosphates[361]. This may reflect the abundance of phosphorus-rich, volcanic rocks in many African catchments, or possible high rates of denitrification in catchment area soils at tropical temperatures.

3.2.3 Heterotrophy in the phytoplankton

Photosynthesis is not the only source of energy and carbon for the phytoplankton. Many species are at least facultatively heterotrophic for small organic molecules such as glucose and acetate, but in most waters such compounds are present in very low concentrations (a few $\mu g \, l^{-1}$). Present evidence indicates that the phytoplankters are unable to compete successfully with the bacteria in most circumstances for these compounds, though Saunders[447] has demonstrated active uptake (as opposed to passive diffusion) of glucose by a blue-green alga, *Oscillatoria agardhii* var. *isothrix* Skuja. as it begins its population growth in water close to the surface of sediments. It is probable that in such highly organic waters as those of sewage oxidation ponds heterotrophy may be a major carbon source for phytoplankton. Here algae of strong heterotrophic abilities (some Chlorophyta and Euglenophyta) dominate the phytoplankton, and dense algal populations restrict their own photosynthesis by self-shading.

3.2.4 Transfers from the phytoplankton compartment—secretion and physiological death

Several processes remove phytoplankton biomass as soon as, or soon after, it is formed. High production and low rates of loss allow biomass to accumulate, but grazing by zooplankton may sometimes reduce the standing biomass to low levels despite very high rates of production. Secretion of organic compounds by phytoplankton may also reduce the rate of build up of biomass by a physiological process. Other losses occur from sedimentation, washout through the overflow or down river, and death through nutrient starvation, parasitism, or both.

Glycollate and peptides are among substances found secreted by phytoplankton[136]. Such secretions are detected in experiments where inorganic ^{14}C has been supplied to phytoplankton in bottles and combined organic ^{14}C is ultimately detected in the water. The secretion could be artefactual—phytoplankton doubtless behave somewhat abnormally in glass bottles—but the phenomenon is so widespread as to make this unlikely to be a complete explanation. Secretion may amount to over half of total carbon uptake and must have some advantage to the organisms, otherwise there would presumably have been selection against it. The compounds may chelate scarcely soluble nutrients (see page 70), or give competitive advantages in the inhibition of growth of other species (see 3.3).

Jassby & Goldman[254] made an estimate of loss rates of phytoplankton in Castle Lake, an infertile lake at 1760 m in the Klamath Mountains of California. They compared observed changes in biomass with the increases that would be expected from the amount of primary productivity measured by the ^{14}C method. Total loss rates were never less than about 20% of the production per day and climbed to 87% in May. Sinking, and grazing by zooplankton were unimportant and physiological death appeared to account for most of the daily losses. Of potential 'natural causes' parasites were not often seen and death from extremes (from the point of view of a particular species) of temperature, light, nutrient scarcity, or toxic substances produced by other species were felt to be most important. This would imply a large daily conversion of phytoplankton to detritus. In other lakes, losses by zooplankton grazing seem most important.

3.2.5 Zooplankton activity
Investigation of the zooplankton is more complicated than that of the phytoplankton. The latter, in a mixed water column is almost uniformly distributed, and although different species have slightly differing nutrient requirements, and may be photosynthesizing at different rates, it is at least possible to obtain a representative sample and to measure the photosynthesis and rate of uptake of nutrients of the community as a whole. The zooplankton, on the other hand, comprises a number of species heterogeneously distributed in the water column and often with markedly different diets, even in individuals of different ages within the same species. As with all animals, zooplankton activity can be described by the relationships: Ingestion = Assimilation plus Defaecation, and Assimilation = Net production (body tissue and gametes) plus Respiration plus Excretion.

Ingestion (feeding)
Production studies of herbivorous and detritus-feeding zooplankters begin with estimates of ingestion rate. The animals either filter small particles from the water (e.g. *Daphnia* spp.) or bite pieces from colonies or aggregates of larger algae or detritus (e.g. *Thermocyclops hyalinus* in L. George, Uganda). In general they seem not to discriminate in an ordered manner between live phytoplankton and dead detritus. This complicates a straightforward account of plankton function! The maximum size of particles which can be filtered depends upon the size of the animal. Burns[55] elegantly showed this by having *Bosmina longirostris* (O. F. Müll) and *Daphnia* spp. feed on a mixture of tiny plastic beads of known size and edible yeast, but size is not the only factor determining whether food is ingested. Some algal species may be rejected since they are of unmanageable shape, or chemically unacceptable.

Ingestion, or feeding rate depends on the volume of water filtered per unit of time and on the food concentration. Burns & Rigler[56] studied this with *Daphnia rosea* feeding on a yeast (*Rhodotorula glutinis*) isolated in culture from lake water. The yeast was grown in a medium containing radioactive (^{32}P) phosphate until it was uniformly labelled. *Daphnia* were allowed to feed in yeast suspensions for 2–5 minutes, after which they were anaesthetized in saturated CO_2 solution, which prevents defaecation on disturbance of the animal. The radioactivity incorporated into the animals was measured and filtering rate calculated as:

$$F \text{ (ml animal}^{-1} \text{ hr}^{-1}) = \frac{\text{Radioactivity per animal}}{\substack{\text{Radioactivity per ml of} \\ \text{yeast suspension}}} \cdot \frac{60}{\text{time (min) of feeding.}}$$

Ingestion rate could then be calculated as filtering rate times the number of yeast cells per ml as cells animal^{-1} hr^{-1}. Filtering rate decreased almost exponentially as food concentration increased from about 1.5–2.0 ml hr^{-1} at 25 000 cells ml^{-1} to about 0.2 ml hr^{-1} at 500 000 cells ml^{-1}; both increasing body length and temperature increases up to about 20°C were associated with increased filtering rates. Increasing food concentrations were met by increasing ingestion rates, up to a maximum of about 100 000 cells ml^{-1}.

These studies were carried out under ideal laboratory conditions, with a food source known to be acceptable to *D. rosea*. When studies were carried out in natural lake water from Heart Lake, Ontario, where *D. rosea* is common, filtering rates were smaller by a factor of 2–3 in mid-June to mid-October, than those predicted from the laboratory studies. Possible reasons for this include a lack of particles of filterable size or the presence in the water of a toxin inhibiting *Daphnia*. Such toxins have been shown to exist, secreted particularly by bluegreen algae[288] but have not been chemically identified. Removal of animals from a lake to measure their filtering rates may itself affect the rates measured. This has led Haney[192] to design a sampler which will enclose *in situ* a volume of lake water and simultaneously will allow labelling of the particulate matter with radioactive yeast.

Material that is ingested is not necessarily digested and assimilated. Some passes through the gut and is rejected as faeces. The faeces also contain some secretions of the gut wall, which represent previously assimilated material. This leads to an overestimate of faeces production, which can be corrected by using a double radioactive labelling technique using ^{14}C and an isotope of chromium which is not assimilated[59]. From measurement of feeding and defaecation rates, assimilation rates can be measured. In many zooplankters, faeces are released as fine particles which immediately mix with the water, where they form part of the detritus compartment. In some copepods, however, the faeces are released packaged in a membrane and can be readily collected and measured. Assimilation rates in animals whose faeces cannot conveniently be collected are measured directly by uptake of radioactive food.

The animals are first allowed to feed on radioactively-labelled food, then they are transferred to a suspension of unlabelled food for a short period to allow evacuation of any remaining radioactively-labelled faeces. The short period minimizes assimilation losses due to respiration (if ^{14}C is used) or excretion (if ^{32}P is used). Assimilation is measured as the amount of radioactivity taken into the animal in unit time.

Various foods are assimilated to differing extents. Some phytoplankters pass out of the gut alive, others dead but only partly digested. Table 3.3 illustrates this. The blue-green algae have a reputation for being poorly assimilated foods, and Arnold[14] has shown that indefinite maintenance of *Daphnia* populations is not possible on certain blue-green algal species. Generalization should be avoided, however, since animal diets are complex and little understood. A given alga may be unable to support *Daphnia* growth in a defined chemical

Table 3.3. Assimilation rates (A) as percentages of ingestion rates (I) of *Daphnia longispina*, *Diaptomus gracilis* and *Cyclops strenuus* when fed different foods under standard conditions. (After Schindler[456].)

Food	A/I × 100 *Daphnia*	A/I × 100 *Diaptomus*	A/I × 100 *Cyclops*
Microcystis (Blue-green colony)	17.9	45.3	8.9
Oocystis (Green colony)	10.5	13.7	8.0
Elakatothrix (Green unicell)	100	31.3	19.0
Gloeocystis (Green colony)	13.6	44.2	18.2
Anabaena (Blue-green filament)	50.8	73.5	25.9
Tribonema (Xanthophyte filament)	68.6	19.9	5.6
Coelastrum (Green colony)	20.8	29.1	6.2
Oscillatoria (Blue-green filament)	25.6	29.7	3.7
Asterionella (Diatom colony)	38.4	20.1	38.0
Cryptomonas (Cryptomonad naked unicell)	91.6	100	18.5

medium, but may do so if suspended in 'conditioned' water—such as that from an aquarium where activity of other organisms has been going on[494]. Presumably certain substances, perhaps dissolved vitamins, are required as well as particulate food.

In L. George, Uganda, the major zooplankter, *Thermocyclops hyalinus* assimilates between 35 and 58% of ingested *Microcystis*, the major phytoplankter present, which is a blue-green algal species[355]. Reports of inabilities of temperate zooplankters to assimilate blue-green algae may reflect the present state of flux in many temperate lakes due to nutrient enrichment (see Chapter 10). This has stimulated growth of previously less common blue-green algae, for which the zooplankters may not yet have evolved suitable digestive enzymes. L. George is not a recently polluted lake, but has had a naturally large crop of blue-green algae presumably for a very long time.

Excretion and respiration

Respiration and excretion by zooplankters release substances back into the dissolved substances pools of the lake, where they become available for uptake again by phytoplankton and bacteria. The importance of respiratory production of CO_2 by zooplankton is probably minor but the excretion of ammonia and urea, primarily the former, and of phosphates, is crucial to maintenance of the cycles of these scarce but essential substances within the plankton. In L. George (Uganda), for instance, it is only by rapid recycling of nitrogen and phosphorus through the zooplankton that dense blooms of blue-green algae can be maintained[155]. The nitrogen and phosphorus compounds would otherwise be lost to the sediments. When released by zooplankters they are taken up so rapidly by the phytoplankton that their excretion can only be detected if the phytoplankton is experimentally removed[154].

Careful experiments have shown that *Daphnia rosea* excretes a mixture of about 90% soluble inorganic phosphate (orthophosphate) and 10% organic phosphate compounds of low molecular weight[401]. Within minutes the orthophosphate is reabsorbed by bacteria associated with the animals and analysis of the dissolved excretion products reveals increasing proportions of organic phosphate compounds.

Calculations based on the plankton of Heart Lake, Ontario by Peters & Rigler[401a] suggest that about 27.4% of the particulate matter in the epilimnion (expressed as its phosphorus content—particulate phosphorus) is ingested by zooplankton each day in summer. Of this 54% is assimilated and during the approximately steady state of the zooplankton population in summer an equivalent amount is daily excreted. Thus 0.54×27.4 or 14.8% of the total phosphorus in the plankton is regenerated daily by excretion and 0.46×27.4 or 12.6% is daily turned into faeces which enter the detritus compartment. Here more of the phosphorus may be released by microbial activity.

Production

Production of the zooplankton is an elusive parameter to measure in natural systems because of simultaneous predation by other zooplankters, insect larvae and vertebrates. It can be measured by estimating assimilation, respiration and excretion of representative animals in the laboratory and by extrapolating the difference found between assimilation and respiration plus excretion to the populations of the lake. If estimates are in terms of carbon or energy, excretion can be ignored. Predation rates can then also be obtained from the difference between calculated production and measured changes in biomass of the zooplankton. An alternative, which avoids some of the uncertainties of extrapolating laboratory measurements to the field, is to use various population parameters to measure production. An example is the study of *Daphnia galeata mendotae* carried out in Base Line Lake, Michigan, by Hall[188]. The method, however, is applicable to any zooplankter, and depends ultimately on finding the birth rates of the animal as the year progresses.

D. g. mendotae reproduces parthenogenetically at rates dependent on temperature, food concentration and other factors. The rate of change of numbers at any instant of time, t, is given by:

$$N_t = N_0 e^{rt} \dots\dots\dots\dots\dots\dots\dots\dots\text{(b)}$$

where N_t and N_0 are the numbers at the end and start of the time period, e is the base of natural logarithms and r is the intrinsic rate of natural increase. r equals (b − d) the difference between instantaneous birth and death rates. Death arises from predation, parasitism, sinking and loss to the outflow, though there is evidence that only predation is very significant.

If by regular sampling in the field the numbers of animals can be established, estimates of r can be obtained for the period t → t + 1, between samplings from integration of the above equation:

$$r = 1/t(\log_e N_{t+1} - \log_e N_t) \dots\dots\dots\dots\dots\dots\text{(c)}$$

r, however, is a composite rate, and to understand the dynamics of the population, and to calculate the total production, b and d must be estimated. b, the instantaneous birth rate, is defined by

$$N'_{t+1} = N'_t e^{bt} \dots\dots\dots\dots\dots\dots\dots\dots\text{(d)}$$

N′ represents a potential population size in the absence of predation and cannot be estimated. Hence b cannot be directly measured. However, a finite approximation to birth rate, B, can be defined as

$$B = \frac{\text{Number of newborn (during interval } t \to t + 1)}{\text{Population size at t}}$$

$$B = \frac{N'_{t+1} - N'_t}{N'_t}$$

and since
$$b = 1/t \log_e \frac{N'_{t+1}}{N'_t}$$

and for
$$t = 1, b = \log_e \left(\frac{N'_{t+1}}{N'_t} - \frac{N'_t}{N'_t} + 1 \right)$$

so
$$b = \log_e \left(1 + \frac{N'_{t+1} - N'_t}{N'_t} \right) = \log_e(1 + B).$$

B can be independently estimated since it is equal to the number of newborn per individual per day. Those about to be born are carried as eggs or embryos by the female until they are released when the female moults.

$$B = \frac{\text{Number of reproductively mature adults } (N_A) \cdot \text{Number of eggs and embryos carried per adult (brood size, } \overline{E}).}{\text{Total population size } (N_t) \cdot \text{Number of days for an egg to mature from production to release (D).}}$$

N_A, $\overline{E}$ and N_t are readily estimated from sampling of the natural population. D depends largely on temperature and is measured in laboratory cultures at a range of temperatures. In *D. g. mendotae* it ranged from 2 days at 25°C to 20.2 days at 4°C. From the lake temperature at the time of sampling, an appropriate D value can be selected for calculation of B. Some objections have been raised to the logic underlying the calculation of b from B[19] but in comparison with errors inherent in sampling natural populations, errors from this source are likely to be negligible.

b and r can be used to calculate d, the instantaneous death rate (r = b − d). Fig. 3.6 gives these data for *D. g. mendotae* in Base Line Lake, together with

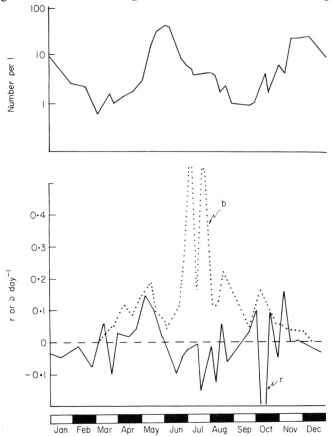

Fig. 3.6. Population changes of *Daphnia galeata mendotae* in Base Line Lake, Michigan (after Hall[188]). Upper graph shows changes in the total population, and the lower one the observed intrinsic rate of increase (r) and the calculated birth rate (b). The difference between the graphs gives d, the instantaneous death rate.

changes in population. From March to early June the population increased and death rates were low. In midsummer the population remained steady at a low level, but birth rates (and therefore production) were very high. Death rates, due to predation by fish were then also high. Although the population was reproducing at such a rate that it could double itself, on average, every four days, predation more or less accounted for all of the production and kept the population at a low level.

Production, expressed as numbers by Hall, could readily be changed into carbon or energy units by suitable analyses of the animals. This example illustrates once more the highly dynamic nature of the plankton. Despite an apparently static population in summer, the turnover, due to predation, was 25–50% per day. In another lake, Crawford lake, Ontario, higher summer mortality of *Daphnia rosea* was concentrated at or just before the hatching stage[413]. This work also gives a sophisticated criticism of the principles on which Hall's work was based.

3.2.6 Detritus and microorganisms

Sources of particulate detritus
The dead cells of phytoplankton, corpses of zooplankton, and zooplankton faeces comprise some of the many sources of detritus present abundantly in the plankton. Other sources include allochthonous particles washed in from the catchment. Bacteria colonize algae and zooplankton even when they are alive and the transition from living compartments to dead detritus is bridged by a progressive colonization of microorganisms. Dynamics of the detritus compartment can thus only be considered along with those of microorganisms and of dissolved organic substances, some of which are produced by the microorganisms and others of which are used by them.

Some algae are far more resistant to decay than others. Gunnison & Alexander[186] found that some blue-green and colonial green algae took more than a month to lose their overall shape when killed and incubated in contact with lake water bacteria. This was despite ready colonization by bacteria of colonies with their often mucose covering[264]. Other species were completely decomposed within a few days. The cell walls of the resistant species appeared to be difficult to digest though the internal cell contents readily autolyse and decompose. Transfer from living phytoplankton to detritus may be accelerated by parasitization. Many chytrid parasites are present, colonizing their host cell as zoospores, and extending a branched rhizoid system within the wall and contents of the host. Sporangia are ultimately produced, often after sexual reproduction, and a new generation of zoospores is released. Myxomycete, protozoon and other Phycomycete parasites also occur. Some are host-specific, whilst others attack a limited range of hosts. They are difficult to study, for few have been cultured and microscopy of a high standard is needed for their investigation. However, it may be that a steady attrition of phytoplankton by an array of parasites makes a significant contribution to the detritus compartment.

Bacteria are attached to the detritus but may also be present as free-living bacterioplankters feeding largely on dissolved organic compounds. Attached bacteria break down detritus, which normally appears as amorphous brownish-coloured lumps, and absorb the soluble compounds so produced. Other bacteria may use the detritus only as an attachment site and may feed on dissolved compounds. Much of the apparently soluble organic matter present in the water may in fact be colloidal and it is not known how this is utilized and by what. Products of detrital decomposition are CO_2 and other inorganic nutrients, which become once more available for algal uptake.

Dynamics of particulate detritus

Saunders[448] prepared 'artificial detritus' by autoclaving a radioactively-labelled plankton community. He resuspended the detritus in lake water and found that it was decomposed at rates of 5–20% per day. About 1% per day leaked out as soluble organic compounds. This was lower than the rate at which phytoplankton secreted organic compounds in the living community, but the generally greater concentrations of detritus meant that the absolute amount of release was about twice that normally released from the phytoplankton per day. Detritus may also be grazed by zooplankton. Assimilation efficiencies (assimilation/ingestion) for *Daphnia* spp. were 2–18% for detritus, 12–52% for bacteria, and 20–88% for phytoplankton. The least palatable detritus came from dead thick-walled green and blue-green algae and that most readily assimilated from thin-walled flagellates. Although detritus was not nutritionally so effective as bacteria or particularly phytoplankton, it nonetheless could form a major part of the energy requirement of the zooplankton on account of its high concentration relative to that of phytoplankton. It was 1.3–16.9 times as abundant in a Michigan lake and frequently comprised nearly half of the assimilation by *Daphnia*. Detritus can be regarded as a food store or larder for the plankton ecosystem, providing a steady food source not directly subject to the vicissitudes of weather, etc., which may affect phytoplankton production.

Dynamics of dissolved detritus

Relationships between microorganisms, particulate detritus, and soluble organic compounds, the dissolved detritus, encompassing, as they do, a wide variety of organisms and substrates, are the least well studied of plankton processes. The technical problems can be illustrated by an account of a method currently necessary to measure the concentrations and turnover times of simple dissolved organic substances in the water. Despite their many origins—secretion by phytoplankton, decomposition of particulate detritus, exudation from animal guts and animal excretion—the levels of such substances are very low, often too low for convenient and precise chemical analysis. The life of a given molecule may be only seconds and the entire pool of some dissolved substances may be turned over every few minutes in summer. Of the wide range of organic substances found in lake waters (amino acids, carboxylic acids, carbohydrates, peptides, vitamins, free enzymes, phenols, etc.) glucose and acetate have been studied the most in relation to their uptake by bacteria.

This seems often to follow the relatively simple relationship:

$$v = \frac{v_{max}S}{K_m + S}$$

where v is the rate at which the substance is taken up, v_{max} is the maximum uptake rate achievable by the organisms, S is the concentration of substance, and K_m (the saturation constant) is the concentration of substance at which half the maximum uptake rate is attained. This equation was first proposed, for laboratory cultures, by Monod[349]. It sometimes applies to natural mixed populations but sometimes does not[253]. When graphed (Fig. 3.7), the equation shows

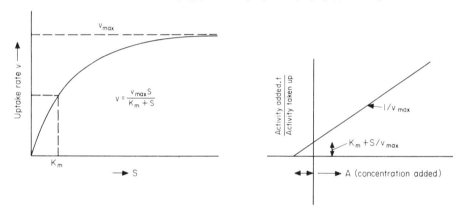

Fig. 3.7. Uptake kinetics graphs for absorption of organic compounds by bacteria. Left-hand diagram shows the rate of uptake in relation to substrate concentration. Right-hand graph shows an example of how the uptake parameters can be derived from appropriate experiments.

an increasing uptake rate until v_{max} is reached. At this point the transporting enzyme systems in the bacterial cell walls are saturated. Increasing substrate levels cannot then result in greater uptake rates.

Lake water is placed in bottles and a series of additions of the substance under test is made with radioactively labelled substance. The total concentration of substance in a bottle is then that which has been added, A, plus the natural concentration, s. The bottles are incubated, suspended in the lake for a known period (t) in which the bacteria absorb some of the substrate. The water is then filtered through bacteria-retaining filters and the incorporated radioactivity is measured.

Rate of uptake, v is given by:

$$v(mg\ l^{-1}\ hr^{-1}) = \frac{\text{Radioactivity incorporated into bacteria } l^{-1}}{\text{Radioactivity added to water } l^{-1}} \cdot \frac{A + s}{t}$$

Rearranged this becomes:

$$\frac{A + s}{v} = \frac{R_{added}t}{R_{bact}}$$

and since $v = \dfrac{v_{max}S}{K_m + S}$

then by algebraic manipulation

$$\frac{S}{v} = \frac{K_m}{v_{max}} + \frac{S}{v_{max}}$$

In the experiment, S is (A + s) and

$$\frac{A + s}{v} = \frac{R_{added}t}{R_{bact}} = \frac{K_m}{v_{max}} + \frac{A + s}{v_{max}} = \frac{K_m + s}{v_{max}} + \frac{A}{v_{max}}$$

$$\therefore \frac{R_{added}t}{R_{bact}} = \frac{A}{v_{max}} + \frac{K_{m+s}}{v_{max}}$$

This is an equation of the form $y = mx + C$, for a straight line. If $\dfrac{R_{added}t}{R_{bact}}$ is plotted against A (Fig. 3.7), the slope of the line is $1/v_{max}$ and the intercept on the y axis is $\dfrac{K_{m+s}}{v_{max}}$. Thus v_{max} can be found from the gradient and used to calculate $K_m + s$. Since K_m has been found by other means to be very small, a slight overestimate of s can be calculated. Derivable also is the turnover time of the substance s/v since if $A = 0$,

$$\frac{s}{v} = \frac{R_{added}t}{R_{bact}}$$

'Turnover time' is the time taken for complete uptake and replacement of the pool in the water. In L. Lötsjön, Sweden, Allen[6] found that on one August day, the concentration of acetate in the water was about 6 μg l^{-1} and was being replaced more frequently than hourly. The total dissolved organic carbon concentration was about 10 mg l^{-1} or about a thousand times as large as the acetate pool. In an Oregon lake, Wright[545] found mean pool sizes for acetate, glucose, glycine and glycollate of 26, 11, 9, and 118 μg l^{-1} respectively and turnover times of 2.3, 2.4, 8 and 26 days. This illustrates the highly dynamic state of some of the dissolved organic matter and how estimates of a whole compartment or sub-compartment can give little idea of the dynamics of individual members.

3.2.7 The dissolved substances compartment
Inevitably the dissolved substances have been previously and frequently alluded to. Along with other compartments there is a wide range of individual member pools, each behaving differently and turning over at greatly differing rates. The least important dissolved substances are those whose pools are large and readily detectable. Such substances—Na^+, K^+, Ca^{2+}, Mg^{2+}, Cl^-, SO_4^{2-}, HCO_3^- for instance—or the ratios of their concentrations one to another, may be important in determining which particular species occurs in a given lake water, but their general abundance suggests that they never limit transfers within the plankton.

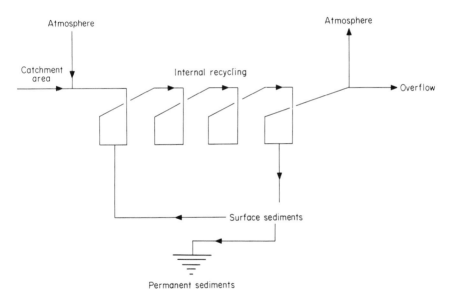

Fig. 3.8. General model of the pathways a substance may take in an aquatic ecosystem.

It is those substances whose dissolved pools are very small compared with their levels in the particulate fraction which are likely to be crucial. By way of integrating the workings of the plankton so far considered, the movements of two such elements, phosphorus and nitrogen, are now discussed. Carbon, although it normally has large pools of inorganic compounds, is also considered, for some of its scarcer organic forms may significantly influence the rates of several processes. Fig. 3.8 shows a generalized flow diagram of the behaviour of any chemical element in a lake. In the present chapter it is the internal recycling that is discussed.

Cycling of phosphorus in the plankton

The phosphorus cycle within the plankton community is shown in Fig. 3.9. The diagram is a tentative one, not least because it is currently very difficult to measure accurately the level of dissolved phosphorus in lake water! The most convenient methods all depend on reaction of inorganic phosphate with molybdate under acidic conditions in the presence of a reducing agent. The acid hydrolyses some organic phosphorus compounds (esters and fulvic acid—metal complexes) to phosphate. The method thus measures a quantity called soluble reactive phosphorus (SRP) of which only part may have been inorganic phosphate in the original lake water[423]. Accurate estimation of inorganic PO_4 is only approached by passing filtered lake water through a column of hydrous zirconium oxide which selectively retains PO_4. The other compartments through which phosphorus moves are those of dissolved organic compounds, and the particulate live plankton and detritus.

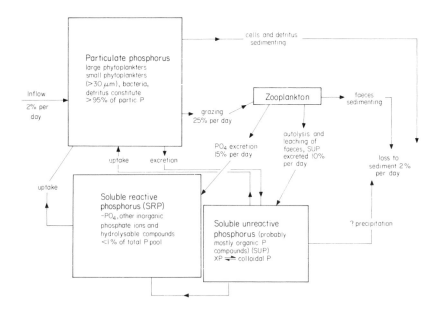

Fig. 3.9. Cycling of phosphorus during a day in summer in a temperate lake. There is relatively little through-flow of water so losses to overflow are negligible. Values are given as percentages of the particulate pool turned over per day and are approximate. (Compiled from various authors cited in the text.)

Filtered lake water may be irradiated with U.V. light or evaporated and the residue hydrolysed with perchloric or persulphuric acids. Organic phosphates are then converted to carbon dioxide and inorganic phosphate which may be detected in a molybdate reaction. The amount in excess of the SRP previously detected is referred to as soluble unreactive phosphorus or SUP. SUP probably includes some previously insoluble and colloidal inorganic phosphates, e.g. of iron. SRP and SUP are determined in water filtered, usually, through 0.45 μm pore size filters. The particles retained on the filters may be digested and dissolved. Their contained particulate phosphorus probably normally accounts for over 90% of the total phosphorus present. Indeed, filtration through 0.01 μm filters or ultracentrifugation reveals that much of the SUP pool is finely particulate[66a]. The SUP certainly has several components. Free phosphatase enzymes perhaps released on autolysis of organisms passing through zooplankton guts as faeces or from bacterial attack on detritus are present in the water[33] and these must mediate some reactions within the SUP pool.

Phosphorus enters the cycle as it enters the lake from the catchment area (see Chapter 1) or sediment release (see Chapters 2 & 5). It leaves it when it is washed through the outflows, or is incorporated permanently into sediments as detritus or precipitates. During the plankton cycle between these events, there seem to be at least two main sets of regenerations of SRP—by zooplankton, previously considered on p. 58, and by bacteria. Information on the latter is meagre, but Lean[283] has shown some of the complexity.

He added radioactive inorganic phosphate to water from Heart Lake, Ontario. Within a minute, half of it had been taken up by phytoplankton and microorganisms. Within an hour an equilibrium had been established in which only about 0.21% of the dissolved phosphorus was present as dissolved inorganic phosphate and 98.5% was in particulate form. By filtering the water through molecular gel filters two further sets of phosphorus compounds were detected; one with a molecular weight of about 250, called XP, constituted 0.13% of the total and one with molecular weights greater than 5×10^6 constituted 1.16% of the total and was called 'colloidal phosphorus'. Both of these would normally be measured in the SUP compartment. XP was released from the particulate compartments probably by bacterial rather than algal secretion, but certainly not through death and decay as the process was too rapid (less than three minutes). XP reacted with 'colloidal phosphorus' with release of inorganic phosphates, but XP did not itself readily hydrolyse to inorganic PO_4. The turnover time for the passage of an inorganic PO_4 ion through the particulate, XP and colloidal P compartments is likely to be only a few minutes in summer[431]. Some of the colloidal P ceases to react with XP after a few days and hence its contained phosphorus becomes unavailable to the plankton.

The overall picture of phosphorus movement is thus one of very rapid cycling and reuse of SRP by the phytoplankton and bacteria. SUP seems not generally available directly to the phytoplankton, but must be metabolized first by bacteria other than those which produce and secrete it. Zooplankton also cycle phosphorus with a rather longer turnover time (days rather than hours) and may also be prime agents, through sedimentation of their faeces, of loss of phosphorus to the sediments. This seems to occur at the rate of a few per cent of the total phosphorus pool per day and implies again (see Chapter 2) that maintenance of plankton production relies on continued renewed supplies of phosphorus from the catchment.

The nitrogen cycle in the plankton

The behaviour of nitrogen in the plankton is as complex as that of phosphorus (Fig. 3.10). There are four main dissolved inorganic pools—ammonium, nitrite, nitrate and molecular N_2. The first three forms are available for uptake by most phytoplankters and may be simultaneously absorbed[34]. Some species, particularly the Euglenophyta[287] seem able to absorb only ammonium, and often other plankters may absorb ammonium preferentially since it is energetically less

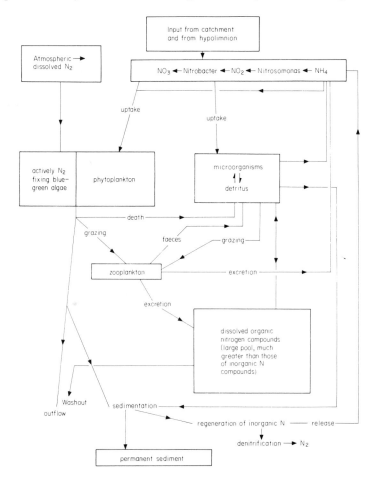

Fig. 3.10. Main pathways of nitrogen transformation in freshwaters.

costly to process in the cell. Ammonium and nitrite rarely have large dissolved pools, however. Even nitrate may be barely detectable in summer and in lakes where the loading of phosphorus is large the rate of supply of inorganic nitrogen compounds may limit the rate of growth of phytoplankton (see section 3.2.2).

The turnover times of the inorganic nitrogen pools are largely unknown. Ammonia excreted from *Thermocyclops hyalinus*, the dominant zooplankter of the equatorial L. George, was measured by Ganf & Blazka[154]. They showed that the ammonia released was rapidly taken up by phytoplankton, with a turnover rate of about 1.5 d^{-1}. Parallel estimates of inorganic PO$_4$ turnover of about twice per day are likely to be underestimates[401a]. The very small pools of inorganic nitrogen in the lake (no nitrate was detected at all) suggest that nitrogen should be cycled at least as rapidly as phosphorus and the turnover rate may be much greater than 1.5 d^{-1}.

Molecular nitrogen can be used only by nitrogen-fixing blue green algae and by certain photosynthetic and heterotrophic bacteria. Until recently it was believed that only blue-green algae possessing differentiated cells called heterocysts could fix nitrogen. Nitrogenase, the complex of enzymes responsible, is inhibited by oxygen, and heterocysts lack that part of the photosynthetic apparatus, photosystem II, which is responsible for oxygen release[131]. Isolated heterocysts have been shown to contain active nitrogenase[481], and there is a high correlation between measured nitrogenase activity and various functions of heterocyst numbers[225]. Nitrogenase is contained in ordinary cells of some blue-green algae which do not have heterocysts[482] though it is active only under anaerobic conditions. A coccoid species, *Gloeocapsa* has been shown to fix nitrogen in aerobic conditions[546] despite a complete lack of heterocysts, and this may mean that nitrogen fixation is more widespread in lakes than presently believed.

In many lakes nitrogen fixation seems not to be an important source of nitrogen. Less than 1% of the total nitrogen income of L. Windermere is provided in this way[224], but production is probably limited by phosphorus supply in this lake. In lakes with high phosphorus loadings, large crops of blue-green algae may account for about half of the total nitrogen income by fixation[225].

Nitrogen fixation is directly measured by supplying experimentally the non-radioactive isotope ^{15}N to a sample of plankton and later measuring, by mass spectrometry, the uptake of the isotope after a suitable period of incubation. The method is tedious and since its introduction in 1943[57] relatively few measurements of natural fixation rates have been made by it. However in 1966, Dilworth[103] showed that the enzyme system capable of reducing nitrogen to amino groups was also capable of reducing acetylene (ethyne) to ethylene (ethene). Acetylene is injected into a small rubber-capped bottle partly filled with a plankton sample. After some time, a sample of gas is withdrawn and the proportions of ethylene and acetylene determined by gas chromatography. The rate of acetylene reduction is believed to be crudely proportional to the potential rate of nitrogen fixation[480] and, with occasional calibrations against the direct ^{15}N uptake method provides a rapid, inexpensive assay which has led to a profusion of estimates of potential nitrogen fixation.

Dissolved organic nitrogen compounds
Blue-green and other algae and bacteria secrete nitrogenous organic compounds into the water where they form part of a pool of dissolved organic nitrogen compounds about which little is known[328, 329]. The pool is also supplied by

excretion of urea by fish, and perhaps with amino acids and urea by some zooplankters[255]. Many different substances comprise the organic nitrogen pool.
Walsby[515], for instance, showed at least twelve separate polypeptides to be
produced by a single species, *Anabaena cylindrica*. In L. Mendota dilute pools
($< 0.01\ \mu M$) of free amino acids have been measured, with serine and alanine
the most prevalent of ten acids examined. A tenfold larger pool of combined
amino acids was simultaneously measured and both pools increased during
decomposition of large algal populations[156].

At least part of the dissolved organic nitrogen pool—that of small molecules
such as amino acids, probably turns over rapidly at rates similar to those of
other simple organic compounds. Bacteria probably account for much of the
uptake because algae, although capable of using amino acids[3] and other
organic nitrogen compounds[35] at high concentrations, do not compete
effectively with bacteria at low concentrations[215]. The vitamins cyanocobalamin (B_{12}), thiamine, and biotin, variously required by some phytoplankton
species, are all dissolved organic nitrogen compounds and are rapidly used once
formed. Pools of up to 8 ngl^{-1} of cyanocobalamin, 400 ngl^{-1} of thiamine and 40
ngl^{-1} of biotin were recorded in the Japanese lake Sagami by use of bioassay
techniques since chemical analyses are insufficiently sensitive[387]. The vitamins
seemed to be secreted by bacteria and some blue-green algae, and were taken up
rapidly by planktonic diatoms.

Dissolved organic nitrogen compounds, although not solely responsible,
may also act as chelators. Chelators reversibly combine with metal ions in such
a way that equilibria are set up in the water between free metal ions and soluble
ion-chelate complexes. As ions (such as Fe^{3+}, Mn^{3+}, Mo^{3+}) are removed from
the complexes by phytoplankton the equilibrium moves in such a way as to
replace them. Such ions are readily precipitated inorganically as hydroxides or
carbonates and retention of them in soluble chelator complexes ensures a steady
supply for the phytoplankton. The peptides secreted by blue-green algae can
act as chelators[139] and it is significant that the two most commonly used
chelators in media for the laboratory culture of algae are the organic nitrogen
compounds ethylene diamine tetraacetic acid (EDTA) and nitrilotriacetic acid
(NTA).

Cycling of carbon in the plankton
The cycling of carbon in the plankton (Fig. 3.11) provides a final example of the
role of dissolved substances whilst also integrating much of the foregoing
information. Photosynthesis, grazing, detritus formation, release of soluble
organic compounds, whether or not they contain phosphorus and nitrogen, are
all processes that can be expressed in terms of carbon transfers and it is unnecessary to repeat discussion of them.

There are usually large pools of inorganic carbon, mainly in the form of
CO_2 and HCO_3^-, and of refractory organic compounds. These pools are probably turned over relatively slowly, because of their size in the case of the inorganic compounds and because of their resistance to decomposition in the case

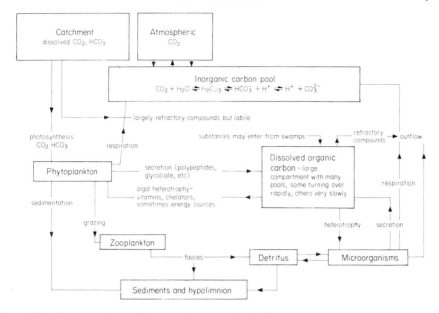

Fig. 3.11. Main interrelationships of carbon transfers in the plankton.

of the refractory organic compounds. These seem to come mostly from the catchment where soil bacteria failed to decompose them before they were washed out by rain. They are often yellow or brown in colour and were found, in Linsley Pond, Connecticut, to be a mixture of phenolic and enolic hydroxy carboxylic acids with a mean molecular weight of 456[462]. They were resistant even to chemical oxidation, formed a pool of about 4 mg l^{-1} and acted as chelators of iron.

The behaviour of the very labile soluble organic compounds has already been discussed in other contexts, so that only the behaviour of the inorganic carbon pools need be now considered.

Inorganic carbon compounds and the plankton
Inorganic carbon exists in lakes in various forms—CO_2, H_2CO_3, HCO_3^- and sometimes CO_3^{2-}. The relative proportions of these forms are governed by the equilibria:

$$H_2O + CO_2 \leftrightarrows H_2CO_3 \leftrightarrows HCO_3^- + H^+ \leftrightarrows CO_3^{2-} + H^+$$

and the state of the equilibria depends partly on [H^+], of which pH is a measure. At and below pH 4.5, bicarbonate and carbonate are unstable and essentially only H_2CO_3 and CO_2 are present. At about pH 10 and above the equilibria move in such a way that carbonate formation is favoured and only very small proportions of CO_2 and H_2CO_3 are present. The absolute amounts of any form depend on the total amount of inorganic carbon present. Addition of bicarbonate, for instance, will increase the absolute levels of CO_2 and CO_3^{2-} though their proportionate relationships to HCO_3^- at the same pH will remain constant. Changes in $[H^+]$ occasioned by other reactions involving the anions of weak acids such as silicate, borate and those of organic acids in the water will also cause adjustments in the carbonate equilibria. Biological activity itself causes marked adjustments.

As phytoplankters withdraw CO_2 or HCO_3^- for photosynthesis the equilibria are shifted towards the left so that H^+ and carbonate, if present, or bicarbonate and H^+ will reassociate. This reduces $[H^+]$ and increases the pH. The pH increase itself alters the state of the equilibria, tending to push them towards the right. The ultimate state is generally one in which the pH rises, often from 6 or 7 to 9 or 10, and the amount of CO_2 available decreases, at least temporarily. CO_2, however, readily diffuses into lakes from the atmosphere and replaces the dissolved CO_2 pools in compensation for the CO_2 incorporated into the phytoplankton cells. Over periods of more than a few hours the atmospheric invasion of CO_2 seems high enough[453] to support at least as much production of algal biomass as can be attained by the incident light available and excessive supplies of all other nutrients. There seems no question, therefore, of a carbon limitation of phytoplankton where production of biomass is concerned. The temporary imbalance caused by changes in the carbonate equilibria during bright, sunny periods may limit the rate of carbon uptake particularly for those species unable to use HCO_3^-. Only one or two species have ever been shown to be able to take up CO_3^{2-} but most lake waters probably never reach a state where both HCO_3^- and CO_2 levels are negligible. The CO_2 depletion which is normally noted by day in fertile lakes whose phytoplankton biomass and photosynthetic rates are moderately high may give rise to marked diurnal fluctuations in pH and gross photosynthesis, though net production is probably little affected except indirectly in very hard-water lakes.

In these (marl lakes) the normal HCO_3^- level is high (3–5 mEq l^{-1}) and even low levels of photosynthesis push the pH from its usual 7.5–8.0 to values where carbonate concentrations increase greatly. Because of the abundance of limestone or chalk or glacial debris derived from such rocks in the catchments of marl lakes, Ca^{2+} and Mg^{2+} are also very abundant. The solubility products of $CaCO_3$ and $MgCO_3$ may thus be exceeded. Resultant precipitation of these carbonates as colloids or larger particles follows. In itself this has little effect on the phytoplankton, but the precipitates seem also to incorporate phosphates and trace elements and also organic compounds, including dissolved vitamins and chelators. Indirectly then the net production of phytoplankton may be kept at a low level by aggravated shortage of key limiting nutrients[529].

3.2.8 Summary of plankton functioning

The plankton is very dynamic. Our picture is one of rapid transfers of key nutrients between the water and the living organisms set against a more stable background of major ion water chemistry. In this way phosphorus and nitrogen can be viewed as reusable catalysts concerned in the process of packaging organic compounds produced by photosynthesis, which then become available for zooplankton energy needs, or are lost to the sediments where they form part of the food of the sedimentary benthic animals (see Chapter 5). Despite the efficiency of the cycling it is not entirely closed. Some phosphorus and nitrogen are lost to the sediments from which they may not return (see Chapters 2 & 5) and a continuous supply from external loading is almost always necessary to maintain plankton production.

There are good reasons why phosphorus should usually be the key limiting factor to plankton production. It is lithospherically scarce (see Chapter 1) and it lacks the vast atmospheric reservoir that nitrogen has. Of course, this reservoir must be regenerated also, and on a world scale, denitrification, the reduction of nitrate and ammonium to molecular nitrogen in the oxidation of organic compounds by bacteria in waterlogged anaerobic soils and sediments, must compensate for the amount of nitrogen fixation. The existence of nitrogen fixation theoretically means that nitrogen limitation of plankton production should never develop[450] and it would not do so if populations of nitrogen-fixing blue-green algae developed instantly when dissolved inorganic nitrogen sources were used up in lakes. However, the lag period while they grow to sufficient size to feed much nitrogen into the plankton cycles is long enough for a nitrogen-limiting hiatus often to intervene.

Whilst the plankton changes hourly in the cycling of substances within the community, the community itself changes on a long-term basis with a seasonal periodicity of different species being reflected week to week or month to month. This change is imposed variously, by changes in the water caused by climate through its effects on temperature, rainfall and hence nutrient loading; by the earth's rotation with its effects of daylength and light intensity, and by internal chemical changes as the community reacts to the external factors. These are the subjects of the next sections.

3.3 Seasonal changes in the plankton

Almost all species which increase their populations in the plankton at some time during the year are ever present in the water as small residual populations. Detection of them may be difficult but an exhaustive search might reveal several hundreds of species of phytoplankton, whilst a count of the more numerous and therefore readily detectable species will amount to perhaps ten or twenty. A few species may form resting stages in the surface sediment and new ones may be brought in from time to time on water birds or by wind or floods.

There is then a great reserve of varied forms, each best fitted to exploit a particular set of conditions in the water, when its population will increase, and each unable to compete in other conditions, when its population will decline. The changing water mass throughout the year, in turn, selects the better fitted species for a particular time, and, in turn, leads to their decline. The result is a procession of overlapping, large populations against a background of small, declining populations.

How can even a few species co-exist in an apparently uniform mixed epilimnion, not to mention the several hundred usually present? Many phytoplankters seem to have rather similar nutrient requirements and Gause's law of competitive exclusion would predict that ultimately only one should persist. For this 'paradox' of the plankton[235] several explanations have been offered[236]. The most compelling is that the conditions of equilibrium necessary for Gause's law to be valid are rarely found in the plankton. The submerged environment changes so rapidly that no one species is favoured long enough for it to exclude the others. The phytoplankton theatre has many understudies for its leading roles and changes the cast frequently.

If changes in weather are a major driving force in determining seasonal periodicity, the least marked periodicity must be expected in lakes at the Equator. L. George, astride the Equator in Uganda, experiences a very constant climate. Incident radiation, although irregularly intercepted by cloud, varies within a range of only ± 13% of the mean, and the water temperature is always about 30°C. There are two dry seasons but their potential effect in determining

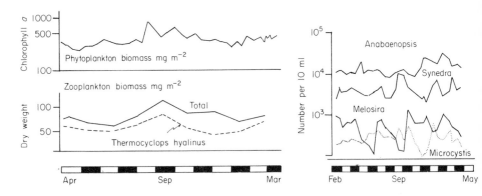

Fig. 3.12. Seasonal changes in the plankton of L. George, Uganda. Details are given of the major zooplankter and four of the most abundant phytoplankters; *Melosira* is a diatom genus, the others are of blue-green algae. (After Ganf & Viner[155] and Ganf[153].)

changes in nutrient loading is offset by mountains in the catchment whose high run-off permits a continuous inflow to the lake[510]. This constancy is reflected in the low diversity of the plankton. Over 99% of the plankton biomass is phytoplankton and of this amount 80% comprises six species of blue-green algae. Only a dozen or so other species have been recorded, compared with hundreds in more seasonally variable lakes. Fig. 3.12 shows the great stability in phytoplankton biomass and species composition, and also in zooplankton biomass, which is dominated by only two species, *Thermocyclops hyalinus* (80% by weight) and *Mesocyclops leuckarti* (20% by weight).

Seasonal differences become more marked the farther a lake is situated from the equator, as do changes in the plankton, although these, even if well described, are infrequently understood. A typical pattern of temperate phytoplankton periodicity is shown in Fig. 3.13. A late winter/spring pulse of several diatom species is overlapped by one of Chrysophyta, largely *Dinobryon* species, as it declines. The onset of summer brings a wide variety of green algae, Cryptomonads, dinoflagellates and, in mid to late summer blue algae develop measurable populations. In autumn diatoms may grow again.

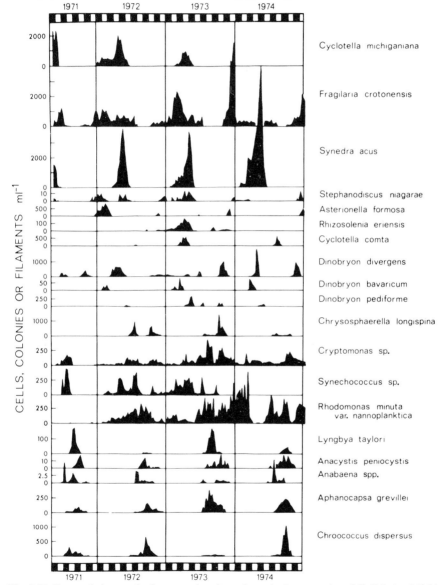

Fig. 3.13. Seasonal changes in the most abundant phytoplankton species of Gull Lake, Michigan, U.S.A., over a period of several years. Numbers of organisms are the means of counts taken at several depths in the water column of this 30 m deep lake. The top seven organisms are diatoms, the next four are Chrysophyta. *Cryptomonas* and *Rhodomonas* are Cryptophyta, the remainder are blue-green algae. (Redrawn from Moss *et al.*[372].)

Some algal growth goes on in winter even under a thick ice cover. Ice is usually transparent unless it is covered by snow, which is opaque. A variety of species grow in winter and physiological investigations have shown some winter algae from polar seas to be adapted to low temperatures and low light intensities[53], though examples are not yet available from freshwater studies. Such species will not grow at 10°C or above. In temperate lakes in winter, even so, growth is not great and the levels of soluble available key nutrients like silicate, phosphate and nitrate are able to build up. Temperatures, day lengths and light intensities increase in late winter and so also do cell division rates of various diatoms whose populations reach maxima in the spring. In L. Windermere, U.K., the diatom *Asterionella formosa* (Hass.) is the first phytoplankter to form a prominent population after the turn of the year. Lund[302] has shown that it is primarily low light intensities which minimize growth in winter for this diatom. Cells are always present in the water and will multiply in winter water if brought into more brightly lit conditions in the laboratory.

From around February onwards the *Asterionella* cells divide and exponential increase brings the population from about 1 cell ml^{-1} in winter to perhaps 10 000 ml^{-1} by late spring[303, 306]. Levels of dissolved nitrate, phosphate and silicate, built up by the inflows in winter all decline during this growth as the cells take them up, and in May or early June the *Asterionella* population suddenly declines. Some dissolved silicate (about 0.4–0.6 mg SiO$_3$–Si l^{-1}) which is required in quantity for production of diatom cell walls, remains in the water, but the cells seem unable to take it up unless small amounts of phosphate are added[232]. On the other hand, addition of silicate without phosphate will also allow some further growth in the water at the time when the population is declining in the lake. Clearly there is a nutrient limitation but the mechanism is complex. As this point is reached the cells seem unable during division to obtain enough silicate to complete the cell walls of their daughters. Weak walls are formed and there may be invasion by bacteria; eventually most of the population is lost to the sediments, perhaps with some regeneration of phosphorus and nitrogen from the cells as they sink.

The major spring diatom growth, which is a feature of many temperate lakes does not always involve *Asterionella*, and does not necessarily end due to silicate limitation, though nutrient shortage of one kind or another is often, but not always, implicated. The genus *Melosira* is common in many fertile lakes and seems to persist in the water only when mixing is vigorous. Its walls are thick, its cells heavy and the onset of stratification, whether inverse under ice during the winter or direct in late spring, leads to sedimentation of the cells. Most phytoplankters are grazed or die in other ways once they reach the sediment, but *Melosira italica* var. *subarctica*[304] in some English Lake District lakes, survives in a quiescent state, with its cell contents contracted, buried in the surface mud during stratified periods in the lake. Once mixing is vigorous enough to disturb the surface sediment it is resuspended, expands its cell contents and divides, so long as light and nutrients are available.

The major spring diatom species may be accompanied, or succeeded by others, particularly those which can take up nutrients at the low concentrations left after the major growth, and continuing low since the inflow of water and hence of new nutrients decreases as the spring and summer wear on. Some indication of how these species may co-exist or replace the major species has come from Titman's work[501] on the interaction between *Asterionella formosa* and a centric diatom *Cyclotella meneghiniana* in Michigan lakes. The relation between growth rates of these species and nutrient concentrations were found to be approximately described by:

$$\mu = \frac{\mu_{max}S}{S + K}$$

where μ is growth rate, S nutrient concentration and K the nutrient concentration giving half the maximum rate (the half saturation constant). The lower the value of K, the more a species is able to grow at low nutrient levels. KP for *Asterionella* was found to be 0.04 μ molar PO_4, and for *Cyclotella* 0.25 μ molar PO_4. This suggests that under limiting phosphate conditions *Asterionella* will tend to compete favourably with *Cyclotella*. On the other hand under limiting silicate conditions the reverse is true, for K_{si} was 3.9 μ molar SiO_3 for *Asterionella*, and 1.4 μ molar SiO_3 for *Cyclotella*.

For each species in turn growth rates will be similar when

$$\frac{\mu_{max}S_{si}}{S_{si} + K_{si}} = \frac{\mu_{max}SP}{SP + KP.}$$

Hence both nutrients are in balanced supply for growth when:

$$\frac{S_{si}}{SP} = \frac{K_{si}}{KP}$$

For *Asterionella* this ratio is 3.9 : 0.04 or 97, so that at Si: P molar ratios greater than 97 in the water, *Asterionella* will be phosphorus limited, and at ratios below 97 it will be silicate limited. For *Cyclotella* the ratio is 5.6, and above and below this the diatom will be phosphorus and silicate limited respectively. When the two species are present in the water together, both will be phosphorus limited when the ratio is greater than 97, and *Asterionella* will tend to survive rather than *Cyclotella*. When ratios are below 5.6, both are silicate limited but, being more efficient at silicate uptake, *Cyclotella* will predominate. But between 5.6 and 97, each diatom is limited by a different nutrient and they should be able to co-exist. Tilman[498] has shown this to be the case. In a lake the ratio experienced will vary according to the weather and the day-to-day nutrient loading and on the growth of other species. The outcome of competition between the two species may, therefore, vary from year to year and from lake to lake where they occur together.

The diatom spring growth is often overlapped and succeeded by Chrysophyte populations, particularly those of *Dinobryon* species. It has been thought that these organisms were inhibited by even moderate concentrations of dissolved phosphate, but it now seems that they have very low KP ($<$ 0.5 μ molar P) values and hence can grow in nutrient depleted water when other organisms cannot[289].

Information on the very complex interplay of populations of the many other species that grow in the summer period rather than in spring is meagre. In summer, the water is much more of a self-contained system than it is in spring when nutrients are coming in with the inflow water. The rapid cycling of phosphorus and nitrogen, and the metabolically induced changes in pH and CO_2 levels caused by high summer photosynthesis make interpretation difficult. Some of the green algae form summer populations because they grow slowly and though division begins in a very sparse background population early in spring, its effects are not really shown until summer[307]. Other species, the Cryptomonads for instance, which require vitamin B_{12} and often thiamine also, may benefit from enhanced bacterial activity and secretion at the higher water temperatures. The blue-green algae of late summer may exploit the microaerophilic zone of the metalimnion (see Chapter 10) since low dissolved oxygen levels seem to favour their growth, and if dissolved inorganic nitrogen becomes scarce then nitrogen-fixing blue-green algae will grow in late summer.

An attempt to account for the major trends in phytoplankton periodicity by means of a model has been made by Lehman et al.[290]. Their rationale was that the growth of various species or groups of species can be approximately predicted from existing knowledge of their physiology, some of which has been discussed above, and from physical and chemical measurements in the field. For instance the blue-green algae, as a group, tend to grow well at higher temperatures than the diatoms and most other phyla; the Chrysophyta seem very tolerant of low phosphate concentrations; the desmids (Chlorophyta) divide slowly and might be expected to build up sizeable populations only over a lengthy period, whereas diatoms seem able to take advantage of abundant nutrients at the end of winter and to build up their populations rapidly. A search of the literature allowed Lehman et al. to decide, for 'typical' members of the blue-green algae, chrysophytes, green algae, diatoms and dinoflagellates, representative values of various growth parameters, some of which are included in Table 3.4. Some generally held beliefs about phytoplankton physiology are quantified in the table—the intolerance of high light intensities, low affinity for nutrients and low sinking rates (owing to gas vesicles) of the blue-green algae for instance.

The relationships of the growth parameters to external conditions such as temperature, light and nutrient concentrations, were quantified in sets of equations also based on available literature.

Small inocula of cells of a diatom, a blue-green alga, a chrysophyte, a green alga and a dinoflagellate were assumed to be present in the winter water with relatively high concentrations of nutrients available. The ensuing wax and wane of the various algae populations could be predicted by manipulation, in a computer, of the equations describing seasonal trends in temperature, light and nutrient loading and the effects of these on nutrient uptake and photosynthesis.

Table 3.4. Some growth parameters of 'typical' representatives of some major algal groups. (From Lehman *et al.*[290].)

	Green alga	Blue-green alga	Diatom	Dino-flagellate	Chrysophyte
Optimal light intensity for photosynthesis (ly min^{-1})	0.045	0.03	0.1	0.2	0.06
Maximum photosynthetic rate per cell (μ moles C uptake cell^{-1} hr^{-1} × 10)	7	1.5	3	15	1.8
Maximum cell division rate per day	3	2.5	3	1.5	1.5
Maximum uptake rate of nutrients (μ moles cell hr^{-1} × 10^9) P	10	1	3	10	1
N	30	10	20	100	10
Si	0	0	700	0	0
Nutrient concentration at which half maximum rate of uptake is achieved (K) (μM) P	1	2	1	3	0.6
N	2	5	1.5	5	0.75
Si	0	0	7	0	0
Sinking rate (m d^{-1})	0.8	0.2	2.5	0.5	0.5

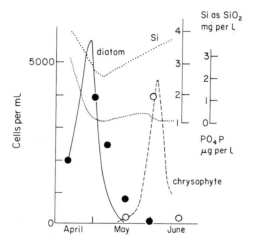

Fig. 3.14. Some results of a simulation model intended to predict phytoplankton seasonal changes. The curves give the predicted changes in silicate, soluble reactive phosphate, a diatom and a chrysophyte population. Superimposed are actual data points for populations of a diatom (*Synedra*, solid symbols) and a chrysophyte (*Dinobryon*, open symbols) in Linsley Pond, a lake comparable to that intended in the simulation. (After Lehman *et al.*[290].)

The model predicted a series of species changes which corresponded generally and sometimes quite closely, to events in a real lake (Fig. 3.14). This suggests that the physiological properties of the algae are major determinants of phytoplankton seasonal periodicity.

The successful twin approaches of analytical field observations and experiment and the subsequent syntheses of such data into predictive models perhaps bias our view of the causative factors of seasonal periodicity by their concentration on the larger and more readily studied species. The smaller plankters are less easily studied for they may form ephemeral populations, readily decimated by grazing and they may also be more dependent on organic nutrients. Some species secrete compounds which, in laboratory culture, inhibit or stimulate the growth of others[274, 288, 415]. The difficulty is that these substances are less readily detected in the rapidly changing field community than they are in the tranquillity of the laboratory culture flask. Here, populations not decimated by grazing can build up to levels where their secretions are sufficiently copious to be readily analysed. There seems little prospect of a ready solution to this problem, but it must be emphasized that although the seasonal cycles of a few species can be accounted for by major trends in physical and chemical factors, those of perhaps several hundreds of others in a lake cannot, and we must seek subtler mechanisms for studying them.

Differential grazing can change the composition of a phytoplankton population and influence seasonal periodicity. Porter[409] enclosed 500 litre samples of lake water in large polyethylene bags which she sealed and resuspended for several days in the lake. From some of the bags she had removed the larger zooplankters (Crustaceans) by previously filtering the water through a 125 μm mesh net. In others she increased the zooplankton population severalfold by adding animals caught with a net from the open water. The major effect of grazers, which included filter feeding herbivores (*Daphnia galeata mendota* and *Diaptomus minutus*) and raptorial herbivores (*Cyclops scutifer*) was to suppress populations of small flagellates and nanoplankters and of large diatoms. Populations of large colonial green algae were not affected or even increased. They were ingested but not digested and may even have benefited from passage through guts. Phosphate released there by digestion of other species was taken up by the colonial green alga *Sphaerocystis schroeteri* as it passed through undigested and emerged growing healthily on copepod faecal pellets[410].

Apart from a size selection of food imposed by the physical ability of the animal to eat it, some species are actively rejected by grazers[115], while others seem particularly sought after such that the animals can distinguish them from others of a closely similar size. Edmondson[117] measured the reproductive rates of several species of rotifers by the egg : female ratio and found high correlations with the abundance of particular small phytoplankters.

The larger phytoplankters are not completely immune from grazing. There is selective predation of large prey by smaller predators when protozoa move into the colonies of the larger green algae and attack the cells one by one with pseudopodia or suction organelles[61]. The protozoa may be very specific in their choice of host, and thus eat species which escape conventional grazing by crustaceans and rotifers. Other groups previously considered to be rarely grazed, such as the larger colonial blue-green algae, have also been shown to be eaten and digested, in this case by raptorial copepods[355]. It seems that all of the algae are grazed at least some of the time.

3.3.1 Seasonal changes in the zooplankton

These depend partly on physical conditions since egg development is temperature dependent, partly on available food supply, and partly on the predation pressure exerted by the carnivorous zooplankton, fish and other vertebrates. Graphs of changes in zooplankton biomass often show apparently irregular fluctuations as a result, but a general increase in population in summer compared with winter is usually apparent in temperate and polar lakes. In L. George, astride the equator, seasonal fluctuations are essentially absent as they are in the phytoplankton (Fig. 3.12).

Because they can assimilate detritus, zooplankters are partly buffered from the problems which might be caused by lengthy shortages of edible algae. Zooplankton populations thus do not always correlate with those of phytoplankton and are sometimes very abundant in the near absence of the latter. These circumstances may also be explained if the algae have all just been ingested.

Zooplankters in temperate waters overwinter as adults or resting eggs. As the water warms and food becomes more abundant their populations increase and many generations are produced during a year. Rather fewer species of zooplankton than of phytoplankton are present in a lake but the potentialities of niche differentiation are great. Their movement in swarms, the vertical migrations which they make and the possibilities of food selectivity mean that several species of zooplankton can co-exist abundantly with minimal competition. Seasonal patterns of relative abundance do exist, however, and their explanation partly concerns the effects of predation. A common shift is from predominance of large species early in spring to smaller species as the summer progresses and seems related to selection by fish for the larger prey which they can see more easily when they begin to feed in spring.

The effects of fish predation have been shown clearly when fish have been newly introduced into lakes where previously they did not live. Fig. 3.15 shows the marked shift in zooplankton species and sizes which followed introduction of the planktivorous fish *Alosa aestivalis* into Crystal lake in Connecticut which previously lacked such a planktivore. The zooplankton community was changed from one of *Epischura, Daphnia* and *Mesocyclops,* which are all usually more than 1 mm in size to one of *Ceriodaphnia, Tropocyclops* and *Bosmina* which are all rather smaller than 1 mm. *Cyclops,* at just under 1 mm, persisted in both situations.

This work led Brooks & Dodson[49] to state their size-efficiency hypothesis to explain why crustacean zooplankton communities at a given time during the year tend to be of rather uniform size range. They believed that all zooplankters competed for the 1–15 μm particulate matter in the water but that the larger

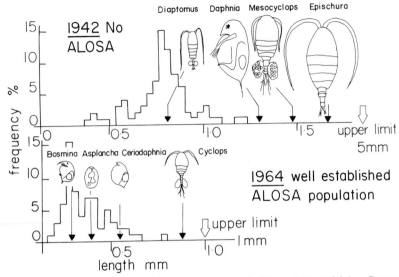

Fig. 3.15. Composition of the mainly crustacean zooplankton of Crystal lake, Connecticut before (1942) and after (1964) introduction of a planktivorous fish, *Alosa aestivalis*. Planktivorous fish had previously not been present. Specimens are drawn to scale and represent mean size for each species. The arrows indicate the size of the smallest mature instar of each species. The effect of the fish has been to replace a community of large species with one of much smaller organisms. (From Brooks & Dodson[49] Fig. 4. Copyright 1965 by the American Association for the Advancement of Science.)

zooplankters competed for it more effectively. Small animals were thus excluded by starvation if large ones were present. Vertebrate predators (largely fish but also amphibians), however, select the larger Crustacea (Cladocerans, Calanoid copepods), and, depending on the intensity of predation, smaller zooplankters (smaller Crustaceans, rotifers) could co-exist up to the state of the complete elimination of the large forms where predation was intense.

Subsequent work has failed to demonstrate any greater efficiency at feeding by larger zooplankters. Predominance of larger zooplankters, when vertebrate predation on them is relaxed, seems to rest on size-selective predation of the smaller plankters by carnivorous zooplankters in the same way that predominance of small forms depends on vertebrate predation of the larger ones[105]. This subject is critically reviewed by Hall *et al.*[190].

3.3.2 Cyclomorphosis of zooplankton

Many zooplankters, particularly Cladocerans, undergo a change in shape as their generations succeed one another through the year (Fig. 3.16). This often takes the form of an extension to the head, the helmet, which is more translucent than the body, sometimes coupled with a reduction in size of the body. The change is not one through which an individual goes but may be phenotypic rather than genotypic. Females are capable of producing a series of forms ranging from non-helmeted to those with extreme helmet development. In winter, when fish predation is low or absent, non-helmeted forms are produced, which, not having to invest energy in helmet production, are able to increase their numbers most rapidly. In summer the balance may turn in favour of helmeted forms when their usually effectively smaller body size (the helmet being almost invisible under water) may make them less vulnerable to fish predation[48].

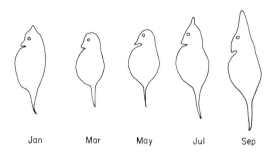

Jan   Mar   May   Jul   Sep

Fig. 3.16. Changes in shape (cyclomorphosis) of *Daphnia cucullata* generations during a year.

The small zooplankter *Ceriodaphnia cornuta* in Gatun Lake, Panama, provides a variation on the theme[548]. Seasonal changes are absent but spatial ones occur. *C. cornuta* exists in horned and unhorned forms, the latter having much less development of certain body processes. The unhorned form has reproductive advantages of a shorter generation time and a greater number of viable eggs produced per female. However, it is selectively eaten by a plankti-vorous fish, *Thyrinopsis chagresi*. The horned form predominates in offshore areas of the lake where, in this case, fish predation is greatest. It is, however, effectively no smaller than the unhorned form, but the production of horns appears genetically linked with the production of a much smaller eye than that produced in the unhorned forms. The larger, black eye of the unhorned forms catches the fishes attention more easily and leads to selective predation.

Other hypotheses have been advanced to account for cyclomorphosis. These are reviewed by Hutchinson[236], and Hebert[206] suggests that the flatness of the helmets may have a role in improving gas exchange in warm water. Helmets are not produced in cold, oxygen-rich water.

3.3.3 Summary of seasonal changes in the plankton

Studying seasonal changes in plankton has the same appeal as following the development of any complicated situation, human or otherwise, and of ponder-ing the outcome. The diversity of phytoplankton species, measured by some index which takes into account the equitability, the distribution of numbers among species, as well as just the length of the species list, increases from winter to autumn, then falls (Fig. 3.17). Some observers have likened this to a similar increase supposed to take place in the succession of communities in 'new' habitats on land—a glacial moraine, a new volcanic island, a cleared forest—others have likened it to the change in land communities which occurred

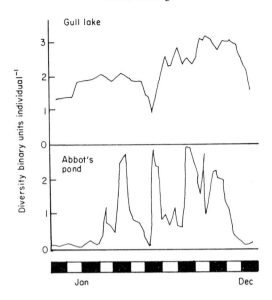

Fig. 3.17. Seasonal changes in diversity of the phytoplankton communities of a deep lake (Gull Lake) and a shallow pond (Abbot's Pond). Diversity indices are on a logarithmic scale and are calculated from the Shannon formula[461], Diversity $= -^i_1\Sigma P_i \log_2 P_i$ where P_i is the numerical proportion of the ith species of the total number of individuals present. (After Moss[363].)

on climatic changes following the retreat of the continental glaciers some 10–15 000 years ago.

In many ways the seasonal periodicity of phytoplankton is better viewed as a regular cyclic change in which a tendency for the community to achieve some sort of equilibrium is regularly frustrated by the change in weather which ultimately drives the change. The phytoplankton seems greatly controlled either directly by climatic factors, through temperature, light and the rate of inflow of nutrient laden water, or indirectly through climatic influence on fish activity, predation on zooplankton and grazing[411]. The increase in diversity experienced in the phytoplankton in summer may thus partly reflect the greater variety of niches available in the water as all of these factors come into play, and partly the simple accumulation of residual populations of those species whose maximal growth was early but whose populations have not yet entirely been reduced to the background level[363].

The community composition of the zooplankton throughout the year, and from lake to lake, is much more a matter of the relation between invertebrate and vertebrate predators and their respectively smaller and larger zooplankton prey (Table 3.2). Rotifers and the smaller Cladocerans are readily eaten by carnivorous zooplankters and by insect larvae, such as those of the Dipteran fly *Chaoborus*, the phantom larva, which migrate upwards at night from the mud. However, they reproduce rapidly by parthenogenesis and meet their losses by high production. The larger zooplankters are invulnerable to invertebrate predation but are readily taken by fish. They reproduce relatively rapidly, but also have evolved strategies to minimize predation by cyclomorphosis or by vertical migration. Many retreat to the dark hypolimnion by day, where fish cannot easily see them. At night they move to the dark surface waters to feed. The very largest zooplankters, some of the Copepods, do not reproduce rapidly and are delectably sized morsels for vertebrate predators. They do have the genetic advantage over parthenogenetic zooplankters of sexual reproduction in each generation, and they may also move extremely rapidly (10–50 times as rapidly as Cladocerans[4]) and with swift changes in direction can avoid the more clumsy of their predators.

For both phytoplankters and zooplankters there are then many possible successful survival strategies even in the apparent uniformity of a mixed water column—an object lesson for those who would wish to impose uniformity of institutions on human societies.

3.4 Further reading

Knowledge of the plankton has been transformed in recent years, and the rapidity of developments has meant that specific references have been widely quoted in this chapter. For more detailed accounts of specific areas, Hutchinson[236] gives masterly accounts of classic work on the plankton, and Fogg[137] is a wonderfully clear account of how work with cultures of algae can aid in the solution of problems such as those of seasonal periodicity. Wetzel[530] is strong on chemical transformations, particularly those of carbon, whilst Rigler[433] comprehensively reviews work on phosphorus. Work on bacteria in the plankton is reviewed in Rheinheimer[426], but is much less copious than that on other plankters. Methodology for determining production in the plankton is collected into Edmondson & Winberg[121], Sorokin & Kadota[468] and Vollenweider[511]

Identification of planktonic organisms can be sought in Donner[107], Edmondson[116], Harding & Smith[193], Huber-Pestallozzi[231], Illies[245], Prescott[414], Scourfield & Harding[458] and West & West[525]. George[162] is a guide to the sources of keys to algae.

CHAPTER 4

STREAMS AND OTHER

EROSIVE HABITATS

4.1 Introduction

It may seem strange, but the mean velocity of water movement increases from the upstream, often torrential stretches, to the lowland reaches of a stream. Stream flow is a complex matter and a single stretch will include parcels of water each moving at very different speeds—lower close to the bed and banks, where frictional retardation, or drag, occurs, and higher in the middle.

As a stream becomes larger, through collection of water from an increasing area of catchment, its discharge, D (the volume of water flowing), and its cross-sectional area, A, increase. These quantities are related by the equation

$$D = A.V,$$

from which it can be seen that V need not inevitably increase if both D and A do. Generally, however, since the proportion of water slowed by friction decreases as the cross-sectional area becomes larger, the mean velocity does increase. At the edges velocity may decrease nonetheless to values much below those recorded upstream, allowing the deposition of silt eroded from the catchment area. In turn, swamps of emergent aquatic plants may colonize this and compound the process. Large streams, or rivers, often have long enough courses for a plankton community (see Chapter 3) to develop and as a river enters a lake, the great increase in A, coupled with a little changed value of D means that V departs from the general rule and decreases sharply.

This chapter is mostly concerned with the upper sections of streams where the flow appears fastest, generally because of the smaller scale of the stream ecosystem. The mean flow rate is lower than in downstream sections but the range of current speeds is narrower, so that flows do not often fall to levels where silt or even the heavier sand and gravel may be deposited. The rocks and boulders in the stream are thus kept clear of deposits, but some may accumulate in pockets on the bed where flow rates fall in the shelter of boulders or other obstacles such as fallen trees.

The plankton and reed-swamp communities of large lowland rivers are fundamentally similar to those of lakes (see Chapters 3 & 6). They differ in that the riverine ecosystem has a larger supply of organic detrital matter kept in suspension and may depend on autochthonous primary production to a lesser extent than a lake ecosystem. A similar parallel may be drawn between upstream ecosystems and their nearest equivalents in lakes, the vigorously wave-washed rocky shores, where they occur. Streams, however, receive large amounts of litter from the catchment area and recent work has shown that this is usually a far greater component of the stream energy supply than *in situ* photosynthesis. The to-and-fro of wave action on rocky shores in lakes keeps them free of organic particles which are washed to greater depths. This leads to a dependence largely on photosynthesis by algae attached to the rocks (epiliths). Such scoured shores have been less thoroughly studied than upland streams but such comparisons as can be drawn will be considered.

4.2 Energy supply in upland streams compared with other aquatic ecosystems

Some of the generalities of Section 4.1 can be illustrated by comparison of three ecosystems—an upland stream, a riverine lake with a high throughput of water and a lake whose flushing time is comparatively long.

The latter is Lawrence Lake, Michigan, U.S.A., a small water body formed originally by the melting of an ice block buried in glacial drift. It is fed partly by groundwater and partly by small streams and has been extremely thoroughly investigated by R. G. Wetzel and his co-workers[530, 534]. Fig. 4.1 compares

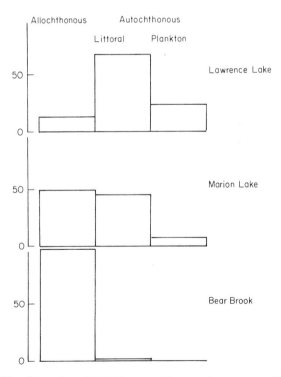

Fig. 4.1. Availability of organic matter to three aquatic ecosystems, expressed as percentages of the total input in each case. 'Littoral' includes aquatic macrophytes (see Chapter 6) and all algal communities (see Chapters 4, 5 and 6) other than those of the plankton.

the various annual organic carbon supplies to the lake and it is clear that autochthonous primary production is predominant. The residence time of water in the lake is probably more than a year and it is perhaps mainly because of the low rate of inflow that the allochthonous component is small. The concentration of organic matter in the incoming water is not unusually low.

The riverine-lake considered in Fig. 4.1 is little more than a widening and deepening of Jacob's Creek in the coastal mountains of British Columbia and called Marion Lake. Its catchment area is about a hundred times that of the lake itself and the climate is wet. The lake basin is flushed many times per year. As a result, the contributions of dissolved and particulate organic matter entering the lake are as high as that of autochthonous production[191].

Thirdly, Bear Brook, in New Hampshire, U.S.A., is an upland stream draining undisturbed hardwood forest which even forms an almost continuous canopy over the stream bed in summer. Allochthonous organic matter comprises by far the bulk of the organic carbon supply to the stream[133]. The only primary producers of any consequence in the stream itself are mosses and these supply only 0·02% of the total organic influx, in contrast to autochthonous contributions of 51·5% in Marion Lake and 88% in Lawrence Lake.

This pattern seems to reflect the relative size of a water body to its catchment —the smaller it is, the more likely allochthonous organic matter supplies from a relatively large area will dominate autochthonous supplies from a small area. The smaller the water body also, the more likely it is that shading from overhanging terrestrial vegetation will inhibit photosynthesis in the water. And the smaller the water body the greater the likelihood that flushing rates will be high enough to prevent phytoplankton development and perhaps the establishment of suitable habitats for aquatic macrophytes.

4.3 The processing of organic matter in streams

4.3.1 The sources of organic matter

It is not an easy matter to account for all the sources of organic matter to a section of a stream, and to balance these inputs with equally good measures of what happens to them. Work of this detail has been carried out on the 1700 m section of Bear Brook, New Hampshire, introduced above[133]. This stream section was ideal for the work since it lies on hard bedrock (sillimanite-zone gneiss) which prevents deep seepage and directs all the rain and snow which is not evaporated or transpired by the vegetation, to the stream and allows full accounting of water movement. It is also a small stream with a total area of about 0·6 ha and a maximum depth of about 60 cm, which eases sampling problems. Furthermore it is surrounded by a uniform terrestrial vegetation which means that relatively fewer sampling sites are necessary than if the vegetation was heterogeneous, and lightens what is necessarily a heavy workload of measurement in any case.

The forest edge overhangs the stream and direct fall of litter was measured by collection in suitable boxes placed along the stream bank. Litter includes not only leaves, but branches, bud scales, flowers, fruits and the exuviae and droppings (frass) of leaf-living animals. Although most litter fall was in autumn it is not negligible at other times of year, for there is a continual turnover of leaves even in summer, bud scales fall in spring and frass is a constant source in summer. Wind and the weight of snow breaks off tree branches in winter. Not only is there a direct litter fall, there is also a contribution blown sideways into the stream along the forest floor, and this was collected in traps placed at right angles to the stream. Rain dripping directly into the stream from the overhanging leaf canopy in summer picks up dissolved organic matter exuded from the leaves and from leaf insects. This throughfall was collected and measured by chemical means.

At the top of the stream section the rate of entry of organic matter carried from upstream was measured in three categories—coarse particulate organic matter (CPOM), greater than 1 mm in size; fine particulate organic matter (FPOM), less than 1 mm in size but collectable by filtration through a glass fibre filter of pore size 1 μm, and dissolved organic matter (DOM) which passed through such a filter. DOM presents few problems if sampling is fairly frequent as its concentration is relatively steady and its total contribution can be calculated easily if the discharge of the stream is known. This was regularly measured at Bear Brook by a V-notch weir and varied from 2–1000 l sec^{-1}. Calibration of such a weir allows discharge to be monitored by continuous, automatically recorded measurement of the depth of water over a V-shaped notch of concrete or wood firmly set across the stream bed.

FPOM and to a much greater extent CPOM present problems of measurement since they tend to come down the stream in pulses related to spates of water passing after thaws and rain storms. In Bear Brook, very frequent sampling was necessary to obtain a reliable measure of FPOM. Even this failed to give a reliable measure of CPOM when it comprised merely the spreading of a 1 mm mesh net across the stream for a timed period every week or two. Fortunately, in a nearby similar stream as part of another experiment, CPOM was continuously collected in a concrete ponding basin built into the stream. This collected continuously all CPOM passing and gave a value some twenty times higher than that obtained by regular discrete sampling.

DOM entered along the length of Bear Brook from the podzolic forest soils bordering it. Samples of this water were collected for analysis from seeps and its volume determined from the difference between the total amount of water entering the stream at the top of the stretch and that leaving it at the bottom. Finally, estimates were made of the photosynthesis of the moss population of the stream (algae and higher plants were said to be absent) by a method based on the oxygen light and dark bottle technique (see Chapter 3). The conversions in energy through animal consumption were estimated from biomass measurements and productivity data from other sites. Though these estimates were very crude (the methodology is discussed later) the pattern of energy flow in the stream would be little altered if they were as much as an order of magnitude in error.

Table 4.1 shows the energy budget which Fisher & Likens have constructed for the stretch of Bear Brook they investigated. Most notable are the high contributions of litter (43·7%), particularly direct leaf fall, and of dissolved organic matter (46·3%). Autochthonous primary production, as mentioned in Section 4.1 was negligible in this stream.

Table 4.1. Annual energy budget for Bear Brook, New Hampshire[133]

Inputs (k cal m⁻²yr⁻¹ (%))			Outputs (k cal m⁻²yr⁻¹ (%))		
Litter fall:			Transport downstream:		
Leaves	1370	(22·7)	CPOM	930	(15)
Branches	520	(8·6)	FPOM	274	(5)
Miscellaneous	370	(6·1)	DOM	2800	(46)
Side blow litter	380	(6·3)	Respiration of		
Throughfall	31	(0·5)	microorganisms	2026	(34)
Transport from			Respiration of		
upstream:			invertebrates	9	(0·2)
CPOM	430	(7·1)			
FPOM	128	(2·1)		6039	100·2
DOM	1300	(21·5)			
Ground water					
DOM	1500	(24·8)			
Moss photosynthesis	10	(0·2)			
	6039	99·9			

The outputs of energy also show some startling features. Firstly, if all the measured outputs (CPOM, FPOM, DOM, and utilization by animals) are added, a total of 4013 k cal m⁻² yr⁻¹ was accounted for, whereas 6039 k cal m⁻² yr⁻¹ entered. The difference, 2026 k cal m⁻² yr⁻¹, must be attributed to the respiration of microorganisms feeding heterotrophically on the organic matter entering the stream, and thus processing about a third of it. The remaining two-thirds were mostly exported downstream. Such streams as this would appear therefore to be relatively efficient processing factories for the large amounts of organic matter they receive, and the next section considers the mechanics of these processes.

4.3.2 Mechanics of processing of organic matter in streams

Microorganisms

Leaf litter is very different, chemically, from the living vegetation it once was. Labile, reusable substances, both inorganic and organic have been translocated back into the perennating organs and what is left is largely cellulose, lignin and resistant carbohydrates, plus substances like polyphenols which may be meta-bolic waste products. Microorganisms have already colonized the litter before it has fallen, and begun the decomposition which, for most leaves, will be com-pleted in the soil of the terrestrial ecosystem.

For litter which falls or blows into a stream a parallel sequence of events occurs. A large proportion of any remaining soluble substances is leached out within a few days, or even hours[244]. The speed of this depends on the leaf species, but up to 30% may be lost in the first 24 hours and deciduous leaves are leached more rapidly than those of conifers. This DOM, together with that washed out of the catchment soils, may be taken up by microorganisms associated with the stream bed, or may be washed downstream and utilized there. Some of it may be precipitated and aggregated into fine particles by apparently physico-chemical processes, and may thus join the FPOM fraction.

As soon as it enters the water there is a chance that the litter will be colonized by aquatic microorganisms, and over a few weeks there is an intricate succession of fungi present in and on the litter. Bacteria do not appear to be important at this stage, though of course they are present. The strains found are casual adherents from the bacterial flora present in the bottom substrata and the surrounding land. They do not appear in any orderly succession, nor is their biomass high compared with that of the fungi[486].

The fungi concerned are largely from a group called the aquatic hyphomycetes, whose spores were first discovered in the 1940s by Ingold. The spores are often tetraradiate in structure (Fig. 4.2), a morphology which seems to favour

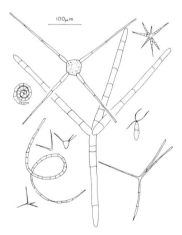

Fig. 4.2. Spores, mostly of aquatic Hyphomycetes, drawn from a sample taken below Sezibwa Falls, near Kampala, Uganda. (Modified from Ingold[249].)

their sticking to the litter when swept against it by the stream flow[249]. The terrestrial fungi already present on the litter as it enters the water seem to play little further role. Suberkropp & Klug[486] have studied the fungi and bacteria which colonize 'new' litter placed in a stream, and, because of the difficulty of studying microbial ecology compared with the activities of larger organisms, some mention of their methodology is useful.

Leaves were removed from the experimental packs (bunches of leaves held in a mass by nylon monofilament line and anchored to a brick in the stream) and discs cut aseptically from them on several occasions over a period of a few weeks. Some discs were homogenized in a blender and aliquots of the suspension produced were plated onto peptone-yeast extract agar medium and incubated at stream temperature for an eventual viable count of bacterial colonies to be produced. Direct counts of bacteria were also carried out on aliquots passed through cellulose ester filters and suitably stained. Fungal mycelium in the discs was stained with lactophenol-cotton blue and examined directly. This gives some idea of biomass, but it is usually impossible to identify the species present. Most fungi are identifiable from their sporing structures so further leaf discs were incubated in sterile inorganic medium which induces sporulation. A measure of biomass was also obtained by determination of the adenosine triphosphate (ATP) content of homogenized discs. ATP is an essential component of living organisms but is rapidly degraded in dead ones. It is estimated by measuring the extent to which it causes light to be produced from a mixture of luciferin and an enzyme, luciferase—the system which allows some dinoflagellates and glow-worms to sparkle.

Two sorts of leaves were studied—white oak (*Quercus alba* L.) and pignut hickory (*Carya glabra*)—in a Michigan stream. Some species of Hyphomycetes appeared on both types, but there were differences in the succession which developed (Fig. 4.3). The biomass (ATP) increased to a plateau after about two weeks then declined after about 12 weeks as the leaves were progressively degraded.

Leaf litter provides only part of the growth requirements of the fungi which colonize it. Although it is rich in carbon and chemical energy, it contains little nitrogen and phosphorus and the fungi must obtain much of their supply, particularly of nitrogen, from the water. The nitrogen to carbon ratio of the colonized litter increases as the fungi build up their biomass[243, 268, 269]. This is important for the next stage in decomposition by invertebrate animals, for the fungal biomass, rich in nitrogen, is a 'better' food for invertebrate animals than uncolonized litter. The speed with which different sorts of litter are consumed largely depends on their food contents as expressed by animal feeding preferences.

The shredders
Mechanical abrasion breaks down some of the colonized (and uncolonized) leaf litter to FPOM, but much of it is chewed by coarse particle-feeding invertebrates, the 'shredders', which bite out the softer parts, between leaf veins for example, leaving the vascular skeleton for later abrasion or consumption[81, 82]. Shredders include insect larvae and nymphs, e.g. Tipulidae, the usually quite large, fat, legless larvae of a family of flies (Diptera); Limnephilidae and Lepidostomatidae, families of caddis flies (Trichoptera), and some stoneflies (Plecoptera). Also included are the larger Crustacea, *Hyallela*, *Asellus* and *Gammarus*. Much work has been carried out on the latter for they are easily maintained in the laboratory.

When *Gammarus* were offered fungally colonized leaves of elm (*Ulmus americana* L.), alder (*Alnus rugosa* (Du Roi) Spr.), white oak (*Quercus alba* L.), beech (*Fagus grandifolia* Ehrh.) and sugar maple (*Acer saccharum* Marsh), they showed a preference roughly in the same order as that in which the leaves support fungi—elm, maple, alder/oak, beech. This same order of preference appears to

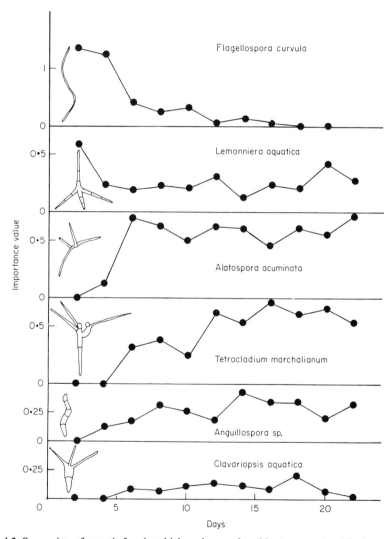

Fig. 4.3. Succession of aquatic fungi on hickory leaves placed in stream water. The leaves were collected as they were shed from the trees and attached to bricks placed in the stream at the beginning of the experiment. The fungi were identified from their spores, which are illustrated and their relative abundance is given as an importance index. This is the sum of the frequency with which each species was noted in a series of replicate samples at each date and the relative abundance of the species, recorded on an arbitrary scale, divided by the total abundance of all fungi. (Redrawn from data of Suberkropp & Klug[486].)

be shared by stonefly and mayfly nymphs[269]. The intrinsic properties of the leaves themselves help determine these preferences, for the order is also maintained if uncolonized litter is offered. Preference is always for colonized over uncolonized leaves, however, and can be influenced by artificial inoculation of particular species of fungi[21, 22]. Part of the reason why *Gammarus pulex* does not colonize particular streams in north-western England is that the available food-grass litter, *Nardia*, a liverwort, and *Tricladium*, the only common aquatic Hyphomycete to grow there, does not meet its dietary needs[538].

The role of fungi in the processing of leaf litter appears similar to that of the sandwich fillings and butter commonly used by us to increase the palatability and nutritional content of modern bread. There are usually several species of shredder present in a stream, and preference for different sorts of 'sandwiches' may explain why they all can co-exist. For example Mackay & Kalff[315] have studied two closely related species of caddis fly larvae, *Pycnopsyche gentilis* McLachlan, and *P. luculenta* Betten in a Canadian stream. *P. luculenta* was found equally in packs of leaves accumulated among rocks and in finer detrital deposits at the stream edges, while 90% or more of *P. guttifer* were found among coarse detritus. The field distribution could be reproduced in laboratory· streams (Table 4.2) when different combinations of whole leaves and finer

Table 4.2. Distribution of two Pycnopsyche species in habitats in West Creek, Quebec, 1968/69 Measurements are of numbers per standard area.

	Detritus	Detritus (+ leaves)	Leaves (+ detritus)	Leaves
No. of samples	18	19	18	20
P. gentilis (mean ± s.e.)	7·7 ± 1·4	8·8 ± 2·0	15·1 ± 3·3	14·3 ± 4·1
P. luculenta (mean ± s.e.)	4·9 ± 0·7	4·6 ± 1·2	2·2 ± 0·6	0·7 ± 0·7

Distribution of two *Pycnopsyche* species in habitats provided in laboratory streams.

	Detritus	Detritus (+ leaves)	Leaves (+ detritus)	Leaves
P. gentilis (%)	10·0	16·5	39·5	34·0
P. luculenta (%)	46·9	31·4	13·6	8·1

detritus are presented to the larvae. These animals use leaves and small twigs not only as food but also for building the cases in which they live. They have similar life cycles and share some food resources, but since these are normally plentiful, competition is minimized. Only when spates of water resulting from the thaw of winter snow wash away much of the habitat provided by the leaf packets, particularly for *P. gentilis*, does there appear to be competition for habitat space and intensive feeding by birds (grackles) on the crowded larvae.

Microorganisms alone can degrade leaf litter, but the process is accelerated by a fifth or more through the action of shredder invertebrates[81], such that as much as $1 \cdot 5\%$ day^{-1} of the litter is converted to animal tissue, carbon dioxide or FPOM. FPOM results from abrasion by water movement and from waste during feeding. It also includes the egested faeces of the shredders. FPOM is additionally colonized by microorganisms as 'new' surfaces are exposed and forms a food source for a second series of invertebrates, the 'collectors', which collect the FPOM by filtration or deposit feeding. Perhaps half of the FPOM they ingest consists of the faeces of shredders[520].

Collectors, scrapers and carnivores
The finer debris that results from shredder and fungal activity, the fine material washed in directly from the catchment area, and particles eroded from the algal communities attached to rocks in the stream-bed, form a rich food source for those invertebrates that can efficiently use it. Most of these organisms are filter or deposit feeders though some scrape it from surfaces, together with the attached algae growing there, where and when the current is slow enough to allow some deposition. Typical scrapers include snails and freshwater limpets (Ancylidae) which rasp at the rock surfaces with toothed organs called radulas, and some caddis fly larvae and a few mayfly nymphs whose mouth parts have evolved stiff bristles with which they scour the rocks.

For deposit-feeding, carried out mainly by burrowing Dipteran larvae (Chironominae, Orthocladinae, Diamesinae) and some mayfly nymphs, pockets of sediment must be able to collect in areas of slack flow, created by the rock configurations or by lodged tree trunks or branches. Such irregularities also allow accumulation of leaf packs and are essential for the overall processing of organic matter entering the stream. Canalization of streams in the interests of flood prevention and improved land drainage (see Chapter 12) removes the irregularities necessary for such pockets to form and reduces a stream's capacity to oxidize organic matter.

Filter feeders show remarkable ingenuity at gathering organic matter from flowing water. Usually they have fringes of fine hairs on the mouth parts or legs (as in some mayfly nymphs, blackfly larvae (*Simulidae*) Diptera) in which particles collect before transfer to the mouth, or they may construct nets (as do some caddis fly larvae) associated with the cases in which the larvae live. Some examples are shown in Fig. 4.4. *Simulium* has large, fan-shaped structures on its head which are held out in fast but smoothly flowing water. Very turbulent flow (which is much commoner in streams) tends to remove particles from the filters as soon as they are caught. *Rheotanytarsus* builds cases of silt particles stuck together with mouth secretions and across projections of these cases it spins fine webs. Periodically it eats and replaces the webs as they become clogged with particles. *Hydropsyche* simultaneously spins two silken threads and with a figure of eight movement of the head weaves these into a net capable of catching the coarser FPOM. The mesh is about 340 μm $\times$ 170 μm and the net is held between twigs and stones, with the larva itself living in a tube behind the net.

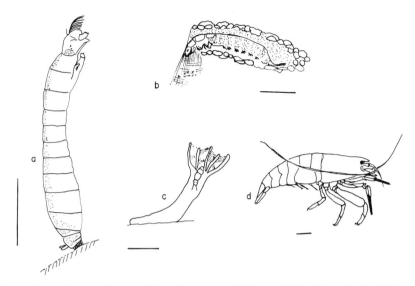

Fig. 4.4. Invertebrate collectors of running waters, (a) larva of *Simulium*—the head fans are held out as the animal itself is trailed horizontally in the current; (b) *Hydropsyche* larva—this insect weaves a fine web at the head of a shelter usually made from small stones, a method similar to that used by (c) *Rheotanytarsus*, also a caddis fly larva; (d) the atyid prawn, *Atya innocous*, whose chelipeds, looking like the tips of artists' paintbrushes at the front of the organism, may be used as filtering fans, or to sweep particles off rock surfaces. Scale lines represent 1 mm for (a), (b) and (c), and 10 mm for (d). ((b) and (c) redrawn from Hynes[242], (d) redrawn from Fryer[146].)

Much finer nets are spun in sandy burrows by *Macronema* species. These are Amazonian caddis flies and direct water through the net by a tube, shaped like a ship's funnel which projects a few cm above the sandy bottom of the stream [446].

Lastly, a group of freshwater prawns has been studied in Dominican streams[146, 147]. They are very important components there of the stream fauna and are often more than 10 cm in length. Some of them have long bristles on a pair of limbs, the chelipeds, near the front of the animal. These bristles are delicate but can be held together like the tip of an artist's wet paintbrush. They can then be used to sweep organic particles from deposits or surfaces and ancillary appendages transfer these to the mouth. Alternatively, the bristles can be expanded into a fan which is held into the current and which, with the help of very fine setules on the bristles, acts as a filtering device. Apart from their intrinsic interest, these prawns provide a reminder that classification of stream (or any other) animals into feeding modes is but a convenience and that the categories are not always distinct. Shredders like *Gammarus* thrive better in laboratory experiments if allowed access to the FPOM of their own faeces as well as to fungally-colonized leaf material, than if given the latter alone!

The food web of upland streams, like almost all food webs, is completed by carnivores, both invertebrate and vertebrate. The invertebrate carnivores include leeches (Hirudinae, Annelida), a variety of insect larvae and water mites (Hydracarina) among others, and just as shredders and collectors may ingest animal material with their predominantly detrital and fungal diet, the carnivores may also have a wider range of food than their name suggests. All the invertebrates are potential prey for fish in the stream, and this leads to a consideration of the productivity of stream invertebrates.

4.4 Invertebrate production in streams

Until relatively recently, before the highly dynamic nature of ecosystems had been realized, the production of stream (and lake) invertebrates was measured as the maximum biomass measured during the year. Samples were taken using suitable grabs, nets or dredges, and the organisms sorted and weighed. Studies on the contents of fish guts, however, suggested that the fish were eating considerably more invertebrates than were ever present in the biomass samples! With hindsight the explanation of this paradox is clear[242]. There is a several-fold turnover each year and the measured biomass, as with that of the plankton (see Chapter 3), represents the balance between production and loss by predation, and in this case also by displacement downstream, called 'drift'. Stream animals are often dislodged and swept into the current where they also become more vulnerable to predation by fish than when they are hidden in leaf packs or between and under stones. Discussion of their productivity must thus start with how they maintain their position in what appears to be, for them, a dangerous habitat.

4.4.1 Drift and adaptation to moving water

For short periods, current speeds in streams may exceed 300 cm s^{-1} and the force of such movement would be sufficient to displace any small animal. Predominantly stony streams must have current speeds of at least 50 cm s^{-1}; for sand to be deposited without finer silt there must be 40 cm s^{-1}, and fine deposits are dropped in currents of less than 20 cm s^{-1}. The force applied even by 50 cm s^{-1} currents is still considerable, yet many stream organisms survive spates in which the mid-water current exceeds this for days. The explanation lies in the heterogeneity of stream beds. There are many small areas of slack flow and the currents experienced by the animals are usually much lower than those measured by current meters in the open water. Even on stones, over which fast-flowing water is passing, there is a layer, a few mm in thickness, where the friction or drag caused by the stone surface reduces the current speed almost to zero and smooths the flow. In this 'boundary layer' many organisms live.

The apparently obvious best way of an animal preventing itself from being washed downstream would be permanent attachment but in fact most stream animals are freely motile. Permanent attachment carries the risk of stranding during dry periods of low flow, and as a result permanently attached organisms, e.g. freshwater sponges, mostly survive on the undersides of submerged rocks. Motility gives flexibility, though with greater risk of displacement.

For some animals, e.g. the mayfly nymph *Rithrogena*, the risks have been reduced by evolution of a flattened body which does not project above the boundary layer on the tops of stones, or which, in the case of other mayfly nymphs like *Ecdyonurus*, allows the animal to crawl under stones. Streamlining, in which the greatest width of the body lies about 36% along its length, confers least resistance to current force and is found in *Baetis* spp. (mayfly nymphs). The long tails of mayflies and stonefly nymphs seem often to act as fins which turn the animal always head on into the current, just as a small boat fares best if headed into the waves of a rough sea.

Some stream animals have suckers with which to cling to the rocks—leeches for example. And the suction pad feet of snails and freshwater limpets are similarly useful. Others, e.g. the water penny beetles, *Psephenus* spp., have friction pads of small movable spines which can be fitted into tiny irregularities of the surface of rocks. Hooks, grapples, and claw-like legs have all been evolved, and some caddis fly larvae make cases with such heavy mineral particles that the case acts as ballast. Some aquatic caterpillars (Lepidoptera) spin flat sheets of silk and attach them to rocks. Under the protective sheets they may form coccoons with low risk of displacement.

Silk is also used by *Simulium* (blackfly) larvae which must represent the doyen of survival in fast-flowing water. *Simulium* is a Dipteran (two winged fly) and like all Dipteran larvae does not have articulated legs. Typically Dipteran larvae have short, fat prolegs and in *Simulium* the prolegs on the hind segment are very much flattened and bear retractable hooks. A proleg on the first thoracic segment is also fashioned into a hook. *Simulium* spins a pad of silk which adheres to the rock and to which it attaches with its hind hooks. It can move by a looping movement, continually producing silk pads and alternately hooking to these with its front proleg and hind hooks. If, despite these attachments, it is dislodged, it has a 'safety line' attached to the body and the pads on the rock, which it can climb up by working it between the front proleg and rough spines on the head, thus regaining the security of its silk pad on the rock.

Despite these adaptations, displacement and drift downstream is common. Most studies indicate that less than 0·5% of stream invertebrates are passively moved downstream each day, but others show higher rates of 3–4% for the fauna in general, and very high rates of up to 43% for particular species under special circumstances[125, 242, 503]. The distances travelled by drifting animals can be measured by suitable placing of nets across a stream. Most animals travel less than 2 m in a single movement but a few as far as 40–50 m.

Drift is a complex process. Not all species, nor even different size classes of the same species, drift to the same degree and the process is partly determined by external factors. Drift of some species is markedly seasonal, peaks of travel coinciding with high current flow, whereas in other cases it may coincide with low current speeds. Diurnal changes may be superimposed[374]. For some species (e.g. *Gammarus*, *Baetis*) drift increases in the period just after sunset, sometimes with peaks late in the night, while in others (e.g. Chironomids) there seems to be no distinct diurnal pattern. Artificial shading of a stream by day, or illumination by night may reproduce a usual day/night pattern, or may not, depending on the species. It may be that some scrapers feed more actively on the tops of stones at night, when they cannot be seen, but hide under stones by day when predators are more active. The risks of dislodgement must then be balanced against the advantages of predator avoidance.

Active movements in contact with the bottom also occur. Usually these are upstream and to some extent counteract the effects of drift, although to the extent of only a 7–10% replacement when drift rates are high. Active downstream movement, however, has also been found for some insect larvae just prior to pupation and emergence of the adults. The adults of some aquatic insects (e.g. some mayflies and stoneflies) tend to fly upstream before depositing their eggs and this too must partly counteract the effects of drift.

The complexity of patterns of drift[125] suggests that it might not merely be a passive consequence of stream living. Some animals are less affected than others (e.g. oligochaetes, beetles, water mites) being proportionately scarcer in the drift than in the stream bed community, and presumably mechanisms to avoid dislodgement completely have had time to evolve if drift has not some adaptive advantage to balance its obvious perils. It results in rapid colonization of new channels opened in a stream of recolonization of stretches denuded by violent spates. It is also relatively high when food supplies are scarce. In an experimental stream, Hildebrand[213] showed that drift of animals feeding on algae attached to stones was high when the algae were scarce, allowing dispersal to sites possibly richer in food. Drift rates of a net-spinning caddis fly (*Plectrocnemia conspersa* Curtis) and a leafpack-inhabiting stonefly (*Nemurella picteti* Klapalek) were exceptionally high (20% d^{-1} and 43% d^{-1}) in a southern English stream when densities of the former were so high (100 m^{-2}) that net-spinning sites were very scarce, and when, in summer, leaf packs for the latter were few. In contrast, drift rates were low in the same period for another stonefly nymph, *Leuctra nigra*, where its food supply, iron bacteria, was very abundant[503].

4.4.2 Problems and methods of measuring the productivity of stream invertebrates
Sampling
As with all animals the basic principles of measuring productivity are contained in the equations

Ingestion   = Assimilation − Egestion

Assimilation = Production of body tissue and gametes
            + respiration + excretion.

These have been used previously in Chapter 3 for zooplankton. Although the present discussion is centred on the invertebrates of rocky streams, much of it, in principle, is applicable to all bottom-living invertebrate communities (see Chapters 5 & 6).

The prime problem in productivity studies lies in the precise determination of the number of animals present and their biomass. Stream bottoms are highly heterogeneous with areas of bare rock, gravel, sand, perhaps some finer sediments, accumulations of leaves, tree debris, and perhaps some mosses or higher plants. Depth varies, pools and shallower riffles can be distinguished. Random sampling, though least biased, is thus less than helpful and random sampling within an approximately uniform area (stratified random sampling) is usually preferable.

Three general sampling methods are available for stony habitats—nets, cores and grabs and placement of trays of cleaned substrata. Most interest has centred on the macro-invertebrates, those more than about 3 mm in size when adult, and the net mesh most commonly used, about 1 mm, reflects this. The net, mounted on a pole, is held in front of the feet of an observer who stands with his back to the current and rubs his feet vigorously on the bottom. The method is called kick sampling and animals so dislodged are swept into the net. A known area may be sampled, or, for relative measurements the sampling goes on for a set time as the observer works his way steadily upstream. Nets are most useful for firm substrata and sediment cores or grabs for soft ones. On sandy and gravelly bottoms one problem is that animals may penetrate to much greater depths (20–30 cm) than are sampled by nets[71]. Usually such deep-living animals are very small ones—microcrustacea and nematodes for example—and this has contributed to relative neglect of them. Artificial substrata, usually comprising cleaned rocks fixed to a tray may be left on the stream bed for several weeks for them to become colonized. A large number of such trays, progressively removed at intervals for examination, allows a picture of seasonal fluctuations to be built up. It is important, of course, that the artificial substrata used closely reflect the natural substrata of the stream stretch under investigation.

A relatively large number of separate samples must be taken for numbers of an animal to be determined with reasonable precision on any occasion. Elliott[126] discusses the statistics of sampling benthic invertebrates in some detail; it is sufficient to say that the complexities of the physical environment, of interactions between predator and prey, and the behaviour of individual species, lead to markedly non-random distributions. Larvae of the caddis fly *Potamophylax latipennis* (Curtis) (an algal scraper) were found, for example, only on the undersides of stones of diameter 11–22 cm in a Scottish river, and even then were grouped in particular parts of such stones[60]. The reasons for this are not known. And although it begs the question of what controls the distribution of the prey, the predators *Plectrocnemia conspersa* (a net-spinning caddis fly) and the active alder-fly larva, *Sialis fuliginosa* Pict., were clearly distributed in relation to prey density in an English stream[214].

Identification of benthic invertebrates, once they have been sampled, may pose problems. Many are larval or nymphal stages of adults which are aerial and short-lived. The larvae of species which are distinct as adults may be barely separable or indistinguishable; other organisms are very small and have never been fully characterized taxonomically (e.g. nematodes). The tendency has been for investigators to concentrate on the populations of the more prominent, identifiable species, and it must be accepted that our understanding of all the processes occurring in streams (and other freshwater habitats) may thus be very incomplete.

Ingestion

The strict concept of herbivores, carnivores and detritivores, plus the convenient category of omnivores for those that did not fit into any of the other three groups appears now to be obsolete. Almost all stream invertebrates are omnivores but usually take a greater proportion of one or other sorts of food at particular stages in their life histories. It is difficult to discover the exact food taken and more troublesome still to apportion the items quantitatively. A usual method is dissection of the gut and examination under a microscope. Resistant remains—spines, mouth parts, carapaces, diatom cell walls—may give clues to food taken but soft-bodied prey may leave no trace and yet have been the prime diet. Certain preferences may be shown however. On a rocky lake shore, Calow [58] determined that a freshwater limpet, *Ancylus fluviatilis* Linn. selected against epilithic blue-green algae in favour of green algae, but took diatoms in preference to both of these. Not only that, but of the diatom genera available, *Gomphonema* was eaten more readily than *Navicula* or *Achnanthes* when food supplies were abundant. The gastronomic qualities of *Gomphonema* which might lie in its shape, its position in the algal community on the stones, or its biochemistry, remain a mystery.

Carnivorous flatworms (Tricladidae, Platyhelminthes) feed by wrapping themselves around their prey (which comprises other invertebrates such as *Asellus*, oligochaetes and snails) and inserting their pharynx into the body of the prey. Digestive juices are secreted through the probe-like pharynx and the liquefied prey tissue is sucked back into the flatworm's gut. There are thus no morphological remains once the flatworm has abandoned it to indicate what the prey actually was. Young *et al.*[547] got around this problem with a serological technique. Potential prey organisms were macerated and the preparations separately injected into rabbits. The rabbits produced specific antibodies in their blood sera. Flatworm species were also macerated and the preparations tested against a succession of specific antisera prepared from potential prey organisms. Precipitation with a particular antiserum indicated the presence of the proteins of that particular prey in the flatworm's guts. Such studies have helped elucidate a detailed picture of flatworm ecology (see later).

Respiration

Respiration of a natural stream animal population is usually calculated from the ambient temperature, biomass of animals, and the rate of oxygen uptake per unit biomass at that temperature, which has been determined in the laboratory. Inevitably the value obtained is approximate but the errors can be reduced by reproducing the stream environment as closely as possible in the laboratory measurements. Stream animals often have much higher respiratory rates than their relatives in still waters, and their respiration rate is also often greatly dependent on the oxygen concentration of the water; often it falls markedly at low to mid saturation levels. Current is also closely involved. Respiration rates may increase with current speed, e.g. in *Rithrogena* and *Baetis* (mayfly) nymphs, *Rhyacophila* larvae (caddis fly), or may not, as in *Ephemerella*, *Ecdyonurus venosus* (Fabr.) (mayfly nymphs[252]). The lowest oxygen concentration tolerated by some stream animals is sometimes decreased at increased current speed, and the general reduction in body and gill movements noted in stream animals compared with similar animals in still waters suggests a dependence on current movement for replenishment of oxygen at the body surface.

Respiratory measurements in the laboratory must therefore be carried out in moving water, for example in oscillating flasks. The internal environment of the flasks must also be carefully arranged. The respiratory rate of *Ephemera danica* (mayfly) nymphs changed by 65% when they were allowed to burrow in sand, their natural habitat, and that of another mayfly nymph, *Ecdyonurus venosus*, by 30–40% when it was allowed to rest on stones rather than on glass[518].

Production

If there was no predation or drift, or death from miscellaneous causes, production could be measured simply by the change in weight of a population over a period of time. However, a very large proportion of the production may be eaten and a more sophisticated approach is needed. The basic data are numbers of an animal per unit area at a given time and the mean individual weight of the animals. It is tedious to weigh all the animals from a large enough sample to be statistically acceptable, so a previously determined relationship between length of animal or width of head and weight is often used. The simplest case is for populations which develop as cohorts from eggs all laid at the same time.

If the cohort begins with N individuals of mean weight w_1 on hatching, the initial biomass will be $N w_1$. At a future time, t, some animals will not have survived, but the remainder N_2 will each be heavier with a mean weight w_2. The total production will thus be that of the survivors, $N_2(w_2 - w_1)$, plus that of those which died,

$$\frac{(N_1 - N_2)(w_2 - w_1)}{2}.$$

The assumption has to be made that on average all the non-survivors were lost half-way through the period, and that weight per individual increased linearly during the period. Sampling at sufficiently frequent intervals reduces the error from this assumption. Expressed graphically, as it was first used (for fish) by Allen[8], the area under the Allen curve gives total production (Fig. 4.5). Number of organisms is plotted on the y axis and mean weight per individual on the x axis. The sum of production of each cohort present gives the total production.

This method works adequately (given well-designed sampling) for organisms with well understood life histories and synchronous reproduction, or for

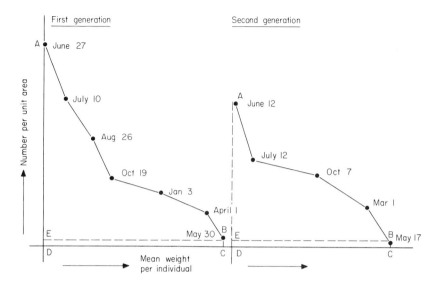

Fig. 4.5. Allen curve method for determining the production of an animal population whose members can be recognized as distinct cohorts, each starting life effectively simultaneously. Numbers per unit area are counted on each of a series of sampling dates and plotted against mean weight of animals on these dates. Area ABCD represents total production; ABE the production of animals which were eaten or died before reproduction; BCDE the production of adults surviving to breed, and ABCD : BCDE the turnover ratio.

Table 4.3. Turnover ratios of selected freshwater invertebrates.
(After K. H. Mann in Edmondson & Winberg[121].)

	Mean annual biomass g wet wt m^{-2}yr^{-1}	Production g wet wt m^{-2}yr^{-1}	Turnover ratio
Annuals			
Glossiphonia heteroclita			
(leech)	0·02	0.105	5·25
Ancylus lacustris			
(freshwater limpet)	0·016	0·052	3·25
Biennials and triennials			
Erpobdella octoculata			
(leech)	0·58	2·04	3·52
Viviparus viviparus			
(snail)	1·69	2·67	1·58
Sphaerium corneum			
(bivalve)	0·49	1·71	3·49
Several generations per year			
Helobdella stagnalis			
(leech)	0·28	1·35	4·82
Gammarus pulex			
(freshwater shrimp)	0·013	0·087	6·69
Asellus aquaticus			
(water-louse)	0·26	0·83	3·19
Chironomid larvae	0·108	0·539	5·0

organisms which can be easily aged, for example by markings on the shell or operculum plate in the case of snails. For animals which reproduce continuously and which cannot be readily aged this method cannot be used. For such creatures, determination of growth rates by serial weighing or measurement of captured individuals in laboratory conditions resembling as closely as possible those of the habitat must be used. Production is then given by multiplying the field biomass by the appropriate percentage increase in weight per day for the contemporary field conditions of temperature and other pertinent factors. An example, for sediment-living invertebrates is given in Chapter 5.

The ratio of annual production to mean annual biomass gives the turnover ratio, or the number of times the biomass is replaced each year. For many species (Table 4.3) this is several times, though for the larger bivalve molluscs (e.g. swan mussels) of muddy habitats it is only once in several years. In either case the biomass alone gives no direct measure of total production.

4.5 Rocky shores in standing waters

The water movement is not predominantly unidirectional but the movement of waves on rocky lake shores creates a habitat superficially similar to that of rocky streams. Little work has been done on the productivity of such shores, but inspection of fauna lists from them suggests that the processing of allochthonous organic matter is not nearly so important on them as it is in streams. The major food base is the epilithic community of algae, bacteria and protozoa which covers the stones; scrapers and predators dominate the invertebrate fauna. This is to be expected since wave action will disturb any loose organic matter arriving from elsewhere and tend to move it, under the action of gravity, down the slope of the lake shore and basin to the sediments at greater depths. What happens there is considered in Chapter 5. A typical fauna list from a rocky lake shore might include nymphs of various Ephemeroptera (mayflies), Plecoptera (stoneflies), Triclads (flatworms), gastropods and limpets, *Asellus*, *Gammarus* and other Crustacea, Trichoptera (caddis) larvae and others. Of course such a shore is never uniform and protected pockets, where sediment may accumulate, leaves lodge or aquatic macrophytes grow, will have a somewhat different fauna characteristic of such conditions.

The rocky shore habitat has been most intensively examined from the viewpoint of what determines the composition of the community of invertebrates. It may best be used here for discussion of how physico-chemical factors, chance, predation and competition all combine to determine the species list and the relative proportions of organisms in it. Firstly, distribution of one group of animals, the Triclads, or carnivorous flatworms, will be outlined.

4.5.1 Distribution of Triclads in the British Isles

Flatworms occur in weed-beds as well as on rocky shores, but are certainly characteristic of the latter. There are about eleven species in Britain but four of them (Fig. 4.6), *Polycelis tenuis* Ijima, *P. nigra* (O. F. Müll), *Dugesia polychroa* (Schmidt) (then referred to as D. lugubris (Schmidt)) and *Dendrocoelum lacteum* (Müll), have been very well studied by T. B. Reynoldson (1968) and his students. The first task was to describe the distribution of the four species in the British Isles. This was carried out with a survey of over 200 lakes throughout Britain which had rocky shores; flatworms were sought by hand on the shores over a timed period necessary to collect 50–100 animals. This relative quantitative method is the best available to date, for flatworms are delicate, readily damaged, and not easily dislodged by kick sampling.

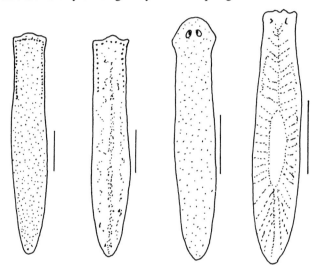

Fig. 4.6. Four species of British carnivorous flatworms (Triclads). From left to right, *Polycelis nigra*, *P. tenuis*, *Dugesia polychroa*, *Dendrocoelum lacteum*. The scale bars represent 1 cm. *Dendrocoelum* is cream-white in colour, the others are dark brown. The *Polycelis* species must be distinguished partly on characteristics of the live organism, particularly the behaviour of the penis[424]. (This figure modified from Reynoldson[423].)

The pattern of distribution found was as follows:
(a)   the total number of flatworms collected per unit time increased directly with Ca^{2+} levels and total dissolved solids concentrations in the water;
(b)   at Ca^{2+} levels lower than 20 mg l^{-1}, *D. polychroa* and *D. lacteum* declined and were usually absent in water with less than 5 mg Ca^{2+} l^{-1};
(c)   at low Ca^{2+} levels above 5 mg l^{-1}, *P. tenuis* was most abundant, but below 5 mg l^{-1}, if flatworms were present at all, *P. nigra* was either predominant or the sole flatworm species present.

To some extent this pattern reflects dispersal of the animals and recolonization of the lakes after the retreat of the last glaciation. Under experimental conditions, with abundant food (slices of earthworm) supplied, waters of the lowest calcium and total dissolved solids concentrations supported breeding populations of all four species indefinitely. Lakes poorest in calcium tend to be at the furthest northern reaches of the British Isles, and in northern Scotland and the offshore islands more lakes than expected had no triclads at all. Part of the triclad distribution pattern at low calcium levels may thus simply reflect a lack of time for colonization to have taken place.

Most of the pattern must be explained, however, in other terms, particularly the change from *P. tenuis* predominance to that of *Dugesia* and *Dendrocoelum* in lakes where all four species co-exist. Predation on triclads by other organisms was found to be unimportant on rocky shores[90]. Of 75 potential predators tested by serological methods only two rocky shore animals, a leech, *Erpobdella octoculata* (L.), and a caddis larva, *Polycentropus flavomaculatus* (F. J. Pictet), ate flatworms, but were shown not to eat enough to alter greatly the population size or balance of species.

Increasing the food supply led to population increases of caged flatworms far greater than those of natural populations in the same waters and it seemed that competition for food might explain part of the distribution pattern. Calcium and total dissolved solids levels, though not directly and causatively implicated in the distribution, are general indices of the fertility of most lakes (see Chapter 1). The production of potential prey organisms and therefore flatworms is likely to be correlated indirectly with levels of these substances as a reflection of increasing fertility of a lake. This leaves the question of the change in balance of species in lakes of increasing Ca^{2+} concentration.

All four flatworms will take a range of foods, but particular and different items seem necessary for indefinite survival of each species. Serological work, examination of gut contents and choice experiments where each species was presented with a range of foods have shown that *Dendrocoelum* prefers the crustacean *Asellus*, which it actively hunts. Of the four species it has the best developed sensory systems and will attach to and coil around a moving nylon bristle the same diameter as an *Asellus* leg[30]. *Dugesia polychroa* eats gastropods, which are spurned by the others in most circumstances. It was thought that the *Polycelis* species specialized on oligochaetes, each taking worms from different oligochaete families, but further work has shown that this is unlikely and that *P. tenuis* takes *damaged Asellus*, attracted to it by oozing body fluids. *P. tenuis* is the most active of the four, even at low temperatures, and such activity is a necessary behavioural trait for a predator which must search for stationary prey. The main food of *Polycelis nigra* has not yet been determined. In laboratory experiments the 'food refuge' of each species had to be provided for indefinite survival either in mixed or monospecific cultures, but *P. nigra* always declined whatever food was given[30]. In natural waters, introduction of *P. tenuis* to locations where *P. nigra* had previously been the sole flatworm species has led to decline (though not disappearance) of *P. nigra* in favour of *P. tenuis*[425].

In lakes of increasing fertility the proportion and abundance of *Asellus* and of gastropods on rocky shores often increases and this seems to explain the coexistence of *Dugesia* and *Dendrocoelum* with *Polycelis* spp. at the higher calcium levels. The whole picture given by Reynoldson and his co-workers also goes some way to explaining the changes in rocky shore fauna noted by Macan[310] within L. Windermere and in a series of lakes of the English Lake District.

Macan listed about 36 species of macro-invertebrates on the rocky shores of L. Windermere, of which 19 were commonly found. The total list included mayfly and stonefly nymphs, caddis larvae, beetles, Crustacea, such as *Gammarus pulex*, *Crangonyx pseudogracilis* (Bousfield) and two *Asellus* species, two gastropods, *Ancylus fluviatis* (Linn.), the freshwater limpet, six flatworm species, including all those discussed above and nine leech species. Their distribution was not uniform among a large number of areas sampled around the lake shore— two extreme sorts of community were found linked by a continuum of inter- mediate ones. At one extreme the fauna had relatively few insects, but molluscs, Crustaceans and flatworms were prominent. At the other, certain stonefly (e.g. *Diura bicaudata* (Linn.), *Nemoura avicularis* (Morton)), and mayfly nymphs (*Ecdyonurus dispar* (Curt.), *Centroptilum luteolum* (Curt.), *Heptagenia lateralis* (Curt.)) were confined. All other species were found throughout the series of sites. The former community type seemed to be associated with shores likely to have been fertilized with effluent from two sewage treatment works (see Chapter 10) and from septic tank and houseboat effluent.

A rather similar pattern was recorded from a series of lakes in the same area. Because of the varying geologies of their catchments and the greater farming activity and human settlement in those of the naturally more fertile areas, these lakes form a spectrum of overall lake fertility (Fig. 4.7). At the least fertile extreme stonefly and mayfly nymphs were most abundant, while certain leeches, gastropods and flatworms reached their greatest abundance (based on a timed collection by hand and net) towards the fertile end.

A tentative explanation of these distributions concerns the interaction of fertility and predation[313]. As the fertility of a lake, or a station in it, increases, so does the productivity of the invertebrate community and a greater population of predators such as flatworms can be supported. There is no evidence *per se* that high fertility discriminates against nymphs of stoneflies and mayflies, and little evidence that they are major prey of the abundant flatworms. However, there is a possibility that their eggs survive less well under crowded conditions than those of animals that are common on such shores. Mayflies and stoneflies deposit eggs which fall to the bottom and catch among the epilithic community of microorganisms on the stones. There they are vulnerable to the continual activity of the scrapers—molluscs and crustaceans—which continually work over the stones. In contrast, the eggs of molluscs are protected in lumps of jelly and those of leeches and flatworms in leathery coccoons; *Gammarus* and *Asellus* eggs are carried on the female's body until they hatch. Doubtless this is only a first level of explanation of what is really a very complex pattern, yet it illustrates some of the factors that must be considered if some pattern of predictive use is to be deduced from the mass of information on community composition that already exists in the literature.

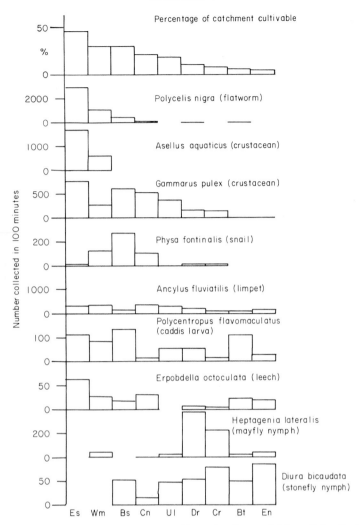

Fig. 4.7. Distribution of animals on comparable rocky shores of lakes of the English Lake District. Histograms show the number of animals of each of nine representative species collected by the same experienced observer in 100 minutes. A measure of the fertility of each lake is given by the percentage of its catchment area which is cultivable. Cultivability in this case largely reflects the availability of easily weathered rocks in the catchment. Es: Esthwaite, Wm: Windermere, Bs: Bassenthwaite, Cn: Coniston, Ul: Ullswater, Dr: Derwent Water, Cr: Crummock Water, Bt: Buttermere, En: Ennerdale Water. (Redrawn from data in Table 37 of Macan 1970[310].)

4.6 Depositional shores and habitats

When current speeds decline in rivers or when shelter from wave action is provided in a lake, various sorts of sediment may be deposited. The continuum of these ranges from gravel, through sands to silts and ultimately to fine, usually largely organic mud. If a bottom of fine sediments lies within the euphotic zone weed-beds may develop (see Chapter 6), but where it does not the sediments support a community of detritivores and their predators discussed in Chapter 5. Between the rocky habitats and the finer sediments of stiller waters lie the fairly vigorously disturbed sandy habitats. Often they have, like gravels, little fauna and flora—the continuous agitation of the sand grinds any colonizing organisms to death and washes away any fine organic particles that might act as food. The least disturbed yet still sandy habitats do have a characteristic set of communities, however, with microscopic algae and bacteria living freely among the sand grains and attached to them. There is also an interstitial fauna of Protozoa, nematodes, small Crustacea, oligochaetes, tardigrades (water-bears) and mites.

Studies of a marine sandy beach of this kind have shown a strong dependence on incoming dissolved organic matter for support of the system. Bacteria growing attached to the sand grains use this organic matter and themselves form the food of polychaete worms and bivalve molluscs, parts of which (the protruding feeding siphons for example) are cropped by young plaice, a flatfish[475]. Such freshwater, sandy beaches as have been studied probably depend more on autochthonous photosynthesis judging by the much higher biomass of algae found attached to the sand grains (epipsammic algae) than is found on marine beaches.

Epipsammic algae (Fig. 4.8) are sometimes of non-mobile genera (e.g. *Fragilaria* spp.) and sometimes of genera which can both attach firmly but

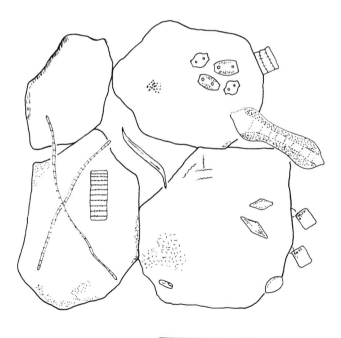

Fig. 4.8. Algae associated with sand grains on submerged sandy beaches. Some are firmly attached (the epipsammon), others move freely over the grains (the epipelon). Scale line represents 1 mm.

reversibly to the grains and also move freely among them. The sorts of sandy beach which have large populations of such algae (largely diatoms) are generally not continuously disturbed by water movements but mixing of the grains occurs from time to time. Diatoms attached to sand grains may thus be buried to depths of several cm where no light penetrates. The motile ones can then detach themselves and move back to the surface, but the permanently affixed must either die or tolerate darkness for several days, perhaps weeks. There is no evidence of much heterotrophic ability[377], but some that they can tolerate both darkness and deoxygenation for several days, while retaining their photosynthetic potential[378]. In contrast, epipelic algae (those that are always free living and move in the surface layers of sand and other sediments) although tolerant of darkness, if buried are less resistant to anaerobiosis. They survive by rapid movement back to the surface. Deoxygenation is usual, even in sands at depths of more than a few cm and in finer sediments below a few mm.

The sediments are of great importance, not only for the organisms which live in and on them, but for the aquatic ecosystem as a whole. Their role is considered in the next two chapters.

4.7 Further reading

There have been two outstanding books on running-water ecology in recent years. Hynes[242] is a supremely readable account, particularly of the stream fauna, and Whitton[536] is a series of reviews which complements Hynes' work. Hynes is stronger on aspects of natural history, Whitton on applied studies. General information on the zoology and natural history of benthic invertebrates is to be had from Clegg[68] and Mellanby[344], and Macan[309] is a thorough review of community ecology and the effects of physical factors, like current, on invertebrates. Methodology is collected together in Edmondson & Winberg[121]. For reviews of the algal communities of streams and shores, Round[437, 438] is a useful starting point. For identification, some valuable works are Edmondson[116], Macan[308], Merritt & Cummins[345], Quigley[417] and Redfern[421]. More specialist keys are the Scientific Publications of the Freshwater Biological Association. A list of these may be obtained from The Librarian, Freshwater Biological Association, The Ferry House, Far Sawrey, Ambleside, Cumbria, LA22 0LP.

CHAPTER 5
THE SEDIMENTS AND
THE HYPOLIMNION

5.1 Introduction

Sediment of one sort or another is inevitable except in the most violent of streams, and even there patches of gravel or sand will collect in sheltered crevices. In the case of more moderate flows, extensive sand and gravel banks may collect in rivers and on wave-exposed lake shores. These habitats are generally poor in species and have been considered in Chapter 4. This chapter looks further along the continuum of current speeds to the levels, in quiet reaches of rivers and in lakes, where fine silts and clays may be deposited, together with organic detritus carried from upstream or produced autoch-thonously. Such is the case at current speeds below about 20 cm s^{-1} and the result is often a richly organic mud teeming with bacteria and invertebrate animals. The latter typically include bivalve molluscs, insect larvae, particularly Chironominae, oligochaete worms, and Crustacea, most of them repeatedly working the sediment through their guts and probably digesting the bacteria growing on the organic detritus, rather than the detritus directly.

If the sediment surface is within the euphotic zone, macrophyte beds and, at the edges, reed swamps, will usually be present, providing habitat space for a great many other species. In turn, these communities support fish and their bird and reptile predators. The macrophyte communities are usually highly productive and are dealt with in Chapter 6.

This chapter is concerned particularly with those sedimentary communities which do not have macrophytes—those below the euphotic zone, which are most extensive in lakes. This habitat is called the profundal zone (Gk–deep) as opposed to the littoral zone (Gk–shore) which is the bottom of a water body which lies in the euphotic zone. Profundal communities are almost entirely dependent on organic matter sedimented from the euphotic zone and exist not only in dark, but also relatively cold and deoxygenated conditions. Even if the overlying water has plenty of dissolved oxygen, the sediment below the surface interface with the water rarely has. The profundal sedimentary benthos, bacteria and animals, is therefore a community of decomposers, providing not only fishfood, but sometimes having also a major effect on the chemistry of the overlying water.

5.2 Sediment and the bacteria which colonize it

There are certainly many bacteria in sediments. No single method is capable of giving an absolute estimate of their numbers, but relative estimates suggest perhaps a thousandfold concentration in the sediments compared with that in the plankton. This is not surprising for the organic detritus from a long column of water is concentrated into only a few mm or less of new sediment each year. The *total* number of bacteria in the plankton of a water column overlying a given area of sediment is probably much greater, however, than the total number in the surface sediment and organic matter is thus usually well processed before it reaches the bottom. The sedimentary bacteria therefore receive relatively more refractory organic material than that present in the plankton. Species which can hydrolyse tough polymers of carbohydrates and other compounds (cellulose, lignin, chitin) are thus readily isolated from sediments.

Refractory organic matter is generally unavailable to animals—they cannot digest it—but they can digest the bacterial cells which are able to grow on it. Benthic invertebrates commonly ingest much sediment, but void a lot of it as faeces able to be recolonized by more bacteria. In using organic detritus as their energy source, bacteria in the surface sediments absorb much oxygen which is replaced by diffusion and by current-induced turbulence of the top few mm of sediment. Below this zone, called the oxidized micro-layer, bacterial activity uses up oxygen faster than it can be replaced and the sediment becomes progressively anaerobic. Some decomposition continues with dissolved sulphate and nitrate being used as oxidizing agents by specialist bacteria which release sulphide, sulphur or nitrogen. When all possible oxidizing agents have been exhausted, most of whatever organic matter remains is preserved in an anaerobic state. There may be some further very slow decomposition by extreme anaerobes such as methanogens which oxidize organic matter with CO_2, producing CH_4 in the process. Many different organic compounds can however be detected in sediments thousands of years old and provide some of the information usable in reconstructing the past history of lakes (see Chapter 8). If organic matter is received at the sediment surface at high rates the community will be able to process it to a much smaller extent before it becomes buried by new sediment and anaerobically preserved. Fertile lakes consequently have more richly organic sediments, other things being equal (e.g. import of organic matter from the catchment), than less fertile lakes.

5.3 Nutrient relationships and the oxidized micro-zone

In the interstitial water of anaerobic sediment inorganic products of decomposition may accumulate, including phosphate, ammonium and silicate. The latter is released by chemical dissolution from the cell wall remains of diatoms. Transition metals, e.g. iron and manganese, are present in their soluble, reduced forms (Fe II and Mn II) and there are generally higher concentrations of many ions than in the overlying water. Some of these may be returned to the water above by diffusion through the oxidized micro-layer (ammonium, silicate), but others are not because they become fixed in the micro-layer in insoluble complexes. Iron, in particular, forms insoluble Fe III hydroxides and oxides in the presence of oxygen and these readily adsorb and precipitate phosphate.

A measure of the state of reduction/oxidation of the sediment surface is the redox potential. If a calomel ($Hg_2 Cl_2$)-platinum electrode is immersed in water or sediment whilst connected into a circuit which also includes a standard hydrogen electrode, electrons will tend to accumulate on the calomel electrode to an extent reflecting the number of free electrons available in the water or sediment. A large number indicates a relatively reducing environment and a large negative potential (the redox potential) is generated on the calomel electrode. This may be measured, in millivolts, by a galvanometer incorporated into the circuit. A well-oxidized medium will have few free electrons and electrons will tend to move from the electrode to the medium, leaving a positive potential at the surface of the electrode. Well-oxygenated waters have redox potentials greater than + 500 mV, whilst anaerobic sediments have negative potentials of sometimes less than − 100 mV.

The oxidized surface micro-zones of sediments in contact with well-aerated water (i.e. with more than 1–2 mg l^{-1} of dissolved oxygen) have redox potentials of + 300 to + 400 mV. It is not until the potential falls to below about + 250 mV that Fe II can exist, and that precipitation of phosphate into Fe III complexes is prevented. Sediments with oxidized micro-zones, which appear as brown crusts over the underlying black (because of the presence of sulphides), anaerobic, deeper layers, are thus efficient traps for phosphate, which, as indicated in Chapter 1, is a very scarce element in aquatic ecosystems.

Oxidized micro-zones occur for at least part of the year on most sediments, the exceptions being those under permanent deep anaerobic water layers such as those of L. Tanganyika (see Chapter 2). In summer, however, the intensity of bacterial activity, even at the sediment surface, may be high enough to use up oxygen as fast as it can diffuse into the sediment. Under these conditions, which need a copious supply of sedimenting organic matter, and therefore generally occur in the more fertile lakes, the redox potential may fall below + 250 mV and the oxidized micro-zone breaks down. Fe II and other transition element ions and phosphate may diffuse freely out into the overlying water. They cannot persist, of course, if this water is oxygenated, but if the lake is stratified and the hypolimnion has become deoxygenated (see Chapter 2), then Fe II, and phosphate, as well as other ions released from the sediment may begin to accumulate in it.

Such accumulation has little effect on the epilimnion however, for the epilimnion and the hypolimnion do not mix during the summer, when shortages of available dissolved key nutrients are greatest in the surface water layers. Internal waves, called seiches, which ripple along the metalimnion as a result of summer gales disturbing the surface of the lake, may result in paring off of pockets of hypolimnial water, but such mixing is minor. In any case, as the Fe II and phosphate are mixed into epilimnial water of high redox potential, the insoluble complexes re-form. The same sort of precipitation must also follow the major seasonal mixing of a stratified lake.

Phosphorus is thus effectively locked away from the surface waters once it has been incorporated into particles which reach the sediment surface, and continual supplies from the catchment area are necessary to maintain the fertility of a lake. Of course some phosphate may leak from the sediment and not be reprecipitated, simply because even ferric phosphate complexes are not infinitely insoluble, but such return is low, as is any exchange between phosphate adsorbed on inorganic particles and the water when sediment is mechanically disturbed. In most lakes the net flow of phosphorus is strongly into the sediments (see also Chapter 3).

Under some circumstances, however, phosphorus may be released to over-lying aerobic water, and not reprecipitated. This may occur in heavily fertilized, shallow water bodies (loaded with sewage effluent, for example) which do not stratify, but in which sedimenting plankton organic matter reaches the surface sediment in quantity (because of the high productivity) and in a relatively labile state (because of the shortness of the water column). Intense decomposition at the sediment surface may then not only break down the oxidized micro-zone but may reduce the redox potential to $-$ 100 mV or less. When this happens Fe II is precipitated as sulphide by the sulphide ions produced by sulphate-reducing bacteria in the sediment. Once all the Fe II has been tied up in this way, phosphate cannot be reprecipitated even in oxygenated water and it becomes available to the phytoplankton. Summer release of phosphate by this mechanism is illustrated in Fig. 5.1 where seasonal cycles of phosphate concentration are shown for two contrasted shallow lakes. In the lake where phosphate accumulated during the summer (Barton Broad) it did so as SRP since the available

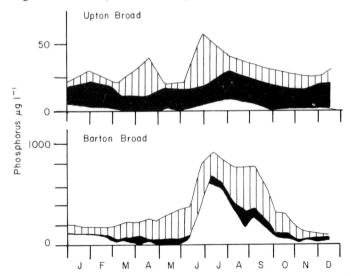

Fig. 5.1. Seasonal changes in two shallow (1–1·5 m) lakes (broads) in Norfolk, England. Upton Broad receives relatively low phosphorus loading, and Barton Broad a very high phosphorus loading from sewage effluent. In the latter there is a considerable release of soluble reactive phosphorus (SRP, see Chapter 3) from the deoxygenated sediment surface in June which persists during the summer period of little flushing. Clear area is SRP, black SUP and vertical lines—particulate P. The upper graph line gives total phosphorus. (After Osborne[390] and Phillips[405].)

nitrogen was then limiting build up in biomass of the phytoplankton. Such releases from the sediment are dependent on the provision of labile organic matter to the sediment surface. Such production needs large external phosphorus supplies and the sediment acts effectively only as a temporary store of these during the spring and early summer before release begins. What is released must first be put there, and the net yearly flow of phosphorus, as in deeper or less fertile lakes, is into the sediments.

Can the release of phosphorus and its reincorporation into the sediment in fertile, shallow lakes ever form a closed cycle in which high summer fertility is maintained irregardless of new supplies of phosphorus from the catchment

Table 5.1. Some chemical characteristics of endorheic lakes compared with exorheic lakes (those with outflows) in East and Central Africa. (Based on Talling & Talling[493], Ganf & Viner[155], Viner[509], Moss & Moss[371] and Milbrink[346].)

		Ca(mgl⁻¹)	Na(mgl⁻¹)	SRP (μg l⁻¹)	Total P(μg l⁻¹)
Exorheic lakes					
Victoria:	inflows	5·4–7·0	11·5	—	—
	lake	5–15	10–13·5	13–140	47–140
George:	inflows	15·6	12·3	79	—
	lake	15–20.2	12·9–13·5	2	412
Albert:	inflows	10–15	—	—	—
	lake	9–16	91–97	120–180	133–200
Mutanda			15·5	18–93	65–110
Mulehe		20·8–21·7	10·8	220	272
Endorheic lakes					
Rudolf		5·0–5·8	770–810	715	2600
Chilwa:	inflows			0–24	—
	lake	7·2–18·0	348–2690	0–7520	—
Abiata		3–10	2870–6375	—	890
Natron		trace	129 000	290 000	—
Nakuru		10	5550–38 000	40–4400	308–12 200
Magadi		10	38 000		11 000
Elmenteita lake		1·1–10	9450	—	2000

area? The conditions for this would require no loss of phosphorus either to the permanent, deeper sediments or to the overflow. Some, often considerable, loss by means of the latter at least is inevitable in all but endorheic lakes, and means that indefinite high fertility cannot be maintained by sediment release. However, endorheic lakes may represent the ultimate case in this series of interactions between water and sediment, for they have no overflows and non-gaseous elements, like phosphorus, cannot be lost from their basins. Such lakes, like Chilwa, Natron, Hannington and Nakuru in East and Central Africa are often shallow, salty and extremely rich in total phosphorus and SRP (Table 5.1). They are also very fertile with dense populations of blue-green algae, like *Spirulina,* and often a spectacular concentration of plankton-filtering birds like flamingoes. Their high phosphorus levels have usually been assumed to have resulted from evaporative concentration like their high levels of sodium, carbonate, chloride and sulphate. However, elements like calcium and magnesium have not become abnormally concentrated (for fresh waters) in them because they are readily precipitated as insoluble salts such as carbonates. Such should also have been the case for phosphate. The inflows to such lakes are not particularly phosphorus-rich (Table 5.1) yet the waters are dramatically so. In some cases phosphate-rich volcanic spring water may be the source, but probably not always. Though no experiments have yet been attempted, it seems likely that in such shallow endorheic lakes phosphorus is continually released from the sediments, once a critical fertility level is reached, as a result of changes in initial phosphorus concentration caused by decreasing lake depth without the counteracting effect of flushing (see Chapter 2). The all year round productivity of these tropical lakes will also reinforce the closed cycle of release and reincorporation by continuous provision of organic matter to the sediment surface.

5.4 Chemosynthetic and photosynthetic bacteria

A wide variety of chemosynthetic and photosynthetic bacteria finds suitable niches in deoxygenated sediments or hypolimnia. Denitrification, the reduction of nitrate successively to nitrite, nitrous oxide and finally to molecular nitrogen, is carried out by some members of several genera including *Achromobacter*, *Escherichia* and *Pseudomonas*. Oxidation of organic compounds by this means may release as much energy as oxidation with molecular oxygen, so long as nitrate is available. *Thiobacillus denitrificans* can release a small amount of energy by using nitrate to oxidize elemental sulphur to sulphate.

Sulphur and sulphides are metabolized by a range of bacteria. Hydrogen sulphide is a product, for example, of *Desulphovibrio desulphuricans* which uses sulphate to oxidize organic matter, and other species form sulphides from the anaerobic decomposition of proteins. Sulphides may be oxidized to elemental sulphur by such sediment-living genera as *Beggiatoa* and *Thiothrix*, while other species oxidize thiosulphates. In the anaerobic depths of sediment, sulphides may react with Fe II, as detailed above, to give the characteristic black appearance of such mud.

Denitrifying and sulphide metabolizing bacteria are to be found in deoxygenated hypolimnia as well as in sediments, and if the sediment surface or parts of the hypolimnion are in the euphotic zone, anaerobic photosynthetic bacteria may also be found. They may form dense populations at the top of the hypolimnion and different species may characterize particular parts of an oxygen gradient in the metalimnion.

These photosynthetic bacteria are essentially remnants from the Pre-Cambrian era, more than 2×10^9 years ago. Then, photosynthesis using water as a hydrogen donor and releasing oxygen had not evolved, but H_2S & H_2 provided reducing power for photosynthetic bacteria, which, in the anaerobic ocean, occupied the niches of present day algae. The anaerobic photosynthesizers are now confined to whatever pockets of illuminated and deoxygenated water are available.

The green sulphur bacteria, or Chlorobacteriaceae, which include *Chlorobium* and *Pelodictyon*, use H_2S and convert it to sulphate or sulphur which is deposited extracellularly. The purple sulphur bacteria or Thiorhodaceae, including *Thiopedia* and *Chromatium* use H_2S or thiosulphate as hydrogen donors and

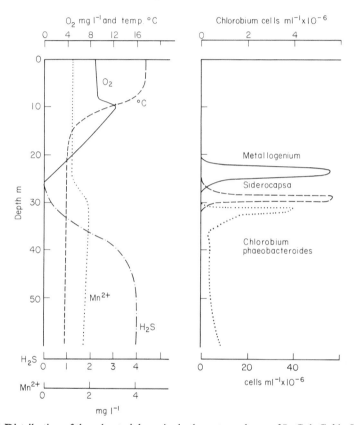

Fig. 5.2. Distribution of three bacterial species in the water column of L. Gek-Gel in September 1970. The lake is meromictic, having a deep and unmixing saline layer from 30–40 m down, which is permanently deoxygenated. *Chlorobium phaeobacteroides* is a photosynthetic bacterium, absorbing light efficiently between 450 and 470 nm (green) and using H_2S as a hydrogen donor. *Metallogenium* and *Siderocapsa* are chemosynthetic bacteria probably using the oxidation of Fe^{2+} and Mn^{2+} as sources of energy. (After Kuznetsov[276].)

deposit the resultant sulphur intracellularly. These organisms may form brightly coloured purple or maroon layers on intensely reducing sediments (for example in shallow pools in forests where much leaf litter constantly enters the water) as well as in deoxygenated hypolimnia (see Fig. 2.10).

Deoxygenated habitats provide excellent examples of niche differentiation on a chemical basis (Fig. 5.2) but as far as the main events in lakes are concerned

the organisms which inhabit them are curiosities, for, with the possible exception
of denitrifying bacteria, they have little influence on the processes going on in
the surface layers where most production takes place. Their interest for biologists
and geologists is considerable, however, because of the clues they may give to
the early history of the biosphere, when it lacked free oxygen. From the point
of view of invertebrate and fish production the importance of the sediment lies
in the timing and degree to which oxygen can reach the surface sediment layers.

5.5 Invertebrate communities of the profundal benthos

A simple comparison between littoral and profundal benthic invertebrate communities recorded at the peaks of their development in the same lake illustrates
just how less diverse is that of the profundal community (Fig. 5.3).

At 2 m water depth in L. Esrom, in Denmark, the littoral macrophytes
and their underlying sediment provide habitat for at least 40 species including
oligochaete worms, triclads, leeches, small and large Crustacea, water mites,
mayfly nymphs, caddis fly and dipteran larvae, gastropod and bivalve molluscs.
A comparable sampling of sediment from under 20 m of water, in the dark
hypolimnion with no macrophytes, revealed only five species—an oligochaete
worm, three Dipteran larvae, and a bivalve mollusc. Exhaustive search of both
habitats would reveal more, rarer species, but would not alter the picture
greatly. The profundal benthos, living in a structurally less complex habitat,
has reduced diversity, but has not necessarily a low productivity for its food
supply may be copious. To some extent the deoxygenation to which the profundal benthos may be subjected may have limited its variety, but even in infertile
lakes with oxygenated hypolimnia the profundal benthos is not species-rich
compared with adjacent littoral communities. Because the hypolimnia of deep
tropical lakes are warm and hence have low oxygen levels even when initially
saturated with oxygen, they rapidly deoxygenate and it is common for there to
be a complete absence of profundal invertebrates in such lakes.

Many profundal communities are dominated by chironomid larvae (Diptera)
and oligochaete worms, with sometimes some small bivalve mollusc species.
Many attempts have been made to classify lakes on the basis of their benthic
invertebrates[41] and although, for the reasons discussed in Chapter 2, these
have been successful only for limited regions or purposes, a general trend in
benthic fauna related to lake fertility can be seen. Well-oxygenated hypolimnia
have a variety of larvae of *Chironomus*, *Tanytarsus* and other chironomid species,
Pisidium spp. (bivalve molluscs), some Crustacea, such as the amphipod *Pontoporeia* and sometimes insect larvae like *Sialis* (alder-fly), but relatively few
oligochaetes. At the other extreme of a sediment covered by anaerobic water for
part of the year, the chironomid fauna is reduced to one or two detritivorous
species, e.g. *Chironomus anthracinus* Zetterstedt, about the same number of
species of larval predators, including *Chaoborus*, the translucent 'phantom-
larva' which migrates to the epilimnion to hunt zooplankton at some stages of

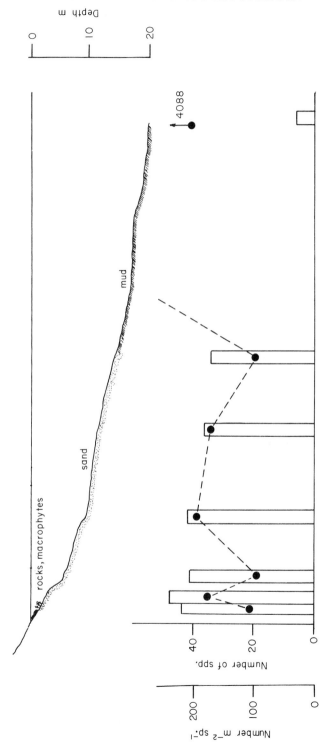

Fig. 5.3. Changes in diversity of the benthic fauna of L. Esrom, Denmark, with depth. Histograms show numbers of species recorded, and the dashed line represents the ratio of numbers of individuals (average over a whole year) per m² to numbers of species at each of several depths. The deepest sediment at 20 m has few species but some of these have very large populations. (Based on data in Berg[31] and Jonasson[26].)

its life history, and a much greater biomass of oligochaete worms such as *Tubifex* and *Ilyodrilus*. Frequently the benthic animals of such lakes are pink or red in colour from contained haemoglobins, though the role that these might play in respiration or survival under deoxygenated conditions is not clear. The profundal benthos of such a fertile lake, L. Esrom, has been studied in some detail[260, 261, 262] and will serve to illustrate something of the biology of these organisms.

5.6 Biology of selected benthic invertebrates

5.6.1 *Chironomus anthracinus*

Chironomus anthracinus in L. Esrom (Fig. 5.4) starts life as an egg mass deposited in May on the lake surface between sunset and darkness, by mated female gnats which dip their abdomens into the surface water film. The egg masses are

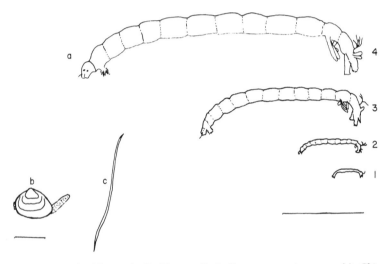

Fig. 5.4. Common detritivores in L. Esrom. Scale lines represent 5 mm. (a) *Chironomus anthracinus*, instars 1–4, (b) *Pisidium*, (c) *Ilyodrilus*. ((a) after Jonasson[260].)

about 2 × 2·5 mm in size when dry but swell a hundredfold when wetted, and are deposited in the lee of beech woodlands on the western shore of the lake. These create the calm conditions necessary for the flight of the gnat, but water currents subsequently distribute the eggs throughout the lake and they sink to the bottom.

By June the eggs have hatched into instars less than 2 mm long. This is the first of four instar stages which precede pupation and emergence of the adult some 23 months later from populations in the profundal. Growth is rapid in June and July and the second instar (3–4 mm long) has hatched into the third in July. The previous growth coincides with a ready supply of food falling from the epilimnion where a major spring population of diatoms has formed (see Chapter 3). The first instar is transparent and probably moves along beneath the sediment surface, swallowing sediment as it moves. The second instar has heavier musculature and has acquired some haemoglobin. It builds a tube, open at the surface and tapering at the base, in which it lives, pumping a current of water past its mouth by undulations of its body. The larvae may also emerge partly from the tube and gather sediment encircling it; this leads to exposure of patches of the underlying dark, deoxygenated sediment until the oxidized micro-zone can reform.

L. Esrom is a fertile lake and the dissolved oxygen levels fall rapidly in its hypolimnion. By July the lower few metres are almost deoxygenated and at around 1 mg O_2 l^{-1} growth of the *Chironomus* larvae stops (Fig. 5.5). This may be because the limited respiration rates then possible produce sufficient energy only for maintenance, and partly because the supply of sedimenting food has

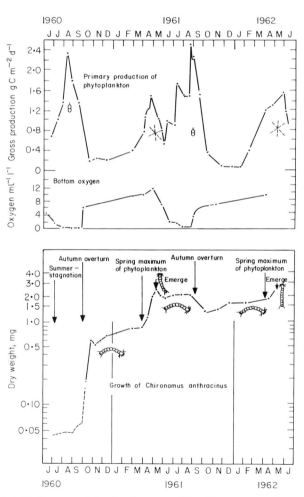

Fig. 5.5. Events in L. Esrom and the development of the *Chironomus anthracinus* population (Redrawn from Jonasson[262].)

changed in quality or amount. In its lengthened journey, delayed in the metalimnion (see Chapter 2), decomposition of labile, energy-rich compounds has been
going on to the advantage of the zooplankton and bacterioplankton, but to the
detriment of the benthos. On the other hand, if growth of *Chironomus* stops, so
does predation on it by bottom-feeding fish, which cannot tolerate the deoxygenation and move to shallower waters. The third instar larvae build a tube,
lined with salivary secretion, which projects a little way above the sediment
surface, and feed on the deposits around the tube. They spread a net of salivary
threads over the mud, to which particles stick, then drag the net down into the
tube to eat it.

   Most littoral animals cannot survive the low or even zero oxygen concentrations that *C. anthracinus* and the related *C. plumosus* L. can[517]. The respiratory
rate of most littoral snails, for example, falls progressively with oxygen content
of the water, whilst that of the *Chironomus* species remains high and constant
over a wide range of oxygen concentrations. The haemoglobin they contain can
store enough oxygen to maintain full activity for only a few minutes, so that it is
unlikely that it helps the animals during prolonged deoxygenation. Animals
beneath the sediment surface are probably at lesser risk from predators than
those at the surface, however, and the oxygen stored by the haemoglobin may
minimize the time that the animal has to spend exposed at the surface, without
loss of activity.

   After the overturn of stratification in L. Esrom in September, growth of
Chironomus resumes and the third instar larvae moult and change into fourth
instar larvae. These are bigger, over 1·5 cm, heavily muscled and bright red in
colour. They produce tubes of sediment particles glued with salivary secretion
which have 'chimneys' projecting 1–2 cm above the mud surface. The role of
the projection is unknown, though it flattens easily and may prevent swamping
of the larva with mud during the autumn and winter when the sediment surface
may be disturbed as the water column mixes. Swamping would scarcely seem to
be a problem for a burrower in mud, however. The height of the chimney, on
the other hand, may be sufficient to reach water only a few mm from the sediment
surface which is much better oxygenated than that immediately in contact with
the sediment.

   Growth of the fourth instar larvae depends on the autumn supply of phytoplankton and detritus reaching the bottom, but is soon reduced as the shortening
days diminish plankton production. It resumes the following spring, but,
because of the pause enforced by deoxygenation during the previous summer,
most of the larvae are not mature enough to pupate and form adults. They must
therefore remain as fourth instars to achieve further growth in the spring and
autumn of the second year. Pupation follows in the succeeding spring when the
pupae emerge from the tubes and float up to the water surface. This happens in
the evening; within about 35 seconds of reaching the surface the pupal skin
splits and the adult emerges, turning from the pupal red to a black colour as it
does so. The adults (imagos) rest on the surface until sunrise when the rising
warmth quickens their metabolism sufficiently for them to fly away and later to

mate and lay eggs on the water. The adult life of the midge is very short—only a few days, compared with the 23 months spent as a larva in the profundal.

At lesser water depths, where summer deoxygenation does not happen, *C. anthracinus* is able to grow sufficiently in one year to emerge the next. The shallow water populations also have higher respiratory rates at any given oxygen concentration, which probably also accelerate development of the larvae. A few of the deep water population do manage to emerge after only twelve months, in the spring following their birth, but they do not contribute any young to the population. Their eggs are mostly eaten as they reach the sediment by the large population of remaining fourth instar larvae.

15.6.2 *Ilyodrilus hammoniensis* and *Pisidium casertanum*

Ilyodrilus is an oligochaete worm (Fig. 5.4(c)), a centimetre or so long when mature. Its development in the profundal of L. Esrom is completed relatively rapidly (within about two months) so that a succession of generations may be found in the sediment. It hatches from egg coccoons, each containing 2–13 embryos and populations of 25 000 worms m^{-2} build up in the summer of some years. The worms burrow through the subsurface sediment, eating each day 4–6 times their own weight of sediment of which they expel most, having digested bacteria from it. They are thus very important sediment processors, causing mixing in the layer in which they live and maintaining the bacterial population in a 'juvenile' state by continual grazing. In the alternate years when *C. anthracinus* exists as a population of fourth instar larvae, the oligochaete population is low, but it increases in the intervening years when these smaller animals seem to compete more effectively with the younger chironomid larvae.

The tiny pea-shell cockle *Pisidium casertanum* (Fig. 5.4(b)) also lives below the mud surface, in burrows parallel to it. Through the openings of the burrows it draws a current of water to maintain its oxygen supply, except of course in summer, when, like the other profundal animals, it is inactive and does not grow. It feeds by sucking in through a tube, or siphon, a current of watery sediment. This is passed over the gills where steadily beating flagella direct a stream of particles to the stomach. The life history of an individual probably lasts more than a year and the young are retained, until fully developed, within the shell of the hermaphrodite adults.

5.6.3 Carnivorous benthos of L. Esrom—*Chaoborus* and *Procladius*

As in all ecosystems the abundance of carnivores in the profundal benthos of L. Esrom is small (about 10%) compared with that of the detritivores. There are two main species, one of which feeds mainly on the plankton and only partly on the sediment community, *Chaoborus flavicans* (syn. *C. alpinus*), and the other, *Procladius pectinatus* also a Dipteran larva, about which relatively little is known. Both species ultimately emerge for brief adult lives in which they lay eggs.

Chaoborus, the phantom-midge, is so called because of its transparent body, punctuated only by its dark eyes and apparently black air sacs at the hind end.

These possibly allow it to regulate its buoyancy when it moves between the epilimnion and the sediment. Eggs are laid in late summer by the adult midges and first instar larvae appear in the plankton in September. Zooplankters, on which *Chaoborus* mainly feeds after seizing them with its prehensile antennae, are reasonably abundant in autumn and the *Chaoborus* larvae quickly pass into their second and third instars. These spend some time in the plankton, particularly during overcast weather when their visibility to prey is least, and part of their time in the sediment. Fourth instar larvae are generally produced the following spring, and these migrate nightly from the sediment to the surface water, where zooplankters are again abundant. During the winter *Chaoborus* spends most of the time in the sediment. Its food then is unknown but it may feed on *Ilyodrilus*. Pupation and emergence are in July.

Procladius is a chironomid species, known to be carnivorous and feeding probably on the smallest *Chironomus* larvae and *Ilyodrilus* juveniles. Its maturer larvae emerge in the spring, while less developed ones may migrate to the shallows, for none can be found in the profundal mud in summer. Their respiration rate is severely reduced as oxygen concentrations decline.

5.6.4 What the sediment-living invertebrates really eat

Organic sediment is, to a large extent, the remains from feeding by a range of terrestrial and aquatic organisms. As a result, it might not be expected to constitute the most nutritious food, and indeed the organic matter by itself seems unable to support the growth of benthic invertebrates. Several studies by McLachlan and his co-workers[340, 341, 342] have uncovered in the invertebrate community of a bog lake in Northumberland a story paralleling that of the feeding of shredders in streams (see Chapter 4).

Blaxter Lough is a shallow basin set in the peat of an extensive blanket bog. Erosion by waves of the peat at the windward edge provides a ready source of organic matter to the bottom of the lake, but a major detritivore, *Chironomus lugubris* Zetterstedt, is conspicuously distributed at the opposite side of the lake basin from the source of peat particles. The eroded peat is washed by water movement across the lake bottom to the leeward shore and as it moves it is broken down into smaller particles. However, although *C. lugubris* does have some selectivity for the size of particle it can eat, this does not seem to be the reason why it does not colonize the areas of the lake where the peat is freshly supplied. Suitably sized particles sieved from eroding or *in situ* peat would not support growth. On the other hand if fresh peat was allowed to become colonized by microorganisms over a few days it would support growth of *Chironomus*, whether in the laboratory or in chambers placed in any area of the lake. The natural distribution of the Chironomids in the lake seems, therefore, to reflect the distance travelled by suitably fine peat particles in order for them to become suitably colonized by bacteria and fungi to be palatable to the animals. The microorganisms absorb nitrogen compounds from the water and by the time they become palatable the peat particles have increased in calorific content per unit weight by only 23%, but have doubled in protein content.

The peat ingested by *Chironomus lugubris* contains a rather greater bacterial than fungal biomass but the balance changes as the microorganisms are digested in the gut. The voided faecal pellets are relatively large, coherent and dominated by fungi. *C. lugubris* is not coprophagic, i.e. it does not eat its own faeces and then digest the new generation of microorganisms which has grown during the interim, as, for example, *Gammarus* (see Chapter 4) seems to do. The faeces are too big for it to eat and *Chironomus* seems to depend on a supply of small, colonized peat particles. Its faeces form the food source of a small cladoceran, *Chydorus sphaericus*. This animal is able to rasp material from the faecal pellets, presumably digesting most of the microorganisms, and producing fine faeces of its own. These are small enough to be ingestible again by *Chironomus*, once re-colonized by bacteria. A rather neat reciprocal relationship therefore seems to exist between the two animals, resulting, with the essential help of micro-organisms, in the ultimate break down of the peat, an initially rather poor food source. It seems highly likely that similar mechanisms and principles might apply to other sorts of sediments.

5.7 Energy flow in the benthos of the Bay of Quinte

The life styles of the various benthic invertebrates are therefore varied, and are influenced by changing environmental conditions. The main energy source of the community is fresh seston (the sum of particulate organic matter, living and dead, falling from the plankton and entering via the inflows) which, by the activity of the animals and bacteria, is converted to more refractory permanent sediment. This is ultimately preserved by the zero oxygen levels below the sedi-ment surface. The benthos is not an isolated community, however, it is modified by predators such as fish, which visit it, just as *Chaoborus* can influence the composition of the zooplankton community through predation[105].

Functioning at depth, the benthos is superficially easy to sample by lowering a grab which scoops up a sample of mud, but elucidation of the processes going on in it is less simple. An instructive study of these has been carried out by Johnson & Brinkhurst[257, 258, 259] in the Bay of Quinte, on the Canadian shore of L. Ontario.

The Bay of Quinte (Fig. 5.6) is long and winding. It is shallow (about 5 m deep) at its inner end, where several towns enrich it with sewage effluent, and opens out, over 100 km distant, into the 30 m deep waters of inshore L. Ontario, where depth and dilution have reduced the fertility of the water (see Chapter 2). At four stations along this gradient the amount of sedimenting material and its fate as it was processed by the bacteria, benthic invertebrates, and fish, was followed. The results emphasize not only the quantity of sediment-ing material as important in determining the productivity of the benthic animals, but also the nature, or quality, of it.

The four stations chosen are shown in Fig. 5.6. In the inner bay, Big Bay station had a benthic community dominated by chironomid species, while the

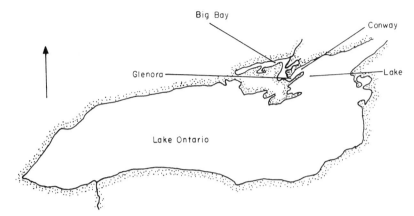

Fig. 5.6. Sampling stations in the Bay of Quinte, L. Ontario.

third station, Conway, had a rich association of bivalve molluscs (*Sphaerium*), oligochaete worms, chironomids and crustaceans. The second station, Glenora, had a community transitional between those of Big Bay and Conway. In L. Ontario, at the fourth station, there was a diverse community of *Sphaerium* spp., Crustacea, and many other species. In general, the diversity of the community increased further into the lake, along with the gradient of decreasing fertility and a gradient of summer bottom-water temperatures, which were above 20°C at Big Bay, but around 10°C in L. Ontario.

The first task was to determine the rate of sedimentation, the rate of supply of materials to the bottom communities. The sedimenting seston was collected in 20 cm diameter funnels fitted into bottles to retain the sediment and suspended about 1·5 m above the bottom. The traps were emptied weekly, as far as possible, and the inorganic and organic parts of the sediment measured. Of course seston is colonized by bacteria while it is still suspended and these bacteria continually decompose it, even in the sediment traps. The rate of this decomposition was found to be, for example, about 5% d⁻¹ between 9 and 14°C and appropriate corrections were applied to give the true rate of sedimentation to the bottom. Table 5.2 gives mean sedimentation rates for the four stations. As might be expected the rates decreased from the inner bay to its mouth, though not steadily, for Big Bay had a much greater rate of sedimentation of organic matter than the others.

Table 5.2. Mean sedimentation rates and community respiration rates at four stations in the Bay of Quinte.

	Organic matter sedimented g m⁻² day⁻¹	Community respiration gO₂ consumed m⁻² day⁻¹	Mean temp. °C
Big Bay	3·01	0·35	17·1
Conway	0·71	0·25	11·8
Glenora	0·29	0·22	10·8
L. Ontario	0·28	0·15	9·1

The overall rate of processing of this material as it reached the bottom was measured as the community respiration, that is the sum of bacterial, invertebrate and fish respiration. The sum of the first two components was measured as the rate of oxygen uptake from water overlying sediment in small cores of the sediment and its community, obtained by a special sampler. The sediment and its overlying water were sealed from the air by a layer of oil and incubated in the laboratory at the temperature at which they had been obtained. Dissolved oxygen was measured chemically in samples taken at the beginning and end of an incubation of several hours. The respiration rates of fish feeding on the benthic community could not be directly obtained. They were estimated from studies elsewhere as about 50% of the total invertebrate production (see below), and this was added to the measured sediment core respiration to give total community respiration. One can see from Table 5.2 that although community respiration decreases from Big Bay to L. Ontario, it does not do so to the same extent as sedimentation rate. This point will be returned to later.

It was then necessary to separate the activity of the benthic microorganisms from that of the macro-invertebrates (those larger than about 1 mm and therefore not the protozoa and organisms like nematodes which had to be included with the microorganisms). This was done by measuring the respiration rates of invertebrates separately. Representative animals of the most abundant species were placed in small jars with a substratum of sand, almost free of micro-organisms, and their rates of oxygen uptake measured at a variety of temperatures. By numerous experiments, Johnson & Brinkhurst were able to derive equations relating respiration rate to the two major factors affecting it, size (as dry weight) of animal and temperature. They did this for all of the major species, and, for instance, the relationship for *Tubifex tubifex* (Müller), one of the oligochaete worms, was:

$$\log_{10}R = 2 \cdot 6126 + 0 \cdot 0455T - 0 \cdot 2654 \log_{10}W,$$

where R is respiration rate in μg O_2 mg^{-1} ash-free dry weight d^{-1}, T is temperature in °C and W is ash-free dry weight (i.e. the organic content) of the animal in mg.

This meant that from routine samplings in which temperature was measured, and animals were counted and weighed, respiration rates for the whole macro-invertebrate community could be determined. It was also possible to find net production, the increase in amount of animal tissue per unit time. This was done by keeping representative animals in small mud cores, freed of other animals by previous heating or freezing, and by measuring by how much their weight increased over several days or weeks. This method has the disadvantage that growth rates may be altered in the absence of competition with other species, but at least gives some indication of the rate. Growth rates were expressed as percentage increases in weight of animal per unit time, and allowed extrapolation to the natural community using the relationship:

Production = Growth rate × biomass.

Production of animals which formed distinct cohorts in their life histories was determined by a variant of the Allen curve technique (see Chapter 4).

Assimilation rates, approximately the sum of respiration rate and net productivity, were then also calculable and these are given in Table 5.3. Biomass of the macro-invertebrates increased towards the outer lake—this contrasts with sedimentation rate and community respiration—and assimilation, production and respiration all reached peaks at Glenora, the second station, and were generally similar or lower at Big Bay and the outer two stations. The high sedimentation rate at Big Bay, therefore, did not support comparably high invertebrate production. The turnover rate of the community, measured by the yearly production to biomass ratio did, however, follow inversely the gradient outwards, probably reflecting the decrease in mean temperature.

Table 5.3. Productivity of the benthic macro-invertebrate community in the Bay of Quinte. All rates in K cal m^{-2} yr^{-1}, and biomass in K cal m^{-2}.

	Biomass(B)	Assimilation	Production(P)	Respiration	P/B
Big Bay	5·45	108·7	74·3	34·4	13·6
Glenora	29·9	368	233	136	7·8
Conway	25·6	142	65·8	75·8	2·6
L. Ontario	38·0	165	51	115	1·3

From all of these data, the diagrams in Fig. 5.7 could be constructed. They show the flow of energy through the sediment community—bacteria, invertebrates and fish—at the four stations, and all quantities have been converted to cal m^{-2} d^{-1} for rates, and to k cal m^{-2} (in brackets) for the standing biomass of the various components. IM is the incoming organic matter, the amount of sedimentation. Community respiration (R_{com}) degrades less than a quarter of it, and animal production less than a twentieth at Big Bay, so that much of it is not utilized (NU) and forms the permanent sediment. Apparently it is not used because it is refractory and difficult even for bacteria to degrade. Its low quality probably reflects partly the chemical nature of the cell walls of blue-green algae which are abundant in the phytoplankton there, but also the large import of more fibrous organic matter left after terrestrial decomposition and washed into the Bay.

The amount of unutilized matter is small at the other three stations and most of the sedimenting seston was degraded by microorganisms (M) or used in animal production. This is reflected in the low organic content of the sediment at the L. Ontario station, 3–4%, compared with 32% in the sedimenting material. The organic content of sediment has, in the past, been used as an indicator of potential benthic animal production in lakes. These studies indicate that it represents only the net result of several processes of accumulation and degradation. These processes are amenable to a general treatment which may have implications wider than those for the Bay of Quinte.

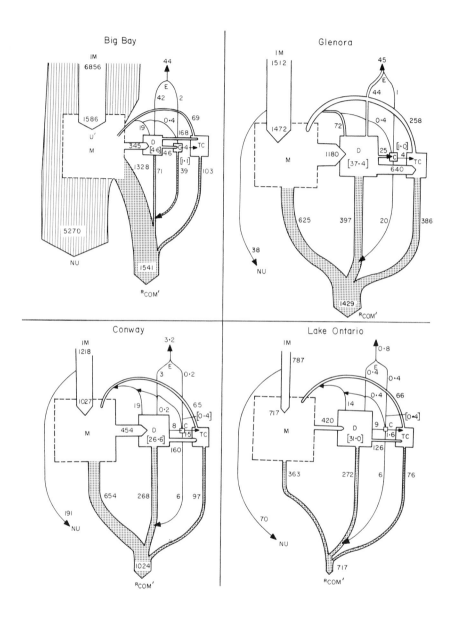

Fig. 5.7. Mean rates of energy flow at four stations in the Bay of Quinte. Boxes represent standing crops and stocks (k cal m^{-2}) and pipes the rates of flow in cal m^{-2} d^{-1}. IM – incoming organic matter; M – microorganisms; U – amount utilized; NU – not utilized and stored in permanent sediment; E – emerging insects; D – detritivores; C – carnivores; TC – top carnivores (fish); R$_{com}$ – community respiration. Stippled pipes show respiratory losses. (Redrawn from Johnson and Brinkhurst[259].)

5.8 General model for benthic production

Several measures of the efficiency of energy use may be deduced for the benthic community. The proportion of incoming energy used by the whole community (microorganisms, invertebrates and fish) is:

$$a = \frac{R_{com} + E}{IM} = \frac{U}{IM},$$

where E is the energy lost in the emergence of adult insects which fly away, and U is the energy used, i.e. not stored in permanent sediment.

The proportion of usable energy chanelled through the macro-invertebrates is:

$$b = \frac{R_D + R_C + E}{R_{com} + E},$$

where R is the respiration of detritivores, D, and carnivores, C. Thirdly, E_{gc}, the net growth efficiency of the invertebrates, is defined as:

$$E_{gc} = \frac{A_D - (R_D + R_C)}{A_D} = \frac{P}{A_D},$$

where A is assimilation and P is production of both detritivores and carnivores.

Table 5.4 gives these quotients for the four Big Bay stations. They show clearly the lower percentage utilization of incoming matter at Big Bay (a = 0·23), but also the fact that the microorganisms at Big Bay take a greater proportion of the utilized material ($(1 - b) = 90\%$) than they do at the other stations

Table 5.4 Efficiency of energy use in the benthic communities of the Bay of Quinte.

	a	b	E_{gc}
Big Bay	0·23	0·1	0·68
Glenora	0·97	0·31	0·64
Conway	0·84	0·27	0·40
L. Ontario	0·91	0·39	0·34

(61–73%). This may be interpreted in terms of the quality of the sedimenting seston. If it is difficult to degrade, a greater investment of its contained energy must be made by bacterial activity to convert it to a form (bacterial cells) usable by the invertebrates.

The product of a and b gives the proportion of incoming energy usable by the invertebrates, and is very low at Big Bay, 2·3%, and 23–35% at the other stations. A general equation for utilization by the animals is:

$$U = a.b. IM,$$

a probably decreases with increased allochthonous import of material (usually of a low quality, refractory nature) from the catchment, while b reflects the cost of processing the material and is low when allochthonous matter, or tough-walled algae, particularly blue-green algae, are present.

The quantity E_{gc} can now be incorporated into the model. Production of macro-invertebrates bears some relationship, c, to utilization of organic matter by the macro-invertebrates:

$$P = c(U)$$

and

$$U = R_{D+C} + E.$$

E, the insect emergence, is generally only a small proportion of the total energy flow and may be neglected so that:

$$P = c(R_{D+C}).$$

Since

$$E_{gc} = \frac{P_{D+C}}{A_D} = \frac{P_{D+C}}{P_{D+C} + R_{D+C}}$$

$$R_{D+C} = P_{D+C} \frac{(1 - E_{gc})}{E_{gc}}.$$

Also, because

$$P = c(R_{D+C})$$

$$c = P/R_{D+C} = \frac{E_{gc}}{(1 - E_{gc})},$$

and since

$$U = a.b. \text{ IM}$$

$$P = a.b.c. \text{ IM}.$$

This is now a relationship which describes how production of invertebrates is related to the supply of incoming organic matter. The two quantities would be directly dependent only if a, b, and c were constants, which they are not—they all decrease as IM increases; a and b decrease for the reasons stated above, c does so because as the import of organic matter increases the rate of deoxygenation in the surface sediment and the water just above it also increases. Animals must then use more energy in obtaining oxygen (in body movements to keep water circulating over the animal's surface or in production of haemoglobin) or, like *Chironomus anthracinus*, have periods when they lie quiescent, respiring, albeit at a low rate, but probably not feeding. This means that a smaller proportion of their energy supply is available for growth. It might be predicted that c should also be low at low levels of import because of the extra activity then necessary in seeking food. The relationship between benthic invertebrate production and import of organic matter to the community may then be much as in Fig. 5.8, with the maximum production at intermediate import levels flanked by minima due to food scarcity on one side, and to low food quality and to environmental changes due to deoxygenation on the other.

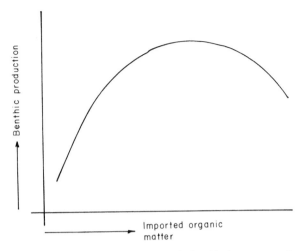

Fig. 5.8. Schematic diagram of a possible general relationship between production of benthic animal communities and the rate of supply of organic matter to them in lakes.

5.9 Further reading

A book has recently appeared of which the subject is largely the biology of the profundal benthos[41]. It also includes a long bibliography and extensive treatment of sampling devices and problems, as does a book edited by Holme & McIntyre[219]. The extensive work carried out on the benthos of L. Esrom can be reached through Jonasson[261], and production in benthic animals is discussed by Edmondson & Winberg[121] and Winberg[539]. Much information on the way that particular environmental parameters affect the benthos is given in Macan[309], and the literature on oxygen tolerance can also be approached with Beadle[24] and Berg & Jonasson[32]. Identification of benthic organisms is often difficult and in addition to works mentioned in Chapter 4, Brinkhurst & Jamieson[42] is a monographic treatment of the Oligochaeta. Such a comprehensive work has, unfortunately, not yet been produced for the Chironomidae.

Deevey[96] is a classic paper on the relationship between lake fertility, hypolimnion deoxygenation and benthic fauna, as is Mortimer[358] on the role of the oxidized microzone. A recent symposium on sediment-water relations is Golterman[177]. Wetzel[530] has a detailed account of the role of chemosynthetic and photosynthetic bacteria. Brock[46] also covers this subject, and more specialist reviews are Kessel[270] and Kuznetsov[276].

CHAPTER 6
AQUATIC PLANT HABITATS

6.1 Introduction

Swamps, wetlands and the reedy or water weed-filled margins of lakes seem to have worried human societies for centuries judging by their reputation in folklore as the homes of trolls, water-witches and the like. For the most part, however, aquatic plant communities are rich and fascinating and offer no greater peril than getting muddily wet to the armpits when wading through them.

It is easy to define where a higher plant dominated community will develop in fresh waters. It will be in shallow enough water to allow net photosynthesis and in habitats where physical stability is sufficient to allow the plants to persist and build up their biomass over what may be a growing period of weeks or months. For most of the aquatic plants (Fig. 6.1) (or aquatic macrophytes, as the

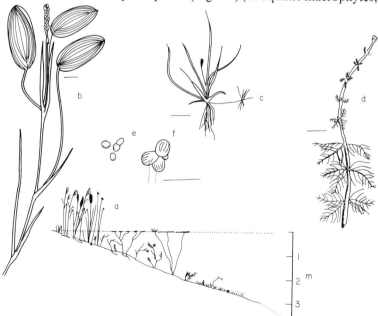

Fig. 6.1. Aquatic macrophyte distribution in a shallow lake (a) with emergent species at the edge, and floating-leaved and submerged species in both shallow and deeper water. (b) *Potamogeton natans* L., a submerged species with floating leaves also; (c) a small submerged species (*Littorella*); (d) a submerged species with dissected leaves, *Myriophyllum spicatum* L., (e) *Wolffia columbiana* Karst., and (f) *Lemna minor* L., two species of duckweeds which float on the surface in quiet water. Scale bars represent 1 cm in each case.

collection which includes the Charophyceae (stoneworts), Bryophyta, Pterido-
phyta and Spermatophyta, is called) this means provision of some sort of sediment and of a water flow sufficient to allow sediment deposition. In man-made
lakes and reservoirs (see Chapter 11), however, with irregularly large water level
changes, rooted higher plants may not survive the unpredictable dry periods and
even a sedimentary shore-line will be bare.

Relatively few macrophytes have colonized fast-flowing streams. These
include some mosses and liverworts, and a flowering plant family, the Podo-
stemonaceae, whose members, with much-reduced flat thalloid bodies can attach
firmly to rock like lumps of green chewing gum. Mostly they are found in sub-
tropical streams. A not diverse, but very important group of macrophytes floats
freely on water surfaces in places where currents and wave action are insufficient
to destroy the plants by smashing them against rocks or casting them ashore.
These include the tiny duckweeds (in the family Araceae), one of whose genera
(*Wolffia*) may be less than 1 mm in size, as well as the much larger floating plants
which have caused economic problems when accidentally introduced into areas
where they were not endemic. The water hyacinth (*Eichhornia crassipes* (Martius)
Solms-Laubach) is the best known, but a fern, *Salvinia molesta* Mitchell, has
been almost as problematic, and both are discussed in Chapter 11. Papyrus
(*Cyperus papyrus* L.) forms huge floating swamps in the Old World Tropics.

6.2 Aquatic plant evolution

There is no precise line at which a lake ends and dry land begins. A continuum
of aquatic plant communities will stretch from the distinctly terrestrial areas
through regions of water-logged soil—fens, bogs, mires (see Chapter 8)—to
those which are rooted in soil over which a layer of free water stands for at least
part of the year—swamps, reed swamps and emergent communities. Such
swamps are typified by such plants as the common reed (*Phragmites australis*
Cav. Trin. ex Steud.) or the reedmace, cattail or bullrush (*Typha* spp.), or by
various Cyperaceae (*Scirpus, Schoenoplectus*). Between the emergent reed
swamp stems where these are not too dense, completely submerged macrophytes
(e.g. *Ceratophyllum, Myriophyllum, Utricularia, Potamogeton, Najas*) are found
together with macrophytes whose leaves, though attached by petioles to rhizomes in the sediment, float on the water surface (water-lilies, *Nymphaea* spp.,
Potamogeton natans L., *Brasenia schreberi* J. F. Gmelin). There may also be
some independently floating plants (duckweeds, *Lemna* spp., *Spirodela polyrhiza*
(Linn) Schleiden *Wolffia* spp., and water cabbage (*Pistia stratiotes* Linn.) in the
Tropics). The emergent plants do not colonize water depths greater than about
1 m, the floating leaved plants extend out in the lake to as much as 3 m, while the
totally submerged species may be found at depths of many metres (see later).

The freshwater macrophytes are clearly a relatively recently evolved group
in which the general trend of plant evolution from water to land has been
reversed. Most flowering plant families in a given area will have one or more

aquatic member, and although there is a proportionately greater representation of Monocotyledons than of Dicotyledons and an under-representation of the rather advanced families of the orchids, grasses and Composites, there is no particular taxonomic grouping of flowering plants which dominates the aquatic flora[239].

The picture is one of an opportunistic colonization of fresh waters by occasional species of predominantly land families able to survive when partially or completely submerged. Sometimes the transition has not been great, for instance in emergent reed swamp plants, but in other cases it has been more fundamental with marked morphological and physiological changes having occurred. In only a few species has the transition been completed to the extent that the entire life-history, including flowering and pollination, occurs underwater. Most produce flowers that emerge above the water surface which are pollinated by the commoner agents (wind and insects) of land plant pollination. The underwater portions, however, have undergone some specialization which affects their distribution in lakes and rivers. These changes become more marked in the continuum from emergent reed swamp plants to those totally submerged. A detailed account of aquatic plant biology is given in Sculthorpe[459].

6.3 Physiological problems faced by aquatic plants

Large, rooted plants cope with several physiological problems underwater not faced by their land relatives, but at least do not have the latter's problems of water supply. Nor, since they grow in a relatively viscous medium, do they need so much support tissue. However, they are often rooted in anaerobic sediment (Chapter 5), the speed of diffusion of oxygen and carbon dioxide to and from their root surfaces is much less in water than in air, and the overlying water absorbs so much light that submerged macrophytes live in an environment of deep shade.

6.3.1 Tolerance of low oxygen levels

The problem of coping with low oxygen levels around the roots has been solved in at least two ways, one structural, the other biochemical. Most vascular aquatic plants are very buoyant because much of their volume (up to 60% in some rhizomes) comprises air spaces, or lacunae (Fig. 6.2). Sometimes in emergent plants (e.g. *Phragmites australis*) they are continuous from the shoots, where they are in contact with the atmosphere through the stomata, to the rhizomes and roots and offer an effective pathway for gas movement. In many species, however, the lacunae are isolated by plates of cells and there is doubt that diffusion of oxygen through such a system would be any faster than through water and sediment itself. Bacteria along the latter route however would scavenge free oxygen and even a restricted pathway through the plant might be advantageous. The lacunae, at least, may help prevent loss of photosynthetically-produced oxygen to the water.

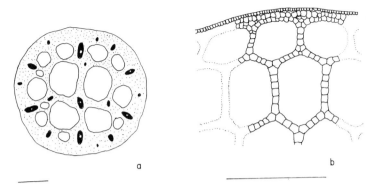

Fig. 6.2. Cross-section of the leaf petiole of *Nymphaea alba* L. (white water lily). (a) Clear areas are air spaces (lacunae), black areas are vascular bundles. (b) Shows a detail of the cross-section of a stem of *Hippuris vulgaris* L. (Mare's tail).

Biochemical mechanisms allowing survival and growth in sediments include tolerance of the usual product, ethanol, of anaerobic metabolism, production of respiratory cytochromes which have unusually high affinity for oxygen[209] and the shunting of the products of anaerobic respiration into substances such as malate which can be accumulated without toxicity at higher concentrations than ethanol can be. Crawford[78] studied species of the genus *Senecio*, some of which grow in waterlogged soils, others in well-drained ones, under experimentally determined different water tables. The dry-ground species died in high water tables as ethanol accumulated in their roots, but the wet-ground species were able to suppress synthesis of the enzyme, alcohol dehydrogenase, which catalyses alcohol formation, and instead produced more oxalacetate, then malate using Krebs cycle enzymes. An enzyme which converts malate back to pyruvate, and hence, under anaerobic conditions to ethanol, was present in the dry-ground species but lacking in those species which survived waterlogging. The wet-ground plants used by Crawford were those of fens rather than of reed swamps or of deeper waters, and the mechanism described has not yet been sought in the latter groups. It would be surprising, however, if it, or something similar to it, was not found in many of them. Some aquatic plants may be able to aid nutrient uptake by their roots by pumping oxygen into the sediment and aerating their own rhizosphere[542].

6.3.2 Inorganic carbon supply
The low rate of CO_2 diffusion into the bulky tissues of aquatic plants as well as the relatively low concentrations of CO_2, particularly at high pH (Chapter 3), could possibly lead to limitation of photosynthesis by carbon shortage. This does not often seem to happen, however, either because light is in even shorter supply or because the plants can use the often more abundant HCO_3 directly as a carbon source. This facility is extremely difficult to demonstrate as it is impossible to create an experimental system where HCO_3^- is present in the absence of free CO_2 or vice versa. The two carbon species are related by ionic equilibria in

water (see Chapter 3) and withdrawal of either by photosynthesis results in changes in the equilibria. However, at pH 4–5 essentially all the carbon is present as CO_2 or as H_2CO_3, whereas at pH 9 the equilibria markedly favour HCO_3^-. Experiments are set up in sealed containers with no gas phase where equal amounts of total inorganic carbon at pH 5 or pH 9 are supplied to replicate plants. If photosynthesis (measured as ^{14}C uptake, or sometimes oxygen production) is significantly greater at the higher pH than at the lower one, then direct bicarbonate use is believed to have occurred[418]. This has been demonstrated in a number of species characteristically occurring in hard, bicarbonate-rich waters, e.g. *Ceratophyllum demersum* L., *Myriophyllum spicatum* L., *Elodea canadensis* Michaux, *Potamogeton crispus* L., *Lemna trisulca* L. and *Chara* spp.[239], while aquatic mosses and plants of more acid, soft waters, *Lobelia dortmanna* L. and *Isoetes lacustris* L. seem to be confined to use of free CO_2. The question is not settled, however, since plants photosynthesizing best at high pH may be favoured by pH for reasons other than inorganic carbon availability and may be able to use free CO_2 at very low concentrations compared with those favoured by low pH. Of course, as CO_2 is withdrawn at high pH the equilibria move in such a way as to raise the pH even further and reduce the CO_2 concentration yet more. Even so there is still a measurable flux of CO_2 through the equilibria and Talling[492] has shown that some algae, at least, can photosynthesize, using free CO_2, at pH's as high as 10 or 11.

In calcium- and bicarbonate-rich waters (HCO_3^- greater than about 2–2·5 mEq l^{-1}) macrophytes and also their attached algal communities often become covered with thick deposits of calcium and magnesium carbonates, known as marl. The mechanism of marl formation has been ascribed to excretion of OH^- ions by the plants as they split HCO_3^- at the cell surface or in the cells to CO_2 and OH^-. The hydroxyl ions are presumed to be excreted at the plant surface. There they combine with free H^+ ions and create a locally high pH, in which the inorganic carbon equilibria favour the formation of carbonate ions. The solubility products of calcium and magnesium carbonates are not high and with ample Ca^{2+} and Mg^{2+} available, precipitates are readily formed. The locally high pH necessary could also be created by the effects of continued CO_2 withdrawal by the plants and need not imply HCO_3^- utilization. Indeed in some highly calcareous lakes, with catchment areas amply supplied with chalk or limestone *in situ* or as glacial debris, carbonates are brought into solution as bicarbonates by the CO_2 content of rain and soil water:

$$CaCO_3 + CO_2 + H_2O \rightleftharpoons Ca^{2+} + 2HCO_3^-.$$

On mixing with lake water of relatively high pH (owing to photosynthetic CO_2 withdrawal) the process may be reversed with inorganic precipitation of a colloidal suspension of carbonates which give the water a bluish-green colour. Examples of such lakes are Lawrence Lake, Michigan[427] and Malham Tarn, Yorkshire[305].

6.3.3 Light

Light availability is unquestionably a very significant factor in determining the distribution of submerged aquatic plants. Aquatic mosses and stoneworts (*Chara, Nitella*) penetrate extremely deeply where the water is clear. In L. Tahoe (California–Nevada) the former are found down to 164 m, and the latter to 64 m[141], though vascular aquatic plants reach only 6·5 m in this very deep, clear lake.

Because of their bulkiness and consequently greater proportion of non-photosynthetic tissue, the compensation point (where gross photosynthesis = respiration) for macrophytes probably lies at higher photosynthetic rates and light intensities than that for micro-algae. Macrophytes may not therefore penetrate to the depth of the euphotic zone found for phytoplankton. The vascular macrophytes, characterized by their lacunae which are lacking in bryophytes and stoneworts, certainly never reach such depths in clear lakes as the availability of light might allow. This appears due partly to the effects of hydrostatic pressure. Experiments in which Gessner[168] applied pressures in excess of 1 atmosphere with mercury manometers to plants of mare's tail (*Hippuris vulgaris* L.) showed that at excess pressures less than 0·5 atmos (equivalent to 5 m of water) growth was almost normal compared with controls, but at 0·75 atmos growth ceased. Work on a variety of other water plants[239] indicates that pressure limits the penetration of vascular macrophytes in lakes to about 10 m, or perhaps a little more in lakes such as Titicaca[504] at high altitude where atmospheric pressure is diminished.

Within the depth zone in which pressure allows growth of vascular macrophytes light exerts an important effect on the overall depth of colonization and on individual species behaviour. The morphology of submerged leaves is similar to that of terrestrial shade plants in some ways. The leaves are relatively thin in relation to their area, comprising only a few cell layers, all of which contain chloroplasts. These may be viewed as adaptations to maximize the use of scarce light energy, and the thinness and frequently extreme dissectedness of aquatic plant leaves may also be devices to minimize diffusion pathways for CO_2 into the leaves.

The overall depth of colonization of macrophytes (both vascular and non-vascular) is often inversely related to the extinction coefficient of the most penetrative light in a rather simple way. Spence[470] has demonstrated this for a series of Scottish lochs (Fig. 6.3), and except for cases where much inorganic turbidity (from glacial melt water, or soil erosion for instance) or organic colour (in peaty catchments) is present in the water the depth of colonization is inversely related to the phytoplankton standing crop and hence to the nutrient loading (Fig. 6.4). The most fertile lakes, with dense phytoplankton populations, are thus likely to have rather restricted submerged plant communities, though emergent vegetation may be prolific at the margins.

A marked zonation of submerged macrophytes is commonly encountered on shelving shorelines. This is at least partly controlled by the photosynthetic abilities of the species concerned[444]. Spence & Chrystal[471, 472] studied a sequence of *Potamogeton* (pond weed) species in a series of Scottish lochs. They found a distinct depth range for each species (Table 6.1) (the light regimes of

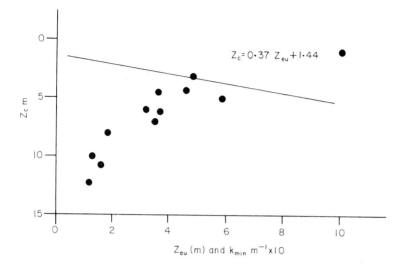

Fig. 6.3. Relationship between maximum depth of colonization, Z_c, of aquatic macrophytes and the minimum extinction coefficient of light, k_{min} (solid symbols), in a series of lakes. The line is that of a regression equation between Z_c and Z_{eu}, the euphotic depth for phytoplankton as represented by the depth at which 1% of the surface light remains. (Based on data in Spence[470].)

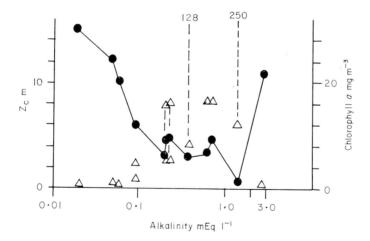

Fig. 6.4. Relationship of maximum depth of colonization of aquatic macrophytes, Z_c (solid symbols), and the size of phytoplankton population (as chlorophyll a mg m^{-3} (triangles)) and the total alkalinity of the water (mEq l^{-1}) in a series of lakes. Z_c bears an inverse relationship to the size of the phytoplankton population. Up to about 1–2 mEq l^{-1} alkalinity, alkalinity gives an approximation to fertility of the water. At the highest alkalinity level of the water, precipitation reactions (see Chapter 3 and section 6.3.2) lead to decreased fertility.

Table 6.1. Depths of colonization of some *Potamogeton* species in Scottish lochs and the extents to which their photosynthetic rates were reduced on transfer from high light intensity (7·08 cal. cm^{-2} h^{-1}) to low light intensity (3.29 cal cm^{-2} h^{-1}).

Species	Mean depth of colonization (cm)	Percentage decrease in photosynthesis
Potamogeton polygonifolius	9	42
P. filiformis	50	59
P. x zizii	120	24
P. obtusifolius	125	2
P. praelongus	190	10

the various lochs were not greatly dissimilar). Leaves of each species were grown at high (7·08 cal cm^{-2} h^{-1}) light intensities under standard conditions in a greenhouse and their rates of photosynthesis measured as oxygen evolution. The leaves were then placed in dimmer light (3·29 cal cm^{-2} h^{-1}) and their photosynthetic rates measured again. Those species which normally grew in shallow water photosynthesized in low light intensities at only a fraction (Table 6.1) of their rates at high intensity, while those normally found at depth maintained proportionately high photosynthetic rates even at low light intensities. The mechanism by which growth is maintained at low intensity was not examined, but in a macroscopic alga, *Hydrodictyon africanum* Yamanouchi, which normally grows in low light intensities, the mechanism appears to be one of reducing respiration through the restriction of some energy-demanding syntheses[419]. One of these is the production of a photosynthetic enzyme, Ribulose diphosphate carboxylase, which is a rather inefficient enzyme but forms a large proportion of the protein content of photosynthetic tissues. In restricting its production of this enzyme, *H. africanum* may reduce its own gross photosynthesis but evidently increases the difference between gross photosynthesis and respiration allowing net growth at very low light intensities.

Other mechanisms, including competition, the nature of the rooting substratum and the behaviour of seeds on dispersal also affect the depth distribution of macrophytes. Seeds of two species of *Potamogeton* in African lakes float for different lengths of time after release and this seems to favour greater depth colonization of the one which floats longest before its lacuna-filled seed coat becomes rotten and waterlogged and the seed sinks to the sediments[473]. Similarly, when L. Chilwa, Malawi, dried out in 1967/68 as a result of below average rainfall for some years, invasion of the remaining mud flat by the predominant *Typha domingensis* of the fringing swamps seems to have been prevented by the inability of the seeds of this species to germinate in the saline mud[227].

6.4 Nutrient supply for aquatic macrophytes

In contrast to phytoplankters, for which nutrient availability seems to be the major factor, with light influencing, productivity (see Chapters 1 & 3), aquatic macrophytes for the most part have a very rich nutrient supply in the sediments in which they are rooted. Experiments using radioactive and other isotope tracers such as ^{32}P and ^{15}N have shown effective uptake and translocation of phosphorus, nitrogen, iron and other ions from the often rather sparse root systems to the shoots[44, 99, 161, 382]. Until recently it had been believed that the root systems were relatively ineffective and that the shoots absorbed most of the necessary nutrients from the relatively much less fertile water. This must indeed be the case for floating macrophytes and for unrooted submerged ones such as *Ceratophyllum demersum* which tend to be most prolific in the more fertile of waters. The sediments are a rich source of nutrients (see Chapter 5) and it seems unlikely that the rate of net production of rooted macrophytes is ever nutrient-limited. Analyses of the ion contents of plants have shown that these are almost always above the minimal levels necessary for nutrient-unlimited growth determined in laboratory tanks. These are about $1\cdot3\%$ of dry weight for nitrogen and $0\cdot13\%$ for phosphorus[167]. The exceptions may occur when either very large biomass levels have already been attained, or, in infertile lakes with sandy substrata, where plants characteristic of these lakes, such as the pipewort *Eriocaulon septangulare* With. and *Lobelia dortmanna*, may have tissue contents only just above, or sometimes slightly below the above-mentioned critical levels which were determined on a range of plants more characteristic of fertile lakes. The critical levels for plants of normally infertile lakes may be lower. A number of experiments[100, 375] indicate that water plants grow better when rooted in organic mud as opposed to sand, but these experiments have generally been carried out on species which do not normally root in sandy substrata. Certainly, it is quite usual to find extensive swards of submerged aquatic plants such as *Littorella* on sand in lakes whose phytoplankton populations are extremely poor because of very low nutrient loading.

6.5 The environment within weed-beds and swamps

Conditions within a well-developed swamp at the margin of a lake or river are very different from those in the adjacent open water. An illustration is provided by studies on L. Chilwa in Malawi. Chilwa (Fig. 6.5) is an endorheic lake (see Chapter 1), though not yet an extremely saline one. It is shallow (maximum depth about 3 m in wet years) and during dry years may dry out altogether. About half of the basin of 2000 km^2 is covered by swamps with *Typha domingensis* Pers. a predominant plant. The open lake water is often turbid with dense blue-green algal plankton and suspended inorganic matter which seems to restrict submerged macrophyte growth. The swamp and open water are thus separated by a more marked boundary than is found in less turbid lakes where a complex continuum of submerged aquatic plants links the two. To the outer fringes of the *Typha* swamp are areas of semi-aquatic grassland, flooded in the summer wet season but without standing water from March to November. The *Typha* swamp itself, except in unusually dry years, is permanently flooded

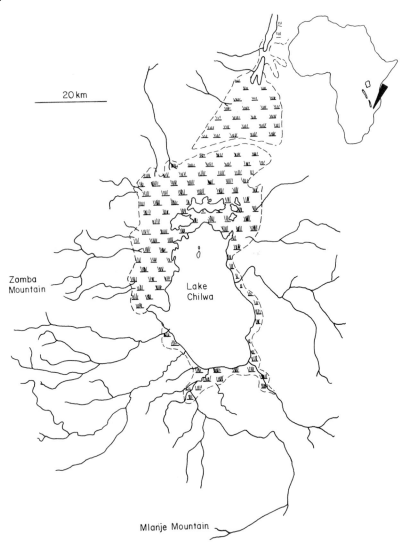

Fig. 6.5. Map of L. Chilwa, Malawi, its inflowing rivers and surrounding swamp.

and the vegetation reasonably homogeneous. During August–September 1975 Howard-Williams & Lenton[229] measured several features of the water and sediment at 24 places in the swamp and at 19 places in the open water. Some of their results are shown in Table 6.2. Notable from these data is the much greater variability, measured by the coefficients of variation (standard deviation as a percentage of the mean) of the swamp compared with the open water of the lake. This is to be expected because of the greater physical structure imparted by variations in the plant growth and because of the much greater effect of the wind over the open water in smoothing out local variations. Clear also is the much lower level of dissolved oxygen in the swamp waters. Again this is due

Table 6.2. Comparison of environmental factors in the open water of L. Chilwa, Malawi, and in the *Typha* swamps fringing the lake.

	SWAMP		LAKE		Level of significance of the differences (F-test)
	Mean	Coefficient of variation (%)	Mean	Coefficient of variation (%)	
Transmission of light (430 nm) through the water (% m⁻¹)	73·2	32·3	38·6	40·8	< ·05
Conductivity (μ S cm⁻¹) of water	1396	66·7	2001	9·9	< 0·01
Saturation of water with oxygen (%)	58·5	97·1	99·8	22·5	< 0·01
pH of water	7·8	9·07	8·9	2·45	< 0·01
Silt content of sediment (g 100g⁻¹)	11·8	53·2	7·2	58·4	< 0·05
Organic content of sediment (g 100g⁻¹)	5·7	49·9	2·8	43·0	< 0·01

partly to the reduced wind mixing, but also to decomposition of the abundant organic material produced in the swamp. Photosynthetically produced oxygen is released to the air from the emergent plants but decomposition of the plant litter takes place underwater where oxygen levels fall and dissolved CO_2 levels increase. This has great consequences for the swamp animals.

In a series of Ugandan fish species, Fish[132] measured the abilities of their blood haemoglobins to combine with oxygen, particularly in the presence of carbon dioxide. Some of the fish lived in open, turbulent water and others in highly deoxygenated swamp waters. The amount of oxygen taken up by the fish blood at different external oxygen pressures was measured by changes in the absorption spectra of the haemoglobin. Fish from open water, such as the active hunter, *Lates albertianus* Worthington had haemoglobin with a relatively low affinity for oxygen, such that O_2 supplied at a pressure of 17 mmHg (about normal atmospheric concentration) was needed to achieve 50% saturation, while in the presence of CO_2 at a pressure of 25 mm (well above atmospheric concentration, but comparable with swamp waters) double the pressure of O_2 was needed for 50% saturation. In the catfish *Clarias mossambicus* Peters, on the other hand, 50% saturation was found at only 6 mm O_2 and CO_2 had a far less depressive action. *Clarias* inhabits swamps and shallow waters and indeed needs atmospheric air to breathe. Its tolerance of CO_2 is crucial since in the intervals between breathing at the swamp water surface CO_2 builds up to high levels in the tissues. The lungfish, *Protopterus aethiopicus* Heck, exhibited this tolerance to CO_2 to an even greater degree. It also is a swamp dweller, and some authors[63] believe that only air breathers among fish can survive away from the edges of tropical swamps. Other fish exist towards the fringes, however, and some gill breathers can use the limited amounts of dissolved oxygen which diffuse into the extreme surface water layers. Beadle[24] gives an excellent review of tropical swamps.

In temperate lakes there is also some evidence that tolerance of low oxygen conditions (or avoidance of them) may be important in allowing some animals to live in swamps and reed-beds and others not. A leech, *Erpobdella testacea* Savigny, a reed-bed species, tested among five other leech species of stream and

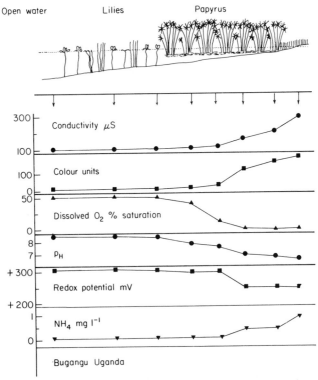

Fig. 6.6. Environmental conditions at the fringe of a tropical papyrus swamp. Decomposition in the stagnant swamp water leads to increased dissolved ion levels, high coloured humic acid levels, and ammonia levels. Oxygen is used up rapidly, redox potentials are decreased and CO_2 production decreases pH. (Redrawn from Carter[63].)

stony habitats[325] was found to be the only species whose oxygen consumption was not reduced by decreasing external concentrations over the range 3 mm–6 mm O_2 even after acclimatization to moderately low oxygen tensions. In a southern Bohemian pond Dvorak[114] studied a transect from the fringing bed of emergent grass *Glyceria aquatica* (L.) Wahlb. to the open water. Similar chemical changes including deoxygenation within the bed were found as described in tropical swamps (Fig. 6.6), and predominant in the fauna of the grassy fringe were snails and the larvae of two-winged flies, Diptera, including mosquito larvae, which were capable of breathing atmospheric air at the water surface.

Although oxygen concentrations have attracted most interest in comparisons of swamps and nearby open waters, many other chemical factors also differ (Fig. 6.6). Some of these are related to aeration of the water, such as redox potential, ammonium and phosphate levels (see Chapter 5), but others depend on the marginal position of most swamps in lake basins. The relatively lower conductivity (a measure of total ion content) of swamp water in L. Chilwa (Table 6.2) is due to initial receipt of rain and river water by the swamp before it mixes with the more saline lake water. The dense swamp delays the mixing. In Ugandan swamps (Fig. 6.6) the reverse occurs. In these lakes, which are not endorheic, the delayed mixing with the main lake allows salts produced by decomposition to accumulate until flushed out by the next spell of heavy rain.

6.6 Measuring the productivity of aquatic plants

Measurement of the productivity of macrophytes presents different problems to those in the phytoplankton, though the same set of basic techniques is used. Macrophytes, particularly the very large emergent reed swamp species present a rather static picture of growth early in the season to a maximum, and then of relative stability for some months as flowering occurs. Early attempts to measure net productivity thus estimated the rate of change of biomass and made the assumption that not much of the production was lost through death or grazing during the season. Recently, using the Allen curve technique (see Chapter 4), where the fate of all the individuals of a particular cohort of new shoots was followed, Mathews & Westlake[337] have shown that in one emergent grass species there was great loss of young shoots by competition and perhaps other agents and that the turnover rate of the biomass may be as much as 3 or 4 times in a year. Rich et al.[427] have demonstrated a similar turnover in a submerged sedge, Scirpus subterminalis Torrey, which produces emergent flowering stalks. Measurement of production by biomass change is thus likely to give serious underestimates if not approached in a sophisticated way. A practical problem of it also is that much of the biomass (often at least 30%) may be underground as roots and rhizomes. Roots are difficult to harvest quantitatively and the sheer bulk of material to be dried and weighed if sampling is to be statistically acceptable lightens the hearts only of laboratory oven manufacturers. The method is also very destructive of the habitat.

On the other hand, enclosure of large plants in gas or water-tight containers so that oxygen release or ^{14}C uptake can be determined also has its problems and enclosure methods have only recently been used for the smaller emergent plants and not yet for the very largest. A community of small Scirpus species has been enclosed[20] in transparent and opaque perspex containers pushed into the sediment, but enclosing both atmospheric and water phases. CO_2 changes were measured by pumping out air samples for analysis with an infra-red gas analyser. Completely submerged plants of Ruppia maritima L. were isolated, still rooted in sediment, in small, water-filled chambers by Wetzel[527] in Borax L., California. Photosynthesis was measured as ^{14}C uptake, for which the same pros and cons apply as to its use for phytoplankton (see Chapter 3). In the bulky tissues of large plants, however, the extent to which fixed ^{14}C is respired and reused may be greater[531]. Oxygen-release measurements in enclosed chambers may also be made, but the lacunae of aquatic plants often retain released oxygen or delay its release so that it may not appear, for measurement, in the water[198].

In an attempt to avoid enclosure of plants in containers, Odum[384] developed the 'upstream–downstream' technique to measure gross production of weed-beds in streams, though it can also be modified to measure community changes in lakes. As in all the available techniques the communities of algae generally found associated with the higher plants (see later) determine that a community photosynthesis (higher plants plus algae) is measured. In the upstream–downstream method a uniform stretch of river perhaps 100 m long is chosen and over 24 hours the concentrations of oxygen are very frequently measured at its upstream and downstream limits. From the difference between the two oxygen curves so obtained, after rephasing to allow for the time it took for water to travel the length of the stretch, the net change in oxygen concentration over the 24 hours can be calculated. It is taken to equal the gross photosynthesis minus respiration, the net effects of diffusion between water and atmosphere, and accrual, the addition of oxygen in seepage water along the banks. Accrual, which cannot easily be measured, is usually assumed to be negligible and is ignored; diffusion is estimated from the oxygen saturation levels and temperatures during the 24 hours; respiration is calculated from the oxygen changes during the night extrapolated to the whole period. Objections to this method, which has been most widely used by Odum's students[187], are legion, as they are to all available methods, but a general picture is now emerging of aquatic plant productivity which is probably approximately correct (Fig. 6.7).

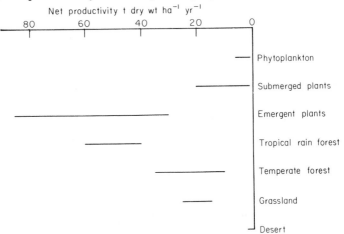

Fig. 6.7. Net primary productivity of aquatic plant communities in comparison with those of phytoplankton and terrestrial vegetation. (Summarized from Westlake[526] and Lieth & Whittaker[292].)

6.7 Productivity of aquatic plant communities

Emergent aquatic plant communities are clearly among the most productive, per unit area, of all the world's vegetation types. This is understandable when it is realized that two of the factors which commonly limit growth on land, water and nutrients, are amply provided in wet sediments receiving supplies of percolation water and 'fresh' silt from the catchment area, while light, which largely limits growth of submerged rooted plants, is also greater for the emergents. Submerged communities are more productive than plankton because of the availability of nutrients in the sediments. Aquatic plant communities must be expected therefore to support more secondary production (including fish production) on a unit area basis than the open water, but their contribution to the productivity of the lake as a whole will depend on the extent to which they are able to develop. This will be fixed by the extent of shallow areas in the lake and on the nutrient loading which will determine the extent to which phytoplankton shading may prevent colonization of these areas (see Chapter 12).

6.8 Weed-bed communities—epiphytic algae and bacteria

Aquatic macrophytes, like all other submerged surfaces receiving enough light, become colonized by dense communities of algae and other microorganisms[5], particularly bacteria and protozoa. Tangled perhaps with marl and mucus secretions of the bacteria, these epiphytic (sometimes called periphytic) communities have a substratum which is not inert. Aquatic macrophytes may secrete organic compounds. This has been demonstrated, using axenic plants (raised from surface sterilized seeds) of *Najas flexilis* (Willd.) Rostk. & Schmidt and ^{14}C tracing by Wetzel & Manny[532], who showed that the growth medium and other environmental factors could result in organic secretion of from 0·05 to 25·3% of the carbon fixed by photosynthesis over a short period. They estimate a mean of about 4% for plants under natural conditions. Much of the organic material secreted was readily oxidized by ultraviolet light, which meant that it was probably metabolically readily available to other organisms. Glucose, sucrose, fructose, xylose and glycine were among the compounds secreted. Such substances can readily be taken up by bacteria and some algae, which Allen[7] has demonstrated with an ingenious apparatus of compartments separated by fine but permeable cellulose ester filters and separately containing ^{14}C labelled axenic *Najas*, and representative axenic cultures of bacteria and algae. The extent to which this secretion may occur in natural situations is hard to estimate but Allen has shown that $^{14}CO_2$ supplied to naturally-rooted emergent flowering stems of *Scirpus subterminalis* in Lawrence L., Michigan does appear in organic form in water close to plants whose epiphytes had been gently scraped off, or in the epiphytes themselves. In the former case, labelled organic matter could still be detected up to 3 m away from the plants after 2 hours.

Provision of such organic matter must help stimulate the growth of the epiphytic community, which in turn absorbs much of the light that would otherwise be available to a submerged macrophyte thus retarding its growth[406]. It seems surprising that macrophytes should not, in the course of their evolution, have acquired chemical or physical means of restricting epiphyte colonization, though, of course, their evolution has been relatively recent and insufficient time may yet have been available. Some large, filamentous algae, e.g. *Spirogyra* and *Zygnema* produce very mucilaginous outer cell walls, which almost entirely prevent epiphyte colonization, but these are the exceptions—vascular plants, mosses and liverworts, and to a lesser extent, stoneworts, are often thickly covered. Hutchinson[239] has recently put forward the idea that an epiphyte community may be advantageous in that it diverts the activity of invertebrate grazers from the host macrophyte itself. Seemingly the macrophytes are much less grazed than their terrestrial counterparts.

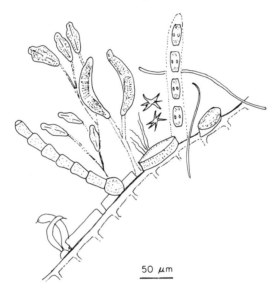

50 μm

Fig. 6.8. A community of epiphytic algae and bacteria.

The epiphyte community comprises a wide variety of algae and bacteria though most emphasis has been on the former. Most phyla of algae are represented, though diatoms are often predominant. Some are attached by stalks or mucilage pads, or by the mucilage secreted through the raphe, which is also an organelle of movement (Fig. 6.8). Some species may not be directly attached but may be embedded in marl or mucilage or may crawl over the surfaces formed by the attached ones. As the community develops, and the weed-bed becomes denser, skeins of filamentous algae may stretch from plant to plant and leaf to leaf and in the quiet water enclosed by this underwater forest, shoals of flagellates move.

The community associated with macrophytes, containing not only firmly attached, but also loosely fixed and free-living organisms is easily disturbed and thus must be sampled delicately. Frequently it is desirable to know its extent per unit area of host plant; determination of host plant area can be made by geometric measurements but these are tedious for highly dissected leaves and calibration graphs relating surface area to wet weight are resorted to. Alternatively, macrophytes may be dispensed with altogether and easily-handled artificial substrata, such as glass microscope slides, or pieces of Perspex (plexiglass), a plastic, may be placed in a weed-bed and allowed to colonize. Use of such devices implies the hope that the colonizing community will be similar to the natural one in production at least and, at best, in composition also. Such hopes are never completely realized—unlike macrophytes, slides are inert, and in different situations quite different results may be obtained in test comparisons. Some reveal evidence of determination of the epiphyte community by the macrophyte itself; e.g. by treating young specimens of three macrophyte species with dilute silver solutions, Prowse[416] was able to free them of epiphytes, then allow them to recolonize, side by side, in the same pond. He found dominance by different diatoms on different macrophytes. Tippett[500] found considerably fewer species on slides than on adjacent macrophytes in a small pond and maximum abundance was found at different times of year on the different substrata. Thirdly, Fitzgerald[135] found that the nitrogen status of a host filamentous alga, *Cladophora*, directly affected the degree of epiphyte colonization.

Despite this evidence of an important effect of the macrophyte on its epiphytes—evidence that is also found for marine macrophytes[194]—many comparisons have shown that communities developing on slides are often similar, at least in their more abundant species, to those on nearby macrophytes[239], and Eminson[128] finds rather similar epiphyte communities on slides and on different macrophyte species in shallow and very fertile lakes in Eastern England.

Of course other factors, such as grazing (see below), help determine the composition of epiphyte communities and Moss[365] has examined these in a set of twenty experimental ponds. The ponds were enriched at three levels with a fertilizer containing nitrogen and phosphorus, and at each level some ponds were stocked with populations of blue-gill sunfish (*Lepomis macrochirus* Rafinesque), while others were not. At the low and medium fertilization levels the ponds retained their initial even sward of *Chara* growing rooted in sediment under clear water, but at the high fertilization level only a few plants of *Chara* persisted and the ponds acquired clumps of *Elodea canadensis* Michx. and some *Potamogeton* species with mats of *Oedogonium*, *Cladophora* and other filamentous algae. There was not a complete coverage of the bottom of such ponds and the reduced biomass of macrophytes was attributed to shading by the much denser phytoplankton populations which developed in the highly fertilized ponds. In the low and medium fertilization treatments the *Chara* bore a diatom community dominated by *Navicula*, *Achnanthes*, *Rhopalodia* and *Epithemia* species, while the *Elodea* in the highly fertilized ponds had a significantly different community with *Navicula*, *Achnanthes*, *Cocconeis* and *Gomphonema* predominant. These

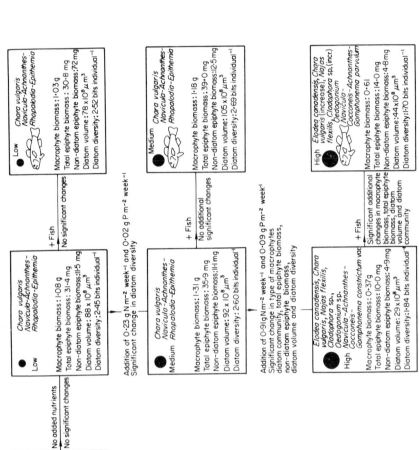

Fig. 6.9. Effects of fertilization and blue-gill sunfish on communities of epiphytic algae and submerged aquatic plants in experimental ponds. Low, medium and high discs refer to fertilization levels; fish symbol indicates treatments where a fish population was stocked in the ponds. The most significant changes occur at the highest fertilization level. (Redrawn from Moss[365].)

differences could be a reflection as much of the very different chemical environments of the ponds, as of the different macrophyte hosts. However, a comparison between the diatom communities on the *Chara* persisting in highly fertilized ponds and on *Elodea* in the same ponds revealed that although water chemistry was the major determining factor, the epiphyte communities on the *Chara* were not so similar to those on the *Elodea* that a complete lack of influence of host type could be ruled out. The fish, which ate invertebrates and which therefore could have had an indirect effect on the epiphytes through their grazers, had no significant effects at low and medium fertilization levels. At high fertilization levels, they caused, indirectly, the marked replacement of one predominant *Gomphonema* (a diatom genus) species, *G. constrictum* var. *capitata* (Ehr.) Cleve, by another, *G. parvulum* (Kütz) Grun., and also allowed greater development of macrophytes and epiphytes. This was attributed to their predation of grazing invertebrates. A summary of the results of the experiment is given in Fig. 6.9.

6.9 Invertebrate communities of weed-beds and reed swamps

Compared with the rocky shore, stream and sediment habitats (see Chapters 4 & 5) the weed-beds of a lake or slowly flowing river present to the animals that live in them a habitat of much greater physical and chemical complexity. Not surprisingly, they support therefore a rich diversity of invertebrates (Fig. 6.10), though for a long period the distinctness of the weed fauna from that of the underlying sediment (see Chapter 5) was not recognized[165]. This has meant that considerably less quantitative work on population sizes, regulation and productivity has been carried out, but that a large amount of the necessary

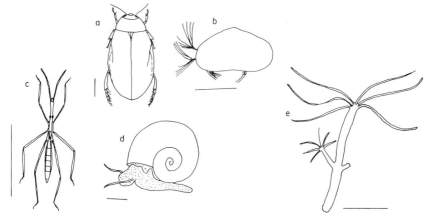

Fig. 6.10. Some weed-bed animals. (a) A hunting predator, the great diving beetle, *Dytiscus marginalis*; (b) an ostracod, which filter feeds in the water among the plants, *Cypris* sp.; (c) a predator which lives on the surface of quiet water and feeds on small invertebrates, the water measurer, *Hydrometra stagnorum*; (d) a grazer on epiphytic communities, the Rams horn snail, *Planorbis corneus*; (e) a lurking, attached predator, *Hydra*. Scale bars represent 5 mm for (a), (c) and (d), and 1 mm for (b) and (e). (After Mellanby[344].)

information on natural history is available. Weed-beds have long fascinated freshwater naturalists and Clegg[69], Macan[309, 311] and Mellanby[344] contain fascinating accounts.

The first problem, inevitably, of studying the weed fauna is sampling. A coarse-weave net swept through the beds misses many of the smaller (< 2 mm) Crustaceans which move among the plants, while a fine net creates so much resistance that many animals can move out of the way. Any net is highly selective —animals like leeches which cling to stems and leaves tend to be underrepresented, while those that normally move around a good deal are overemphasized. Like all other samplers, nets destroy the complex physical distribution of the animals and deprive the observer of all information but a species list and a biased estimate of numerical proportions.

The Birge–Ekman dredge (Fig. 6.11) samples an acceptably constant area of weed-bed by compressing the weeds and then cutting them against the bottom between powerful spring-loaded jaws when these are released by operation of a catch. It is likely to underestimate animals which can move away during its descent. Macan[311] has invented a rotary cutter comprising two concentric cylinders with teeth at their lower edges which can move against one another when the cylinders are counter-rotated by the observer. The device is gently lowered into the vegetation and the teeth operated when it reaches the bottom. A cylinder-shaped section of vegetation is thus cut out and is retained in the cylinder by a plug of peat or sediment at the bottom and an airtight disc screwed to the top. Even with this sampler, however, there must be considerable disturbance and merely the emptying of it destroys the spatial relationships of the several plant and many species of animal which are part of this underwater forest.

Heterogeneity in weed-beds is as much, or more, of a problem for quantitative sampling as it is for other bottom habitats, and many replicate samples must be taken for a statistically acceptable estimate of the population of any of the organisms to be obtained. In an attempt to avoid this problem Macan & Kitching[314] have used 'artificial vegetation' made of polypropylene rope woven into a coarse mesh base weighted with flat stones, to study the weed fauna of a small lake in northern England (Hodson's Tarn). The polypropylene rope is less dense than water and floats upright, giving, with 8 cm strands, a reasonable approximation to the physical structure of the rosette-like *Littorella uniflora* (L.) Ascherson, and, with 45 cm lengths, to *Carex* which fringes much of the tarn. The mats of artificial vegetation are of such a size as to be retrievable with a pond net inserted under them, and the animals can be washed out in a bucket. Although there was some bias in the composition of the communities contained in the artificial '*Carex*' compared with those in the natural vegetation, there was sufficient similarity to make this a useful technique, and it allowed demonstration of some valuable facts. Firstly, the density of artificial leaves determined the sizes of many animal populations, the total number of animals increasing markedly with thickness of vegetation. Secondly, the length of the 'leaves' was important. In an experiment with '*Carex*' where different lengths were used,

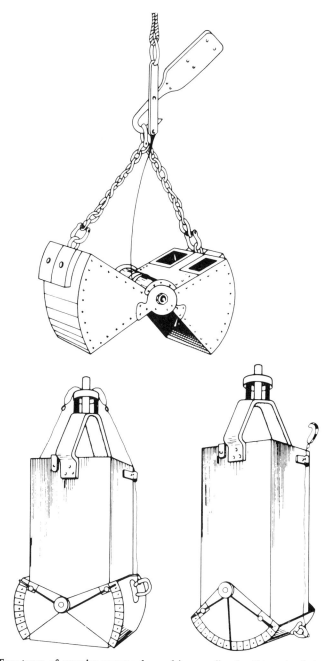

Fig. 6.11. Two types of sampler commonly used in sampling benthic animals in sediment or weed beds. The Pedersen dredge is most useful when the weed is sparse or where soft bare mud is to be sampled. The Ekmann dredge cuts into the weed with its spring loaded jaws which are released by means of a weight (messenger) sent down the line when the sampler is in position. Brinkhurst (1974) from which this figure is redrawn gives details of the performance and relative merits of these and other samplers.

larvae of *Leptophlebia, Cloeon* (Ephemeroptera) and *Gammarus* (Crustacea) were much less abundant when the 'leaves' were shortened from 48 to 8 cm, thus indicating a preference by these animals for the upper parts of the long leaves and hinting at the distributional complexity which other samplers destroy.

Many organisms seem to feed on the periphyton of plants, which, since it comprises a mixture of epiphytic algae, bacteria, protozoa and detritus, must provide a rich diet for these omnivores. It seems to provide a food at least twice as rich in amino acids as underlying sediment in one lake such that chironomid larvae are encouraged to move from their winter sedimentary habitat to feed on the stems of *Typha* coated with periphyton in spring[334]. Other periphyton feeders include mayfly larvae (*Leptophlebia vespertina* (L.) and *Cloeon dipterum* (L.)), freshwater shrimps (*Gammarus pulex*), snails and caddisfly larvae (Trichoptera) all of which normally cling to stems and leaves. Partly clinging, partly swimming and then sometimes filtering detached fine material is a much greater variety of small Crustacea (Cladocerans, Cyclopoid and Harpacticoid Copepods, Ostracods) than is ever found in the open water of a lake[467], and some larger feeders on fine particles, such as sponges (Porifera), are found attached to the plants.

Carnivores in the weed-beds may conveniently be divided into lurkers and hunters[311], though the categories may not entirely be distinct. The former attach themselves to plants or remain stationary among the leaves waiting for prey to pass near. *Hydra*, a small (2–3 mm) coelenterate extends tentacles armed with stinging cells with which it is able to immobilize small prey like Entomostraca (Crustacea), which accidentally brush against them. Leeches (Hirudineae, Annelida) attach by means of a basal sucker and, depending on the species, attack either passing invertebrates, water birds or fish, to which they may attach themselves for some time, loosening themselves from vegetation to do so. Other lurkers include the larvae of damsel- and dragon-flies (Odonata) which are fully mobile but which also wait for their prey. They have a hinged lower mouthpart or labrum which is folded back under the head, but which can be snapped forward to seize the prey with two teeth borne at its tip. Other lurking carnivores include the larvae of the alder fly *Sialis* (Megaloptera) and a stonefly (*Nemoura cinerea* (Retzius)), though stoneflies in general are very unusual in weed beds. Yet others include some caddisfly larvae such as *Phryganea*, whose heavy cases, made from pieces of reed leaves or roots, hamper movement and ensure that only slow-moving prey, or eggs laid on the vegetation are taken. Members of another Caddis family, the Polycentropidae, weave funnel-shaped nets among the leaves and stems and lie in wait at the apex for the prey to become entangled.

Lurking allows a predator to remain concealed, not only from its prey, but also from its own predators, such as fish, but has the disadvantage that prey organisms may very often not pass sufficiently close. Hunting reverses these features, giving greater exposure to predators but readier availability of prey. The quiet water often to be found among emergent vegetation allows those hunting insects which live on the surface tension film of water to build up their populations. These include various bugs (Hemiptera) such as the water strider (*Gerris*), and the water measurer (*Hydrometra*) which has a long head with five stylets which it uses to spear small Crustaceans swimming just below the surface. The water boatman, or backswimmer (*Notonecta*), in common with other of the surface hunters, feeds on insects which become trapped in the surface tension film upon alighting on the water. It hangs from the surface tension film by the tip of its abdomen, moves with paddle like hind legs and seizes its prey with its front and middle legs before impaling it on its sharp beak-like mouthparts. All water bugs suck out their prey. The water beetles (Coleoptera), both as adults and as larvae, are voracious underwater hunters; the adults are able to replenish bubbles of air underneath their wing covers, at the water surface. With this rich supply of oxygen their activity is enhanced. (It is notable that those carnivores confined to lurking are unable to use atmospheric air and are dependent on the more restricted supply of dissolved oxygen.)

6.10 Biology of weed-bed animals

The distribution of individual species within a weed-bed and the ways in which their niches are separated, particularly among the omnivores, provides a rich area for the evolutionist. Relatively little is known. The simplest investigations of effects of temperature and dissolved oxygen, for example, have been carried out infrequently, partly because the complex and daily changing environment of the weed-beds defies simple correlation with experimentally-determined reactions of the animals. Competitive interactions have also been but little studied[309]. Some hint of the sorts of factors which might determine at least the spatial relationships of different species comes from two examples of work done by Macan[311] in Hodson's Tarn, and the first example concerns dragon-flies. The tarn (Fig. 6.12) is surrounded by emergent vegetation of *Carex* and *Eleocharis*, with some *Potamogeton natans* L., with its floating and submerged leaves, and areas of the fully submerged *Littorella*. The deeper water in the middle has beds of *Myriophyllum*, which is submerged though small inflores-cences emerge a few centimetres above the water surface in summer. The nymphs of two species of dragonfly (Fig. 6.12), particularly the younger instars, show marked differential distribution among the vegetation. *Enallagma cyathigerum* (Charp.) is found largely among the *Myriophyllum* though it is not unknown around the edges, while *Pyrrhosoma nymphula* (Sulz.) is confined to the shallow margins. This distribution may reflect little of the differential activities of the nymphs but much of the egg-laying habits of their parents. *Pyrrhosoma* adults fly in copulating pairs among the emergent *Carex* leaves and alight on one of them or on a stem of *Eleocharis* or floating leaf of *Potamogeton*. The female feels underwater with the tip of her abdomen for contact with a stem or leaf and inserts her eggs into the plant tissue as far down below the surface as she can reach without wetting her wings. The pair then fly away and eggs are laid else-where. In the open water, for some reason, the *Myriophyllum* inflorescences

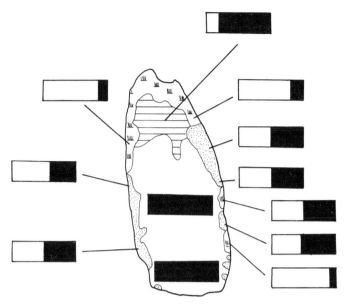

Fig. 6.12. Proportions of the larvae of two dragonfly species in Hodson's Tarn. Main vegetation types are shown as: grass symbols, emergent vegetation; horizontal lines, *Potamogeton natans*; stipple, *Littorella*; white area, *Myriophyllum*. In the bar diagrams one species of dragonfly, *Enallagma*, is represented by black, the other, *Pyrrhosoma*, by white. An explanation is given in the text. (Redrawn from Macan[311].)

appear somehow unattractive to *Pyrrhosoma*, but dense *Carex* impedes the nuptial flight of *Enallagma*. This latter species requires less encumbered air through which to fly and mating pairs land on the *Myriophyllum*, or on stems in the more open areas of the tarn margin. The female starts to walk down the plant into the water, while the male, having copulated, seems not to like getting wet and flies off. The female walks down for as much as a metre, laying eggs in the plant tissue at intervals as she descends. Eventually, on releasing the stem to which she clings, she is buoyed back to the surface by a bubble of air contained between her wings. Oviposition habits seem to explain much of the differential distribution of the larvae of the many species of mosquitoes also[309].

The second example illustrates the possible effects of fish predation on distribution[312]. One year Hodson's Tarn was heavily overstocked with brown trout (*Salmo trutta* L.), a predaceous fish. Subsequently, certain animals, including the tadpoles of frogs and toads, disappeared altogether, and the numbers of active, surface-living, hunting invertebrates, such as the beetles and back-swimmers also declined. Previous to the fish-stocking, one bug *Hesperocorixa castanea* (Thoms.), abounded in the shallow water both among the *Carex* and the *Littorella*. In the first year after stocking it appeared in the fish stomachs and subsequently among the *Carex* but not over the *Littorella* swards. Nor was it then taken by the fish. Apparently sufficient cover was provided only by the *Carex* in the presence of the fish.

The study of weed-bed animals has progressed little beyond the stage of careful observation and in the future, perhaps because of the fascinating detail that emerges from this, the techniques now being applied to stream animals (see Chapter 4) may weld the information into a more complete picture of function. For the moment, the naturalist holds sway and the reader should avail himself of the opportunity to read what a master of this field has to say about it in Macan[311].

6.11 Food webs among the weed community

Very complex food webs must exist within the weed-beds. The aquatic plants themselves are often considered not to be eaten very much. Certainly there would be selective disadvantages for organisms which severely reduced the plants which might be important to them in providing cover from predators. The turnover rates of several times a year among submerged macrophytes (see earlier) testify to some activity however and Gajevskaya[152] has reviewed the evidence.

The Neusiedlersee, a large lowland lake in eastern Austria, has wide reed beds and Imhof[246] has summarized work on food webs and production in beds dominated by the common reed *Phragmites australis*, among which grows *Carex* and other emergent plants and submerged bladderwort (*Utricularia vulgaris* L.). The latter produces between 100 and 200 Kcal (net) m^{-2} yr^{-1}, the suspended algae in the water among the reeds about the same amount, the epiphyte communities about 1000 Kcal m^{-2} yr^{-1}. However, these quantities are dwarfed by the net production of 15000 Kcal m^{-2} yr^{-1} of the reed itself. Despite production of so much plant matter, less than 100 Kcal m^{-2} yr^{-1} appears as net production of animals feeding directly on the plants. These include various stem mining and sucking insects and some gall-forming flies. The latter have a greater effect in inhibiting growth and preventing flowering than in direct grazing.

Detritus-eating animals appear more active, and perhaps ten to twenty times as much energy flows through snails, the water hog-louse, *Asellus aquaticus* (Crustacea, Malacostraca), and chironomid larvae, feeding on dead stems and leaves when they collapse, and on periphyton, than flows through direct plant feeders. Even so only a small proportion, perhaps less than 6% of the net production of plants is eaten by animals so that much must be either laid down as peat (see Chapter 8) or decomposed by microorganisms.

There must be some good reason for the emphasis at least in temperate reed beds on detritus feeding, particularly since these habitats have among the highest primary productivities of the world's vegetation (Fig. 6.7). Possibly it is a response of organisms requiring energy at all times of the year so that the vulnerable larval stages might be passed through as rapidly as possible. It might also be that the bacteria and fungi colonizing the litter provide a source much richer in nitrogen than the live plants, such that much less energy must be invested by the animal in obtaining its supply of necessary elements (see Chapter 4).

6.12 The relationships of swamps and reed-beds with the open water of a lake or river

Although sometimes a fringing swamp may appear to have a distinct boundary with the open water of its lake, the two habitats may have much influence upon each other. Water moves from the swamp into the lake as catchment water flushes previously static swamp water into the lake, carrying with it particulate organic matter as well as its particular chemical properties. Occasionally, on-shore winds may drive lake water into the swamp also[229]. Even completely submerged aquatic plant beds may contribute biologically active substances to the open water and they may provide cover or oviposition sites for animals who spend most of their time in the open water.

Swamps certainly alter the chemical composition of water passing through them[226]. The extensive El Sudd 'an area as large as England, a desolate and sinister maze inhabited only by hippopotamuses, crocodiles, strange water birds and lower forms of life, the air plagued with mosquitoes and other disease-carrying insects, an endless expanse of transient pools and ephemeral streams amidst the stagnant ooze and drifting forest of floating vegetation'[459], lies on the White Nile in southern Sudan. The sulphate content of the river water is markedly reduced as it filters through[488]. Presumably anaerobic sulphur bacteria in the deoxygenated waters (see above) utilize this ion as an oxidizing agent in respiration.

Some authors[226, 335] believe that swamps may make a net contribution of nutrients to the open water when the plants die back at the end of the grazing season. A 'net' contribution is here understood as being a net increase in the nutrient supply (other than organic carbon fixed by swamp photosynthesis) which would be supplied to the lake by the catchment in the absence of fringing swamps. This is difficult to believe since a swamp which is neither extending nor receding will require a constant pool of nutrients for its own growth. It receives 'new' nutrients from the water flowing into it, but some of these become incorporated into the partly undecomposed peat which is laid down (see Chapter 8). Compared with a direct supply of nutrients from the catchment area, supply to a lake through a swamp might thus be expected to be reduced unless the swamp plant roots are able to dissolve inorganic mineral particles washed in with the inflow water from the catchment. A nutrient budget for a bed of water-cress, *Rorippa nasturtium–aquaticum* (L.) Hayck[80] showed no net releases of ions by the vegetation that could not be accounted for by experimental error. And for tropical papyrus swamps, Gaudet[160] has shown a net loss of several ions to the underlying peat, and Brinson[43] has shown very little nutrient release from a swamp forest community.

Displacement of swamp water by heavy rain storms seems often to cause fish-kills in tropical lakes as open-water fish are suddenly engulfed in de-oxygenated water[211]. Less dramatic influx, however, which is more usual, may result in a zone of high animal production and diversity around the junction between open water and swamp. Rzóska[441] gives an account of the shrimps

Cladocera, bugs, Hydracarinids, molluscs and fishes which form a characteristic swamp fringe fauna on the R. Nile as it passes through the Sudd. Presumably, the food base for this community is the fine particulate organic matter exported from the swamp. In a different way the organic litter washed out from the L. Chilwa swamps is essential to certain mud-living Chironomid larvae in that it forms a suitable substratum for the making of the larval burrows, which the lake mud itself does not[339].

Shallow lakes, rich in submerged vegetation sometimes have rather clear overlying water and little phytoplankton despite often moderately high nutrient loading. In some cases this may reflect processes, in very calcareous lakes, of precipitation of phosphorus and iron as calcium and carbonate compounds. The open-water phytoplankton are thus nutrient stressed while the rooted, aquatic plants can exploit mineral nutrients in the sediments. In less calcareous waters this cannot be the entire explanation and Hasler & Jones[203] and Fitzgerald[135] have implied that inhibitory substances might be secreted by the macrophytes. There is no direct evidence yet for this, though macrophytes certainly can release organic compounds (see earlier). Control of many events in the open-water plankton community by substances produced in the littoral zone of weed-beds and sediments in shallow water is advocated by Wetzel whose book[530] should be consulted for a full statement of the case.

6.13 Further reading

Three recent major books, Good et al.[179], Hutchinson[239] and Sculthorpe [459], have been devoted entirely, or almost entirely, to aquatic plants, and contain lengthy reference lists. Two earlier works by Gessner[169, 170] cover much European literature, and Beadle[24] is comprehensive on the biology of tropical swamps. Haslam[199] covers river plants, and Howard-Williams[228] is a very useful review of swamp ecosystems. These compilations make further listing of individual papers unnecessary. Useful floras specifically for the aquatic plants are Fassett[130] for east and central N. America, Mason[336] for western N. America and Haslam et al. 1975 for Great Britain.

CHAPTER 7
THE NEKTON—FISH AND
OTHER VERTEBRATES

7.1 Introduction

'Nekton' is derived from a Greek word meaning something that moves, and in the sense used here includes all those animals which actively move between the communities and habitats described in Chapters 3–6. They may spend much of their time in one community, but are not confined to it and may migrate extensively along waterways, even between the sea and freshwaters. The nekton mostly comprise vertebrates, including cyclostomes (lampreys), fish, amphibians, reptiles, birds and mammals, including man.

Such animals are often studied alone. Their relatively large size has made them ready subjects for natural historians; even so many details of their habits and life histories are still unknown. Much less is known about how these animals influence freshwater ecosystems, as opposed to the reverse case. Through predation they may affect other plant and animal populations far removed from them in the food web, and they provide also a further link with the catchment area. Water birds and amphibious mammals like the hippopotamus (*Hippopotamus amphibius*) may import nutrients to a lake from areas outside the hydrologic catchment, while human populations, by removing marsh products such as reeds (*Phragmites* spp.), may greatly modify the shoreline region. The role of man and his fisheries is considered in Chapter 9, and other effects human populations have on waterways in Chapters 10–12. In this chapter, the non-human nekton is surveyed, with an emphasis on fish, for which most information is available.

7.2 Freshwater fish biology

Fishes are very diverse. They range from the tropical lungfish, able to breath air and to aestivate in mud during droughts, to the salmonids whose tolerance of oxygen concentration and temperature is narrow, but some members of this group migrate between the sea and freshwater. They include omnivores like the roach (*Rutilus rutilus* L.), eating a range of foods from invertebrates to plants, and species such as *Rhamphochromis macrophthalmus*, part of whose diet of fish includes the eyes of prey too large to be tackled in their entirety. Adult size ranges from the 12 mm of a Philippine goby to the (sometimes) 5 m of the European catfish, or wels (*Siluris glanis* L.). Variations on the basic fish body shape encompass bottom-living flatfish and the torpedo shapes of fast-swimming piscivores (fish-eaters). Some fish show spectacular features—the leaf fishes have evolved camouflage and look like dead, brown leaves lying on the bottom; eyeless fish live in cave streams, and the Amazonian piranha work in shoals to clean rapidly the flesh from large carcasses. Small fish have often evolved anti-predator devices, for example spiny pectoral fins which lodge in a predator's throat. A peculiar attraction of one of these small fish species to urine is reported to lead to its painful invasion of the urethras of people bathing in some South American rivers.

The greatest number of freshwater fish species lives in the Tropics. This may reflect partly greater growth rates and shorter generation times in warm waters, which may permit faster evolution and partly the long history as permanent waterbodies of many deep, tropical lakes. The fish fauna of islands such as Great Britain is depauperate, for glaciation eliminated pre-existing lakes and the time

Table 7.1. Some major orders of freshwater fish.

Order	Freshwater spp./total spp.	British spp.	Temperate U.S. spp.	Tropical African spp.	Notes
Lepidosireniformes	5/5	0	0	4	Lung-fish. Air breathing
Polypteriformes	11/11	0	0	6	Bichirs. Partly air breathing
Acipenseriformes	15/25	1	18	0	Sturgeons. Primitive, partly cartilaginous skeleton. Anadromous*
Semionotiformes	7/7	0	7	0	Gars
Mormyriformes	101/101	0	0	101	Elephant fish—snout is often elongated and proboscis-like
Clupeiformes	25/292	2	5+	2+	Herrings, shads. Often plankton feeders with long gill rakers
Salmoniformes	80/508	14	70	0	Pike, salmonids, ciscoes, coregonids, grayling, smelts Believed ancestral to many other orders. Often anadromous
Cypriniformes	3000/3000	20	209	180	Cyprinids ('coarse fish'). With siluriformes constitute 73% of freshwater fish. Carps, etc.
Siluriformes	1950/2000	1	37	345	Catfish. Sensory barbels on head. No scales
Atheriniformes	500/827	0	186	0	Killifish, guppies
Perciformes	950/6880	11	55	700+	Extremely diverse, perches, centrarchids, cichlids; spiny fins
Anguilliformes	15/603	1	?10	0	Eels, often catadromous**
Total	6683	50	597	1332	

Total freshwater fish = 6851, Total fish 18,818
* Breed in freshwaters, but spend part of life cycle in sea   ** Reverse of anadromous

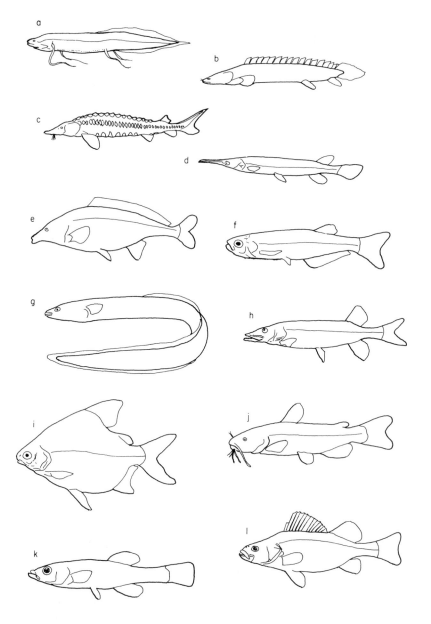

Fig. 7.1. Representatives of some important orders of freshwater fishes (see also Table 7.1). The fish are not to the same scale. (a) Lepidosireniformes, *Protopterus*, an African lung-fish; (b) Polypteriformes, *Polypterus*, the bichir; (c) Acipenseriformes, sturgeon; (d) Semionotiformes, *Lepistoseus*, garfish; (e) Mormyriformes, elephant fish; (f) Clupeiformes, the denticle herring, *Denticeps*; (g) Anguilliformes, freshwater eel; (h) Salmoniformes, *Esox lucius*, Northern pike; (i) Cypriniformes, an African Citharinid; (j) Siluriformes, a north American catfish; (k) Atheriniformes, a cyprinodont killifish; (l) Perciformes, *Perca*, perch. (Redrawn from Nelson[381].)

period afterwards when the waterways of Britain were connected with those of mainland Europe before sea levels rose to isolate the islands was short. The fish faunas of continental North America and Asia are less rich than those of the Tropics but much richer than those of temperate islands. Table 7.1 lists and Fig. 7.1 illustrates some major freshwater fish orders with some indication of their distribution, and an account of six contrasted species (Fig. 7.2) will illustrate the basic biology of fishes.

7.2.1 Eggs and fry

The trout (*Salmo trutta* L.), the Nile perch (*Lates niloticus* (L.)), a 'tilapia' (*Sarotherodon niloticus* (L.)), the walleye (*Stizostedion vitreum vitreum* Rafinesque) and the Chinese grass carp (*Ctenopharyngodon idella* (Val.)) have been chosen for discussion here because of the range of their diets and zoogeography. Trout are carnivores, eating largely benthic invertebrates; they are indigenous to Europe, N. Africa and W. Asia, but have been introduced to suitable waters elsewhere—notably the cool upland streams of the otherwise warmer areas of

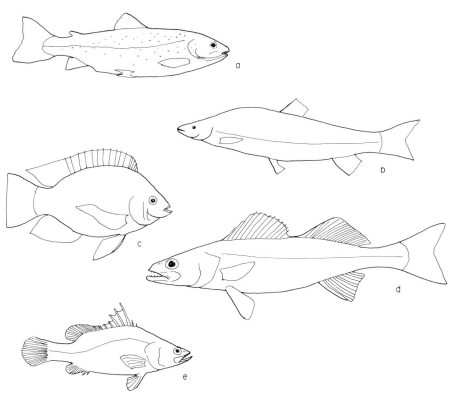

Fig. 7.2. Six fish species of contrasted biology (see text). (a) Brown trout, *Salmo trutta*; (b) Chinese grass carp, *Ctenopharyngodon idella*; (c) 'Tilapia', *Sarotherodon niloticus*; (d) Pike-perch, *Stizostedion lucioperca* (the walleye is a similar though stockier fish); (e) Nile perch, *Lates niloticus*.

the once British Empire, Kashmir, the Kenya Highlands and New Zealand—for their sporting qualities. Nile perch are voracious piscivores in the R. Nile and its associated, or once associated, great lakes Albert and Rudolf. They are valued food fish, being large (specimens weighing over 45 kg are common) and have also been introduced to other African lakes (e.g. Kyoga and Victoria). The tilapia in question feed on fine bottom detritus and on phytoplankton and are also native to the Nile watershed but have been widely introduced to other African lakes and rivers. Walleye are N. American piscivores which feed on zooplankton when young. Some comparison will be made between walleye and their close European relatives, the pike-perch, *Stizostedion lucioperca*[332]. Lastly, the grass carp is an avid feeder on submerged and even emergent aquatic macrophytes and is endemic to the R. Amur and other areas of eastern Asia.

All fish hatch from eggs, mostly released into the water for their development. *Sarotherodon* species are exceptions, for once the eggs are externally fertilized the female gathers them into her mouth for brooding. Under such close protection relatively little investment in numbers of eggs produced is required. *S. niloticus* certainly protects the young fish in this way when predators approach, but may merely guard them as eggs in a shallow depression scraped on the sandy bottom by sweeping movements of its tail.

Trout also carefully excavate a nest or redd, but in a suitable stretch of running water where water flows freely through the interstices of a gravelly bottom[144] and silt is not deposited. The female turns on her side just above the bottom and vigorously flaps her tail up and down. The current so created disturbs the gravel and stream-flow moves it slightly downstream so that after about an hour a depression about 8 cm deep and 25–30 cm in diameter is excavated or 'cut'. The female moves into the redd and the male moves alongside her; eggs are extruded and fertilized after a behavioural sequence with a quivering of the male's body and characteristic posture of the female with her head held up and pelvis downwards. The fertilized eggs sink into the gravel interstices and the female then covers them with gravel displaced from just upstream where she may make a second redd. The young fish (fry or alevins) hatch in up to 5 months at 2°C or less than a month at 12°C, but temperatures much higher than 12°C may be lethal.

The Nile perch and grass carp take little or no care of their eggs and must produce very large numbers of them to ensure ultimate survival of enough young to maintain the population. The eggs of both species are planktonic and probably slightly denser than water. The difference is minimized however by incorporation of oil globules in the former and a water-filled cavity between the egg membranes in the latter. While *S. niloticus*, which guards its eggs in the nest, may produce only a few thousand, the Nile perch releases several millions to the open water.

Walleye also take no care of their eggs. They are scattered onto a large area of gravel in well-oxygenated water having been fertilized during release from the female. Because of dilution of the sperm (milt) in the water the fertilization rate is variable, sometimes 100%, sometimes only 3%. The walleye's European relative, the pike-perch, is similar in morphology, but quite different in spawning behaviour, and illustrates the differences in behaviour which may evolve in a population separated spatially. The male pike-perch excavates a nest on a muddy organic bottom in such a way that the roots or rhizomes of aquatic plants are exposed. The eggs are sticky and adhere to the exposed roots. The female guards them and fans water over them with her tail—this may increase the rate of survival in a habitat which is generally unsaturated with oxygen. The percentage of eggs fertilized is also consistently high for they are confined within a relatively small area.

Many fish in temperate regions lay eggs which adhere to stones (e.g. minnows, *Phoxinus phoxinus* L., barbel, *Barbus barbus* L., and gudgeon, *Gobio gobio* L.) or to weeds in lowland rivers and shallow lakes (e.g. roach, carp, *Cyprinus carpio* L., bream, *Abramis brama* L.) and often these eggs are unguarded. With limited resources available for egg production there has been a tendency for large numbers of small eggs to be produced by those species which do not protect their eggs by either hiding or guarding them, and for smaller numbers of larger eggs to be produced by those that do. In the latter case the individual probability of survival is doubly enhanced for a larger egg contains more yolk for sustenance of the fry. On the other hand, the spawning requirements of the 'protective' type may be more stringent and less easily available and the larger egg, with its smaller surface area: volume ratio may require greater external concentrations of oxygen to enable fast enough diffusion into it. The egg size/number relationship is well shown for some British fishes (Table 7.2) but is complicated in the case of the walleye and pike-perch.

Table 7.2. Relation between size of eggs (mm) and number produced per kg of female body fresh weight in British fish species. (After Varley[508].)

	Egg diameter	No. produced kg^{-1}
Salmon (*Salmo salar* L.)	5–7	1100–3080
Trout (*Salmo trutta* L.)	4–5·5	1100–2640
Charr (*Salvelinus alpinus* L.)	4–4·5	3080–4840
Grayling (*Thymallus thymallus* L.)	3·2–4·0	6600–9900
Pike (*Esox lucius* L.)	2·5–3·0	22 000–44 000
Perch (*Perca fluvialis* L.)	2–2·5	ca. 220 000
Tench (*Tinca tinca* L.)	ca. 1·2	ca. 605 000
Carp (*Cyprinus carpio* L.)	0·9–1·2	up to 1 210 000

The eggs of the pike-perch, which are guarded, are smaller (0·8–1·5 mm diam.) than those of the walleye (1·4–2·1 mm diam.), which are not. Furthermore, the walleye produces only 30 000–65 000 eggs kg^{-1} bodyweight, compared with 110 000–260 000 in the pike-perch. This reversal of the expected trend is probably an adaptive response to the contrasted habitats in which these fish live. The walleye inhabits well-oxygenated waters, the pike-perch is successful in stagnant or slow-moving productive waters with low oxygen levels and high turbidity, although under experimental conditions adult walleye can tolerate lower oxygen levels than pike-perch. Nonetheless the smaller size of pike-perch eggs must enhance gas exchange in a nesting habitat which is within highly organic sediments, and the large number of eggs produced, even though guarded, may be a necessary ploy in a fertile habitat usually characterized by Cyprinid fish which prey on eggs to a greater extent than the species co-existing with walleyes.

7.2.2 Feeding

The fry which hatch from the eggs do so after varying periods of development, dependent on temperature, among other factors. Trout spawn in autumn, walleye between March and June, the tropical Nile perch and 'tilapia' probably at most times of year, the grass carp in rising flood waters in spring when increased turbidity may camouflage the eggs somewhat against predators. The times at which hatching occurs, however, have become adjusted through natural selection to the periods in which food is available for the fry. Adults may overwinter without feeding for they can build up fat stores during the summer. The yolk carried in the egg, however, is only large enough to keep the fry for a very short period and any increase in the amount of yolk produced per egg would mean a corresponding decrease in the number of eggs produced. On balance, therefore, selection has produced systems in which the maximum number of fry are hatched in the spring and summer in temperate regions, when suitable food for the fry is most plentiful.

Small fish can eat food only in small portions; large fish can eat much larger items, and the investment of energy in finding large items is proportionately less than that needed to find the same amount of food in smaller units. The diets of fish thus often change markedly as the fish increase in size. Trout alevins dart up from their shelter in the gravel to take small chironomid larvae and Crustacea in the weeks after their yolk is used up. When they have grown to about 4 cm they station themselves in the water column, facing upstream and space themselves about 8 cm apart, defending their 'water-space' against neighbours by aggressive darts at intruders. They then feed on invertebrates drifting (see Chapter 4) downstream past them. As they grow larger they continue to take drift organisms—insect larvae, molluscs, Crustacea, and also any suitable terrestrial animals falling into the stream. These might include Diptera, aphids, woodlice, beetles, snails, spiders, even an occasional mouse for the largest trout, but the terrestrial contribution to the diet is usually much less than that of aquatic organisms. As a trout ages it will also actively hunt food, particularly if the prey moves along the bottom, and may take fish fry and larger fish when it is bigger than 30 cm in length. Trout have relatively wide mouths and many backwardly directed teeth which efficiently hold prey once it is grasped. Other fish species feeding in mid-water, but on smaller items like zooplankton, have a much narrower mouth, protrusible into a lengthened tube with which they suck up prey, like a vacuum cleaner. The tropical elephant snout fish (Mormyridae) (Fig. 7.1) are good examples.

The adult Nile perch is a voracious feeder even on other piscivorous, but smaller, fish, like the tiger fish (*Hydrocynus vittatus* Cast.). When it is younger,

however, it takes invertebrates. The fry, 0·3–1·35 cm in length, feed on planktonic Cladocera (apparently in preference to copepods) in L. Chad and inhabit
shallow, weedy areas[222]. At about 20 cm they begin to take larger invertebrates
—a bottom-living prawn (*Machrobrachium niloticum*) and snails—and some
small fish. As they grow they take more and more, larger and larger fish; they
themselves grow to several metres in length. The Nile perch is a very active
pursuer of prey. It has a large head, a widely gaping mouth and serried ranks of
backwardly directed teeth; the dorsal and anal fins are situated well back (Fig.
7.2), giving the fish a powerful tail thrust which, as in the trout, allows bursts of
high speed enabling capture of prey, which may itself dart rapidly away.

Newly-hatched walleye begin to feed on plankton—large diatoms, rotifers,
nauplii of crustaceans. They form schools which may ensure a greater survival
rate of the fry, which are only 6–9 mm long, than if they moved individually.
Progressively, they eat larger zooplankters, adult Cladocerans and copepods,
then a mixture of zooplankters and such open water insect larvae as those of
Chaoborus (see Chapter 5). Finally they become piscivorous, seizing their prey
then manoeuvring it until it can be swallowed head first. This is probably
because some favoured prey species, e.g. yellow perch (*Perca flavescens* Mitchill),
have spiny pectoral fins which would lodge in the throat if the prey was swallowed
tail first. Walleye have rather elongate gill rakers (the strips of bone which
protect the delicate gills from damage by large particles pulled in with the respiratory water current) and it is on these that spiny fins could catch. The pike-
perch does not have such elongate rakers and is able to swallow at least some of
its prey tail first. From the point of view of the prey, spiny fins have some
advantage, for in the time taken for a fish, seized usually at the tail, to be man-
oeuvred into a head-first position for swallowing, there is a greater possibility
of escape. Spines also may be used by a fish to avoid capture if they can be
jammed into rock crevices. Other fish mimic twigs and gnarled pieces of wood,
e.g. the S. American catfish *Farlowella* and *Agmus*[300].

The grass carp becomes predominantly vegetarian at lengths above about
30 mm. When fry, it eats rotifers and crustaceans, occasional chironomid larvae
and perhaps some filamentous algae; between 17 and 18 mm in size it takes more
chironomids and fewer of the smaller zooplankters are eaten, and by 27 mm
higher plant food becomes prominent in the diet, and although thereafter it
unavoidably takes invertebrates associated with macrophytes, the grass carp is
well adapted to a plant diet. This is unusual, for relatively few fish eat water
plants.

Like other Cyprinidae the grass carp has a toothless mouth, but it has strong
and specialized projections of the pharyngeal bones, which line the region
behind the mouth, just before the entrance to the oesophagus. These pharyngeal
teeth lie in sockets in the pharyngeal bones. The row on the upper bones has
two small teeth on either side, and on the lower bones are four or five teeth with
serrated cutting surfaces on each side. In older fish these become flattened with
both cutting and rasping surfaces. This change is related to the fact that young
fish eat only the softer submerged macrophytes, whereas older ones can tackle

more lignified emergent macrophytes as well. The teeth tear and rasp the plant food into particles 1–3 mm in diameter. Only the cells which are rasped and ruptured are digested and half of the food passes out, undigested, as faeces[212]. The pH of the gut secretion is quite high—7·4–8·5 in the anterior part, around 6·7–6·8 in the rectum, and such levels are not particularly noteworthy. They provide a relative measure for comparison with those of *Sarotherodon niloticus* which, perhaps because of its extremely low gut pH, between 1·4 and 1·9, is one of the few fish yet examined that can digest even the blue-green algae[353].

Fry of this tilapia may feed on insect larvae—chironomids and *Chaoborus* —but the fish soon turn to an algal diet. In different lakes this may be epiphytic or periphytic, e.g. in L. Volta (Ghana) *S. nilotica* eats the weft of algae loosely hanging from submerged dead trees in this man-made lake (see Chapter 11), and bottom detritus is also taken. In L. George, Uganda, which has a very dense blue-green algal population (see Chapter 3), much of which rests on the bottom but is circulated into the water column by regular diurnal mixing, *S. niloticus* takes the blue-green algae, but whether in the water column or from the bottom is not clear.

Feeding makes little use of the teeth or jaws. Food is sucked in with the respiratory current and entangled with mucus secreted by glands in the mouth, and then it is passed back into the pharynx. The food is thus not filtered out, as it is in many other plankton feeders, by fine projections on the gill rakers. The first lot of food taken during the day is not well digested but as the stomach secretion falls below pH 2 digestion takes place and 70–80% assimilation of the blue-green algae *Anabaena* and *Microcystis* has been noted in laboratory experiments[356]. Assimilation from natural phytoplankton populations in L. George was lower, about 43%, but this is still considerable in view of the long held belief that blue-green algae are indigestible as live cells[211]. *S. niloticus* shows some selectivity in its food. A comparison between the percentage representation of different foods in the gut (r) compared with that in the external environment (p) may be used to calculate an electivity index $r - p(r + p)^{-1}$[251]. In L. George, *S. niloticus* showed a positive selection for *Microcystis*, *Lyngbya* (blue-green algae) and *Melosira* (a diatom) but discriminated against the smaller species, *Anabaenopsis* (blue-green algae) and *Synedra* (diatom).

7.2.3 Breeding

The changes associated with the onset and completion of spawning are major events in a fish's life. The gametes alone may constitute a quarter of the body weight and the energy demands in producing them and in accomplishing the act of spawning may be very great. A 'spent' fish (one that has just spawned) is weak and more vulnerable to predation, and the extensive migrations which some fish undertake before spawning may have exhausted them so much that bacterial and fungal infections are common. Nonetheless, most fish spawn in several successive years after reaching maturity.

Related to studies of the biology of spawning are those of speciation. Distinct species cannot produce fertile, hybrid offspring. Populations of the same species can and do. In fish, however, as in many other groups, there is a continuum of intermediate states as sub-populations, separated geographically or by slight differences in behaviour within the same watershed or even lake, which can produce fertile offspring if brought together in the laboratory, but do not interbreed in nature. In such cases, a decision where to draw the line between one species and another is a difficult, indeed probably undesirable, exercise.

Trout are a case in point. Many local races, separated in discrete catchments, are found and their appearance, even their habits, may be very different. Sea trout spend some of their lives feeding in estuaries and migrate upstream to freshwater to breed. Brown trout live entirely in freshwaters and migrate upstream from lakes into rivers to breed. Both are *Salmo trutta* L. and brown trout imported to New Zealand and the Falkland Islands found their way into estuaries and became sea trout. Sea trout confined artificially entirely to freshwaters will breed and acquire the silvery adult colour of brown trout.

Among brown trout, races vary greatly in size and colour. Those from peaty streams are often dark coloured and those from large lakes are silvery. An Irish race, the gillaroo, has a yellowish back with large, brown spots, a golden belly with pink tints, and large scarlet spots around the lateral line. The lateral line is a mucus-filled canal running along each side of a fish with which it can detect disturbances in the water through pressure variations. The races of trout are separated by geographical barriers and, in time, may differentiate into distinct species. Differentiating species in other lakes may be separated by subtler breeding barriers, including the adoption of complex behavioural patterns, which, if not completed in a particular manner or sequence by both sexes, will prevent successful breeding. Particular gene combinations, linked to genes determining the behaviour pattern are thus preserved intact.

Breeding occurs at different ages in different fish, presumably at a time which represents a compromise between age and the chance of failure to breed through early death, the food supply and size of fish and its ability to cope with the energy demands of breeding. Trout first breed when they are 3 or 4 years old, walleye from 2 or 3 years and *S. niloticum* at only a few months. This reflects to some extent the higher water temperatures and lack of seasonal food scarcity in most tropical freshwaters. *S. niloticum* also exhibits what is known as arena behaviour[149]. In L. Rudolf (Kenya) it has been observed to move, before breeding, into shoals that are circular and about 5 m in diameter. For a short period the fish move at a spacing of about 30 cm around the shoal. The function of this behaviour is unknown but it may be a mechanism by which the fish perceive the size of the local breeding population, and by which, through endocrine changes the number of eggs laid is regulated. This same species may also breed when it is very small. In the main part of L. Albert (Uganda, Zaïre), *S. niloticus* reaches a size of 50 cm and breeds when it has attained 28 cm. In the Bukuku lagoon, a part of the lake now completely isolated by a sandbar and forming a 600 m triangle, the fish breeds at 10 cm and achieves only 17 cm at most. This appears to be a phenotypic effect, not a genetic difference and seems to be a response to the extreme environment of the lagoon, which is very saline and must dry out almost completely from time to time. A fish able to mature earlier and therefore produce more frequent generations may thus have a greater chance of survival than one whose breeding cycle is longer than the survival of its environment.

The role of breeding rituals has been indicated above. Examples are provided by the European three-spined stickleback (*Gasterosteus aculeatus* L.) and the African *Haplochromis burtomi* (Gunther), a cichlid fish. Sticklebacks are silvery-brown, small (5–8 cm) and eat small invertebrates. Sometimes they winter in estuaries. In spring, the males leave the mixed shoals and take on breeding colours—bright blue and red underparts and a translucent appearance to the scales on the back. Each male chooses a small territory on the substratum, a sandy or silted bottom, and defends it against other males. Within the territory, a depression is excavated by the male sucking up the sand or mud and expelling it some distance away. Strands of fine weed or filamentous algae are collected and formed, with the help of a secretion from the kidney, into a tunnel-shaped nest about 5 cm long and broad, which lies in the depression.

At this stage the male becomes responsive to swollen, gravid (bearing ripe eggs) females, though not to thin, spent ones. If a suitable female enters the territory, the male first creeps through his nest and then performs a 'courtship' dance in which he zigzags towards the female, then turns towards his nest. This may not attract the female, though she may swim nearby adopting a characteristic posture with the head up. The male may then dart at the female with his spines raised and usually she will then be led down to the nest. By inserting his snout the male points at the nest entrance, then backs off and swims on his side with his back towards the female. She appears to inspect the nest opening, and then may retreat to the water surface. The whole courtship process may be repeated several times before the female enters the nest, with something of a struggle, for the opening is narrow. Eventually her head is well in and her tail sticks out of the entrance. The male then puts his snout against the base of her tail and quivers violently. The tail begins to rise as the male continues quivering and when it is raised high the female releases a stream of 50–100 eggs into the nest. When the last egg is laid, the female now much thinner, rushes out of the nest as the male bursts in and releases a cloud of spermatozoa quickly over the eggs. Thereafter the female is chased away, the male prods the fertilized eggs deep into the nest, adds sand to reinforce and camouflage it and wafts water over the eggs with his fins. He remains guarding the fry until about ten days after hatching when his breeding colours have also faded.

Spawning behaviour in *Haplochromis burtomi* in L. Tanganyika (Fig. 7.3) is equally complicated. Little is known of the preliminaries of courtship but at its culmination the female lays a very small batch of quite large eggs on the lake bed while the male courts attendance. When she has laid the eggs the female quickly turns and scoops the unfertilized eggs into her mouth. Then the male sweeps past the female, displaying his anal fin as he does so. On it are light, circular markings which, against the darker fin background, resemble eggs. The

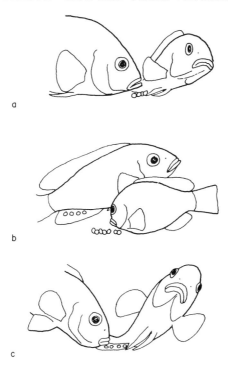

a

b

c

Fig. 7.3. Spawning in *Haplochromis burtomi*, a fish species of L. Tanganyika. (a) The female has laid a batch of eggs. (b) She turns and takes them into her mouth before they are fertilized. (c) The male sweeps past the female, displaying the 'egg-dummies' on his anal fin. In attempting to collect the dummies the female takes in spermatozoa released by the male and fertilization of the eggs takes place in her mouth. (After Fryer & Iles[149], based on a film made by G. H. Wickler.)

female is apparently deceived by these and moves to pick up what she thinks are eggs. In doing so she sucks in water from near the male's genital aperture from which he has just released sperm and the eggs are fertilized in the female's mouth. The process is repeated several times until her mouth is full of fertilized eggs where they are protected until hatching.

7.3 The distribution of fish

Like all other freshwater organisms, fish form a series of intergrading but often reasonably distinctive communities or collections. What determines why a particular set of species is found in a particular lake or river, at a particular time? As with other organisms, at least three general factors must determine this— considerations of evolution and biogeography, the tolerance ranges of the species to many environmental factors, and the action of predation and inter-specific competition.

Because fish, with few exceptions, cannot travel overland, colonization of new waters is difficult. Sometimes eggs may be carried attached to water weed entangled on waterfowl and high floods may sometimes interlink river systems so that fish may pass between them. Interpretation of fish community composition must therefore take into account the possibility of accident. Absence of a species from a waterway may not mean that it could not live there, merely that it has not yet arrived!

Much work, particularly with reference to pollution control (see Chapter 10), has been carried out to determine the laboratory tolerances of fish, particularly to dissolved oxygen levels and temperature[266]. This has given some general understanding of the roles of these factors in determining fish distribution, but just as the potentialities of environmental tolerance can be confounded by the accidents of biogeography, so can they be modified by the competitive effects of other fish species. The interplay of these factors will be considered in a variety of situations.

7.3.1 The British fish fauna

The British fish fauna is, without doubt, depauperate. Comparable areas at similar latitudes on mainland America and Eurasia have many more species. The reason is that ice covered all but Southern England for many thousands of years until 10–13 000 years ago. Only those fishes tolerating very cold water, such as the charr (*Salvelinus*) and whitefish (*Coregonus*) are likely to have persisted in the remaining southern English rivers. Until about 7000 years ago, and for a period of up to 6000 years after the ice finally began to melt, there was a land connection, drained by tributaries of the present R. Rhine, between Britain and mainland Europe. Sea levels, rising as melt water poured out of the polar glaciers, flooded the land bridge around 7000 years ago, but not before about 20 species of fish had reached southern England through the Rhine connection. These included the grayling (*Thymallus thymallus* L.), pike (*Esox lucius* L.), cyprinids like the tench (*Tinca tinca* L.), bream (*Abramis brama* L.), bleak (*Alburnus alburnus* L.), roach (*Rutilus rutilus* L.) and others, and the spiny rayed perciforms, perch (*Perca fluviatilis* L.) and ruffe (*Gymnocephalus cernua* L.). Fish which can spend some of their life cycles in the sea or estuaries also moved in from refuges presumably on the warmer, southern coasts of Europe after the glaciation. These have included the salmon (*Salmo salar* L.), trout (*Salmo trutta* L.), eel (*Anguilla anguilla* L.), sticklebacks (*Gasterosteus aculeatus* L., *Pungitius pungitius* L.), smelt (*Osmerus eperlanus* L.) and sturgeon (*Acipenser sturio* L.). A number of marine species which penetrate estuaries and sometimes lowland rivers (flounder (*Platichthys flesus* L.), shads (*Alosa* spp.), mullets (*Chelon* spp. and *Crenimugil labrosus* Risso) and burbot (*Lota lota* L.), should also be included in this group. *Inter alia* they are the only fish that have naturally colonized Ireland and the offshore north western islands of Britain since the glaciation.

About 42 British fish are indigenous, but apart from some 10 naturally occurring hybrids, the complete list numbers 55. The additional 13 are fish introduced by man for sport (e.g. the Rainbow trout, *Salmo gairdneri* Richardson), food (e.g. carp), aesthetics (goldfish, *Carassius auratus* L.) or scientific interest. A late Duke of Bedford introduced the wels and pike-perch from mainland Europe and they now breed in limited areas.

An inspection of the distribution maps of the indigenous British fish and of those introduced which have had several centuries and the help of the 18th and 19th centuries' canal systems to exploit much of their potential range, shows four general patterns (Fig. 7.4). The first includes fish confined to the generally upland north and west, like the salmon, charr and whitefish (i.e. schelly, powan, gwyniad, *Coregonus lavaretus* L.). The second is of fish with a distribution over much of the island, though this may be biased towards one end or the other—the trout, pike, minnow, eel and perch. In the third group are species largely confined to southern and eastern, that is lowland, England. Typically these are cyprinids—carp, gudgeon (*Gobio gobio* L.), tench, bream, bleak (*Alburnus alburnus* L.), rudd, roach, chub (*Leuciscus cephalus* L.) and dace (*Leuciscus leuciscus* L.), and the percid ruffe. The last category includes predominantly marine species like the flounder and burbot, which have a coastal distribution.

Although it is certain that there are sites in northern Britain where fish of group three could survive and breed, the overall patterns of groups 1–3 can be understood in terms of the physical tolerances and spawning requirements of the fishes. In the north and west the hilly country has streams and rivers with stony or gravelly channels. They are generally cool and saturated with oxygen. For the most part the lakes are infertile, with well-oxygenated hypolimnia, gravelly shores, sparse, submerged macrophyte beds and unproductive benthic communities. Of course, exceptions can be found to each of these general characteristics, but that is the overall picture. In contrast (and again in caricature), the southern and eastern rivers and lakes are silted, warm and weedy. Although well-oxygenated by day, they may become seriously undersaturated at night.

Data for representatives of groups 1–3 are shown in Table 7.3 and it is clear that the requirements of the groups coincide reasonably well with those conditions likely to be found in the waters of their distribution areas. The data also form a basic level of explanation for the often observed change in predominant fish species in a single river which alters its character greatly between its upland stretches and its lowland flood plain. A sequence from salmonids (trout, white fish, and grayling) to pike, perch, and cyprinids like chub and dace, to cyprinids of the warmest, least oxygenated waters, tench and bream, may be noted under ideal conditions. Such data, however, do not explain how species are distributed within, for example, the area of Southern England, or why one particular species of a group predominates in a particular lake or river stretch.

The southern English freshwater fish are notable for the universality of their diet, and much overlap in their food is recorded. Table 7.4 shows that most of the common lowland fish share a diet of invertebrates, and that this overlap extends also to individual groups or species of prey. All of these fish may co-exist but no data are available to show exactly how the food components of the

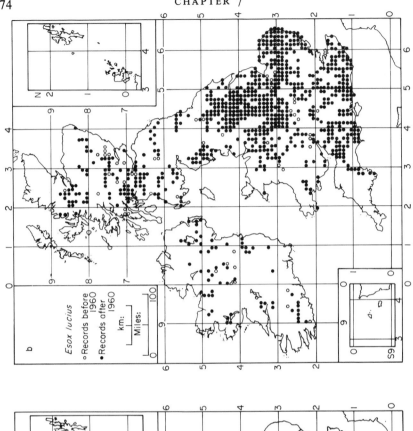

b

Esox lucius

○ Records before 1960
● Records after 1960

km:
Miles:
100

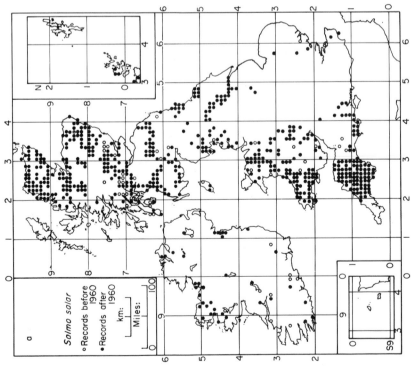

a

Salmo solar

○ Records before 1960
● Records after 1960

km:
Miles:
100

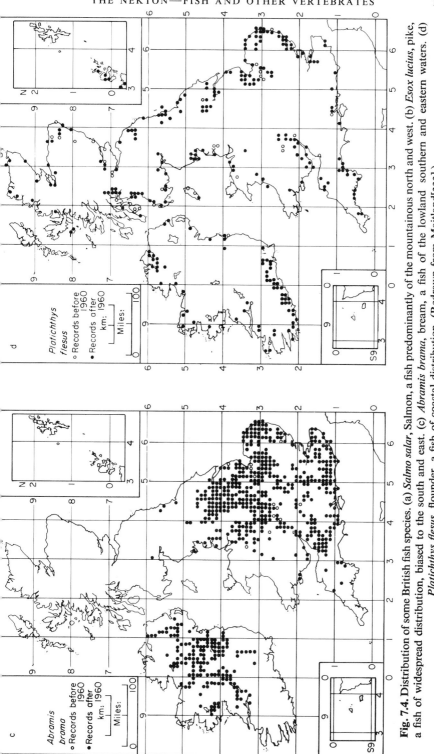

Fig. 7.4. Distribution of some British fish species. (a) *Salmo salar*, Salmon, a fish predominantly of the mountainous north and west. (b) *Esox lucius*, pike, a fish of widespread distribution, biased to the south and east. (c) *Abramis brama*, bream, a fish of the lowland southern and eastern waters. (d) *Platichthys flesus*, flounder, a fish of coastal distribution. (Redrawn from Maitland[322].)

Table 7.3. Data on some British fish species (modified from Varley[508]).

Fish	Optimal growth temp. °C	Upper lethal temp. °C	Spawning temp. °C	Usual dissolved O₂ requirement mg l⁻¹	O₂ level survived at 16°C for 7 days*	Place where eggs are laid	Maximum sustained swimming speed m s⁻¹
GROUP 1 (N & W upland)							
Salmon (fry)		23					
Salmon (adults)		32–34	2–6(−10)			Gravel	8
GROUP 2 (wide distribution, biased to N)							
Trout	7–17	26·5	(→10)	10–16	38**	Gravel	4·4
Grayling		24·1		7–10		Attached to stones	
GROUP 2 (biased to S)							
Pike	14–23	29	10+	5–6		Attached to macrophytes	0·45
Minnow				10–16		Attached to stones	
Perch	14–23	32·8	10+	7–10	1·2	Among macrophytes	
GROUP 3 (southern & eastern)							
Roach	14–23	33·5	10+	5–6	0·7	Attached to macrophytes	
Tench	20–28	35·2	15+	0·7+	0·6	Attached to macrophytes	0·47
Carp	20–28	37·0	15+	0·7+		Attached to macrophytes	0·47

* Downing & Merkens[110]   ** Brook trout

Table 7.4. Diet of fish species in a lowland river system, Shepreth Brook and R. Cam, Cambridgeshire, 1939–41. (After Hartley[1971].) Results are given as % occurrence of a given food in the stomachs of a large sample of each species, taken at various times and seasons.

FEEDER	Fish	Molluscs	Insects	Crustacea	Plants and filamentous algae	Diatoms	Emphemeroptera (mayfly nymphs)	Trichoptera (caddis larvae)	Adult beetles	Simulium	Chironomids	Gammarus and Asellus	Copepods and Chydorus
Pike	85		9	6									
Trout	13	7	48	21	3		26·5	84	43			43	
Eel	19	5	24	44	5		13·5					57	
Gudgeon	12	57	24	6		1·5				5·5		18	
Bullhead		1	58	39				6		9	61		16
Stickleback		2	51	24	14	8		13			21·5	40·5	
Perch	10		30	60							42		30
Loach			65	35									
Roach		19	36	7	30	7				6	14·5	8·5	
Dace		9	53	6	22	10		30	8	8	12	5·5	
Minnow			43	9	21	23					26·5		

niches are separated. Dietary overlap is common in fish of temperate regions and some attempt to explain how it may be sustained has been made for three N. American sunfish (see below). The seasonality of temperate climates means that individual foods are available only periodically and hence specialization would be disadvantageous. Live plant material is generally unimportant in temperate fish diets. This may reflect the even greater seasonality of its availability compared with that of invertebrates. This explanation is not entirely convincing, however, since at least one avid weed-feeder is known from temperate Asia, the grass carp, and this fish, in common with other cyprinids, feeds little in winter anyway.

7.3.2 Fish distribution in tropical Africa

Tropical Africa has undergone many geological and climatic changes. Yet none of them has been so devastating and widespread for the freshwater fauna as the glaciations which removed completely the freshwater habitat from much of the temperate land surface. African fish communities have thus had a long period of development in which there has been an extensive series of fresh waters, and speciation has occurred to a high degree. Fossil remains show a relatively uniform fish fauna in the Miocene period, about 20 million years ago[24]. The African land surface was then one of great undulations with shallow basins separated by low ridges and plateaux. During the wetter periods the basins were likely to have been connected through the upland swamps on the plateaux where their head waters originated, and fish could sometimes migrate from one basin to the next. Thus a group of fish including the Nile perch, *Sarotherodon galileus* (Artedi), *Citharinus citharus* (Geoff.), and the air breathing catfish and lungfish, *Clarias lazera* Val. and *Protopterus* spp., which was then present, is still widely distributed from the Nile valley in the east to the Chad basin in the centre and in rivers of West Africa. This group is called the 'Nilotic' or 'Soudanian' fauna.

About 20 million years ago there began a series of earth movements which raised, by about 1000 m, a belt of land, 500–800 km wide in eastern and northern Africa. Then, between 12 million and 2 million years ago, the plates of the earth's crust began to separate in a line running southwards from the near East in what is now Israel and the Red Sea down into the heart of East Africa, ending just north of the Zambezi valley. This created the Great Rift Valley. The preceding uplift had cut across the previous generally east and west drainage, separating much more distinctly than before the river basins to either side of the uplift. The rift valley, with its uneven floor, created deep lake basins, again well isolated from their neighbours. Accompanying and following the rift movements were other vertical movements of the earth's crust which led to the formation of the relatively shallow depression on the upland between the east and west bifurcations of the rift valley in which L. Victoria now lies (see Chapter 2). Further movements eventually tilted this depression so as to link its waters with those of the R. Nile about 30 000 years ago. Volcanoes produced numerous craters and lava flows which ultimately bore small lakes or dammed rivers to form larger ones, like L. Kivu. A whole series of new lake basins in hillier country was thus superimposed on the previously subdued ridge and depression topography. The basins of L. Chad and of the Zäire (Congo) River changed relatively little and up to between 7000 and 15 000 years ago a series of lakes and swamps cut across the area which is now the Sahara desert to retain the ancient link between the Nile and W. Africa.

Isolation of populations so that they cannot interbreed with similar neighbouring ones leads to progressive divergence and evolution of new species. This is one reason for the rich fish fauna of tropical Africa. Further isolation of some of the basins created by earth movements and volcanic activity has come from the large climatic changes of the last 25 000 years. There have been cooler and wetter periods, the pluvial periods, in some way related to the polar glaciations and warmer, drier periods when lake levels have fallen, and basins previously drained by outflowing rivers have become endorheic (see Chapter 1). 8000 years ago L. Rudolf (see Chapter 2) drained into the Nile, but has been cut off since the dry period which began 5000 years ago. The increased salinity which accompanies closure of a lake basin has also been a stimulus to speciation in creating an environment which makes extreme demands on the physiology of a freshwater organism.

The larger lakes (Malawi, Tanganyika, Victoria) have provided permanent bodies of freshwater for a very long time. Movement upriver of fish from other areas has been hindered by rapids and major waterfalls on their outflows in the cases of Victoria and Malawi. In them have evolved a large number of endemic species (i.e. those confined to the basin) not only of fish but also of other animals. The cichlid fishes, a group of the spiny finned Perciformes, have been extremely successful and predominate in these lakes. The number of species, and the proportion which is endemic, is greater the longer the period of isolation. For example, of 214 species of cichlids in L. Tanganyika, 80% are endemic, but of 37 in L. Rudolf, which was recently connected to the Nile, only 16% are endemic.

The cichlids are widespread in the less isolated lakes, like Chad, Albert and Rudolf for example, but, as in the rivers, other groups are equally or better represented. The catfishes (Siluriformes) in particular are common in the rivers and are part of an extremely diverse fauna in the Zäire (Congo) basin which perhaps reflects the very great extent as well as age of the basin.

The African fish fauna, then, is much richer, for reasons of time and earth history than that of the British Isles. There has been long enough for evolution of a much greater specialization, for example in diet and in the anatomy of the head and mouth necessary to exploit efficiently particular food sources. The more muted seasonality in the freshwater, if not the terrestrial, environments of the Tropics has also meant that particular foods are almost always available. An unspecialized diet, with its implications of potential inter-specific competition, which seasonality has forced on many temperate fishes, is found among tropical fishes, but generally where the environment undergoes unpredictable changes, like irregular drying out. Where the environment is very stable, or changes in a regular way much specialization is found. This can be illustrated by comparison between some of the fishes of L. Malawi, which is stable and relatively constant, L. Chilwa, which irregularly dries out, and river flood plain habitats which change annually but predictably.

Fishes of L. Malawi

In L. Malawi there have evolved 'species-flocks'—groups of species which are very closely related, yet distinct. Many of them are related to the cichlid genus *Haplochromis* and similar flocks occur in L. Victoria and Tanganyika. The range of diet is no greater than elsewhere, but the specialities within the range are much narrower, presumably allowing a very efficient use of food resources. This can be illustrated by considering members of a species flock which feeds on blue-green and other algae which form a felt on rock surfaces[149]. The local African fishermen recognize, unwittingly, the close taxonomic relationships of these fish by calling all by one name, 'mbuna'.

One species (Fig. 7.5) *Pseudotropheus tropheops* Regan has small close-set teeth, like a small file, with an outer row of bicuspid tipped ones and inner rows

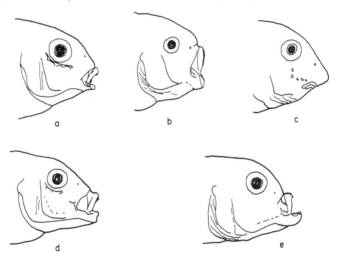

Fig. 7.5. Heads of some of the 'mbuna' of Lake Malawi (see text). (a) *Pseudotropheus tropheops*; (b) *Petrotilapia tridentiger*; (c) *Labeotropheus fuelleborni*; (d) *Pseudotropheus zebra*; (e) *Genychromis mento*. (Redrawn from Fryer & Iles[149].)

of tricuspid teeth. It rasps the algal felt from vertical or steeply sloping rocks and bites off the rasped mass with large conical teeth at the edges of its mouth. In this way it removes even firmly attached filamentous forms. *P. zebra* (Blgr.), on the other hand, is larger and more mobile than *P. tropheops* and this may give some separation of niche to add to the subtle difference in diet. This species has similar outer teeth, but longer and more spaced tricuspids at the back. It opens its mouth against the rock and combs the algae so that the looser species only are removed. A third species, *P. livingstonii* (Blgr.) has similar dentition and diet to *P. zebra*, but inhabits relatively deeper water.

A closely related genus contains *Petrotilapia tridentiger* Trewavas whose teeth are long, curved and flattened at the tips into a spoon-like shape. The whole tooth is flexible. When the mouth is opened against the rock the teeth comb the looser algae, as do those of *Pseudotropheus zebra*, but *Petrotilapia tridentiger* can feed on both vertical and horizontal rock surfaces, 'standing on its head' to do so in the latter case. Lastly among this group of algae scrapers is *Labeotropheus fuelleborni* Ahl. with its rigid, strong jaws strengthened in the plane of forward movement, and a mouth set on the lower side of the head. It hovers over horizontal rock surfaces then moves in to chisel the algae off with its teeth.

It would be difficult to find finer differences, based on attachment of the algae, angle of rock surface and water depth, than these, but rock-scraping fish in L. Malawi have also evolved into scrapers of other surfaces. *Genychromis mento* Trewavas scrapes the body scales from a sluggish bottom-living fish, *Labeo cylindricus* Peters. Though this is its main diet it perhaps belies its evolutionary origin by occasionally rasping epilithic algae, and points to future possibilities by biting pieces out of the fins of other fish. More specialized scale scrapers and fin biters are known also from the lake, and scale scraping is accompanied sometimes by a mimicry that allows the scraper to move unrecognized among shoals of its prey. Fryer & Iles[148] make some general observations on the adaptations of fish to the long-stable environments of the African Great Lakes. The risks of environmental catastrophe there are negligible; population fluctuations are small and reproduction is a sparing process in which relatively few eggs are laid but carefully looked after. Continued separation of species, probably by smaller geographical barriers in the first instance and breeding behaviour subsequently, has led to an efficiency of food exploitation which is referred to as selection for K characteristics. K is the term in the logistic equation of population growth which represents the maximum or saturation population.

The fish fauna of L. Chilwa

Quite different characteristics have been selected for in lakes where environmental stability is not guaranteed. A number of African lakes exist under hydrologically precarious regimes, where a few years of lower than average rainfall may lead to drying out of the lake. Such conditions have led to emphasis on non-specialization, on the ability to change habitat from lake to river, on rapid reproduction to replace a population decimated by drought or the increase in salinity and lowering of oxygen levels which may accompany it, and on flexibility in diet. It is of little value for a fish in such a lake to rely entirely on, for example, felts of algae on the marginal rocks when unpredictably (for the fish) the water level may fall, leaving the rocks high and dry and a muddy bottom the only source of food.

Such an environmentally unstable lake is L. Chilwa, close to L. Malawi but much smaller and therefore much more vulnerable to climatic changes. It was between 96 and 128 km long when David Livingstone described it in 1859 and was approaching that size in 1978. In the interim, however, it has been reduced from time to time to a dry mud flat, the latest instance being in 1968. Raised beaches up to 35 m above the present lake level testify to a connection more than 9000 years ago with the Rovuma River and drainage to the sea. The lake now lies in a closed basin with at most about 4 m of water depth. Bottom sediment is readily stirred into the water by wind, leading to deoxygenation; water temperatures near the shore may exceed 30°C and the salinity of the water may fluctuate from 0·3–16·7°/$_{oo}$ (sea water about 33·5°/$_{oo}$). During a drying phase (they seem to have occurred several times in the last century) the phytoplankton switched from a moderately diverse blue-green diatom and green algal flora to a dense population of *Spirulina platensis* (Nordst.) Gom, which is characteristic of saline lakes, and snails and other littoral invertebrates were left stranded around the shoreline.

Compared with the hundreds of fish species to be found in a comparable area of L. Malawi, seventeen are recorded from L. Chilwa and its streams and swamps[150], with only three being at all common in the open water[271]. These are *Sarotherodon shiranus chilwae* (Trewavas), a cichlid sub-species endemic to L. Chilwa but closely related to *S. shiranus shiranus* (Blgr.) Trew. which is more widespread in east central Africa, a catfish *Clarias gariepinus* (Burchell), and a cyprinid *Barbus paludinosus* Peters. Both these latter fish are widespread species and indeed, other than the sub-species mentioned above, none of the Chilwa fish are endemic to the lake. Its past vicissitudes have not favoured speciation.

S. shiranus chilwae is an unselective feeder on bottom-living algae, detritus, higher plants and zooplankton. It is highly tolerant of low oxygen levels, surviving at 0·6 mg $O_2 l^{-1}$ and at temperatures as high as 40°C for brief periods. It will cope with alkalinities (bicarbonate plus carbonate) of up to 60 mEq l^{-1}. Extreme conditions during drying phases, particularly deoxygenation, do kill this fish, but it can survive in streams and stream-fed pools in the swamps surrounding the lake. These may persist even when the main basin is dry.

S. shiranus chilwae breeds when it is quite young and small—in its second year—and spawns on wave washed, sandy, littoral areas up to three times during the November–March wet season when conditions are likely to be most favourable for survival of the young. (The Great Lakes species flocks tend to spawn throughout the year.) The genus *Sarotherodon* (*Tilapia*), as a whole, exhibits similar features of flexibility and it is not surprising that it is widespread, even where extreme conditions are found. Where hot spring water flows into the highly saline L. Magadi the lagoons have been colonized by *S. grahami* (Blgr.) which is able to live permanently at 39°C. The stunting and early reproduction of *S. niloticus* in a lagoon at the edge of L. Albert and of other *Sarotherodon* spp. elsewhere appears also to be a reflection of their ability to survive extreme and irregularly changing conditions by rapid replacement of populations decimated by catastrophe. Hybridization, with its genetic advantages of producing new variation in changing habitats, is also common among the genus[148].

Except that it is a generalized feeder, mainly on zooplankton but also on plant material and algae, less is known of the second abundant Chilwa fish, *Barbus paludinosus*. It may breed in its first year and after a period of catastrophic drying in the lake, survivors re-entering from the swamp pools may breed more frequently than in a 'normal' year. It too seems very much a 'generalist' as is the catfish, *Clarias gariepinus*. It was the last species to survive in the lake during a recent drying out, and the first to recolonize afterwards. *Clarias* has a leathery, relatively impermeable skin, the ability to aestivate in mud, and to migrate over wet land using its spiny pectoral fins as levers. It is highly tolerant of deoxygenation, having accessory air breathing organs in the gill cavities. These comprise stiff, vascular, bushy projections to each of which is attached a small muscle capable of erecting it in air. The projections are covered in fine folds, or villi, and glands secrete a mucopolysaccharide over them which probably protects them from drying out. When *Clarias* spawns, it produces a very large number of eggs, up to 180 000 per fish and can spawn in its first year. Its diet also shows the expected flexibility of Chilwa fish—although it is piscivorous *Clarias* takes also zooplankton, bottom invertebrates and higher plant material.

The available 'habitat volume' of L. Malawi has been carved into many individual niches but that of L. Chilwa can support only a few 'larger' niches, for flexibility within these niches is necessary to complement an equally changing habitat.

The next instance to be considered, that of seasonally flooded river plains in the Tropics is of interest because it is comparable to that of temperate habitats. There is an unfavourable period but this is regular (i.e. predictable) in occurrence. For temperate fish the annual less favourable period is the winter; for flood plain fish it is the dry season reduction in habitat as the water retreats from the surrounding land to the river bed.

Flood plain fish communities

The flood plains of many African rivers are very large because the land surface is often very flat; the ratio of land seasonally flooded to permanent river area varies from 25 : 1 to 3 : 1. The rivers regularly flood because of the seasonality of rain in their often large catchment areas and the water may take weeks to move downstream. Because of this the flood at a particular place may occur in what is nominally the local dry season, and since tributaries, themselves often large, may contribute much water, the pattern of flooding downstream may be complex, though in general regular from year to year.

During the drying out period (drawdown) water is confined to the river bed and to isolated pools on the plain. The fish appear generally to be in poor condition[300]; they lose fat reserves built up previously and food is relatively scarce. The reduction in area of water reduces the areas of aquatic plant beds and benthic habitats, and at the same time forces fish from a large area into a smaller one. Food competition and predation are high. Many fish die and few breed during this period. The exceptions are some zooplankton feeders such as, in Africa, *Microthrissa* and *Pellonula*. Zooplankton cannot build up large populations during the flood because of the diluting effect of the incoming water, but do so as the discharge falls. Food for planktivores is thus relatively more plentiful in the low water period.

The river fishes respond to increases in the river water level, though the mechanism is unknown. As the river rises they migrate up small tributaries and out onto the riverine grasslands as these flood. Cyprinids, Characoids, Siluroids and Mormyroids (Table 7.1) are predominant in African rivers. They spawn on the flood plain almost as soon as they can reach it. The eggs usually ripen quickly and may hatch in only two days or so. This means that the fry are produced just in time for the burst of algae, plant and consequently invertebrate productivity which quickly follows flooding. The previous terrestrial vegetation and the droppings left by animals which grazed on it quickly rot, releasing nutrients and sometimes causing deoxygenation for a time. Measurements are few, but it is likely that many of the river fish can tolerate low oxygen levels—some catfish (Siluroids) are potential air breathers, and the lung fish are members of the flood plain fauna.

The fish find abundant food and cover from predators; populations increase and fat stores are laid down, often as ribbons along the intestines. The fauna is diverse—a short section of the R. Niger exhaustively fished during engineering operations necessary for the building of a large dam revealed 82 species, and though it is very large the Zaïre R. may have up to a thousand species.

Water levels fall as the flood passes and the fish are forced back to the main river or are trapped in pools which may eventually evaporate in the approaching dry period. It seems that some predators may anticipate this and move back to the tributary mouths just before their prey. During the drawdown both bird and piscivore predation is high and local commercial and subsistence fisheries have their most productive period. Those fish that survive in the river or flood plain pools do so either by using up their fat stores, switching their diet from, for example, aquatic plant seeds and benthic invertebrates to zooplankton (*Alestes*, *Citharinus*, some *Sarotherodon* spp.), aestivating in mucous coccoons (lungfish) or burrows (some catfish) in the mud, or as drought resistant eggs (some Cyprinodonts).

Thus there are many features of this habitat that are reminiscent of the environment of high latitudes, and the diversity of fishes is high. There seems some reason, given sufficient time uninterrupted by catastrophes like glaciation, to suppose that temperate habitats should have as diverse communities, area for area, as those in the Tropics. Insofar as microorganisms like the algae, which are easily distributed as spores, in droplets and on water birds, are concerned, the diversity of temperate communities almost certainly is as great as that of tropical ones. The scale of algal habitats is smaller than that of fish, however, and the micro-waterscape may provide far more physical settings for algal niches than the macro-waterscape does for fish. The seasonally cold water of very large temperate lakes might also preclude fish speciation to the extent that it has occurred in the African Great Lakes. This is because certain food sources, like water plants and the eggs and larvae of other fish, would not be available all year around and hence some flexibility in diet would be essential. On the other hand this limitation applies also to tropical flood plain communities and L. Baikal, a large lake in the southern U.S.S.R. situated in the southern part of the north temperate zone, has a very diverse fauna rich in endemic species. Whether or not its fauna is as diverse as those of the tropical Great Lakes is impossible to say. The complete extent of the fauna of Baikal or of the African Great Lakes has never been recorded and in any case a comparison between a sample of one lake in one case and a mere handful in the other would be statistically suspect. What is certainly true, however, is that the diversity of temperate waterways, depleted by glaciation, could certainly be much higher than it is at present.

7.4 The roles played by nekton in fresh waters

The significance of nekton organisms to an aquatic ecosystem depends on their wide ranging movements. They may move into several parts of the system modifying the prey communities by selective feeding. Movement within a large river system or between land and water may also mean importation of nutrients as excreta which would not otherwise reach the water.

7.4.1 Predation by fish

The potential effects of predation by vertebrates on other communities are very great and can be illustrated by simple experiments in which fish are introduced to, or excluded from, small ecosystems established in tanks or ponds. Many suburban gardens have such an experiment (though uncontrolled) when a delight in goldfish leads to the keeping of such large populations in small ponds that zooplankton populations are completely eaten by the fish. The result is a dense phytoplankton population which prevents the fish being seen!

Controlled small pond experiments usually go on for a short period in a small system. They cannot explain the intricacies of what happens in a larger, more heterogeneous system where predation may be thwarted by access of the prey to shelter in weed-beds or other cover, and in which invasion by predator-resistant species might be a response to heavy predation[330]. They do, however, give some clue to the sorts of processes involved and they have parallels in certain semi-natural habitats or unusual situations. An experiment set up in tanks on the roof of San Diego State College, California is typical[233].

Six replicate systems were set up in 30 cm deep × 2 m diameter pools. To each pool was added a 3 cm layer of sand, tap water to a depth of 20 cm and a litre of dried alfalfa pellets as a source of organic matter and nutrients. A small sample of plankton from a nearby lake and an inoculum of *Daphnia pulex* (De Gear) were added, and the pools left for a few weeks. Then fifty 3–5 cm fish, *Gambusia affinis* (Baird and Girard) were added to each of three of the ponds, and observations were made on the water chemistry, plankton and benthos; some of the results are shown in Table 7.5.

The fish readily ate the larger zooplankters *Daphnia pulex* and *Chydorus sphaericus* (O. F. Müll) which developed significant populations only in the

Table 7.5. Effects of *Gambusia affinis* on pond ecosystems. (After Hurlbert *et al.*[233]).

	Without fish	(Significance P = o·1, rank sum test)	With fish
Zooplankton (No./10 litres)			
Daphnia pulex (2 Dec.)	92 ± 56	S	0·3 ± ·47
Daphnia pulex (3 Feb.)	1837 ± 868	S	0
Total rotifers (2 Dec.)	1134 ± 426	S	5927 ± 1911
Total rotifers (3 Feb.)	32 ± 21	N	4·3 ± 3·2
Phytoplankton (cells ml⁻¹ × 10⁻⁶) $(cells\ ml^{-1} \times 10^{-6})$			
Coccochloris peniocystis (10 Jan.)	0	S	116·5 ± 26
(2 Feb.)	0	S	219 ± 7·5
Colonial algae (2 Feb.)	78·9 ± 55	N	52 ± 37
Macroscopic bottom algae (g pool⁻¹)			
Spirogyra sp. (6 Feb.)	312 ± 132	S	29 ± 22
Chara sp. (6 Feb.)	24 ± 7·6	N	31 ± 8·6
Phosphorus levels (μgP l⁻¹)			
Soluble reactive phosphorus (3 Feb.)	9·7 ± 5·0	S	0·33 ± 0·1
Soluble unreactive phosphorus	18·4 ± 2·3	S	54·7 ± 31·2
Particulate phosphorus	12·4 ± 3	S	271 ± 26
Total phosphorus	40·5 ± 9·4	S	326·2 ± 9·8
Benthic invertebrates (N/cm² bottom sediment)			
Chironomid larvae (3 Feb.)	26 ± 3·1	S	0
Oligochaetes (*Chaetogaster* sp.) (7 Feb.)	14 ± 3·3	S	0·7 ± ·5
Insects collected as pupae or imagos at water surface between (5 Nov. & 7 Feb.)	486 ± 131	S	0
Extinction coefficients m⁻¹			
425 nm (blue)	5·1		63·8
680 nm (red)	0·83		12·0

fishless ponds. The less preferable, smaller rotifers increased to greater levels in the presence of fish, though they were also abundant in the absence of fish. Predation on rotifers by the larger zooplankters in the absence of fish is a likely explanation (see Chapter 3).

In the ponds with fish the reduction in zooplankton grazing allowed a very dense population of a unicellular blue-green alga, *Coccochloris peniocystis*, to develop, which markedly reduced light transmission and the growth of *Spirogyra* sp. (a filamentous green alga) on the bottom. The fish also readily ate the benthic invertebrates, either in the bottom sediments or as they emerged, and probably took rotifers as other food supplies became short. At least, this is one explanation for the eventual decline in rotifers. Significant too was the apparently greater mobilization of phosphorus in the ponds with fish, conceivably through the exploitation by the fish of the bottom fauna and release in fish excreta of SRP, which became available to the plankton subsystem. This may, of course, not have been sustained in the long term as the phosphorus would be progressively locked in inorganic precipitates rather than the readily decomposable alfalfa pellets initially added.

The fish thus led to a reduction in the diversity and structural complexity of the system. This has happened also in larger man-made ponds created for the watering of game animals in Wankie National Park, Rhodesia[519]. The trampling of the larger animals churned the bottoms of the shallow pits and prevented any macrophyte development. The only cover for invertebrates on the bottom was the fibrous cellulose layer derived from the dung of large herbivores like elephants; the omnivorous catfish, *Clarias gariepinus* (see p. 183) was present in some ponds but not in others and there were no other fish species. In ponds with *Clarias*, the invertebrate fauna, dominated by beetles (Coleoptera) and bugs (Hemiptera), was sparse with only 13 species compared with 53 species in ponds lacking *Clarias*. Total numbers and biomass were similarly contrasted. Such dramatic differences result from the extremely simple physical structure of the ponds, where prey have little cover from a predator. Most lakes and rivers are more complex and the effects of predators less obvious, though nonetheless profound.

Much evidence now exists (see Chapter 3) for switches in the zooplankton population even in large lakes, caused by fish predation on the larger animals in preference to the smaller ones, but effects on the benthic fauna have been less widely studied. In quarter acre experimental ponds Hall *et al.*[189] found that bluegill sunfish (*Lepomis macrochirus*) Rafinesque (Centrarchidae) selected the larger, more abundant invertebrates such as *Chironomus tentans* Fabr., dragonfly larvae, and the amphipod *Hyallela azteca* (Saussure), though the presence of fish did not alter the total benthic production. Invertebrates which could hide in the dense macrophyte cover of the ponds increased in production. Similar effects have been noted in Hodson's Tarn, a small lake in the English Lake District where Macan[313] introduced brown trout. The fish took the eggs and tadpoles of frogs and toads (Amphibia) and the surface-feeding bugs and beetles, but were then forced to hunt in weed-beds for prey. Their effect there was much less marked for the macrophyte beds were sufficiently dense for prey populations to maintain themselves despite predation. Production studies were not made but it is clear that fish predation leads to compensatory mechanisms in complex habitats such that the dramatic changes noted in small scale experiments are not usual.

Such compensatory mechanisms may include niche changes in the fish which permit several related species to co-exist despite a food source common to all. An example comes from studies on American sunfish by Werner & Hall[522, 523]. Three species, the bluegill, pumpkinseed (*Lepomis gibbosus*) and the green sunfish (*L. cyanellus*), are commonly found in shallow waters in central N. America. 900 individuals of each species, in their second year of growth, were added separately to three replicate experimental ponds, 29 m in diameter and 1·8 m deep. In each was a fringe of emergent *Typha* and a submerged macrophytic vegetation of *Chara* and *Potamogeton* with some areas of bare sediment.

In a fourth such pond 900 individuals of each of the three species were confined together. Throughout a summer, fish were regularly sampled and their gut contents examined to determine what they were feeding on. At the end of the summer the entire populations were removed and weighed. Results are shown in Table 7.6.

Table 7.6. Food taken by three species of sunfish co-existing or living alone. Results are expressed as % occurrence of various food types in stomachs examined. Figures outside brackets are for each species living alone, inside brackets for when all three species were stocked together.

	Bluegill	Pumpkinseed	Green sunfish
Invertebrates living in vegetation	61(15)	4(5)	43(40)
Bottom-living invertebrates	10(15)	12(34)	23(12)
Open water zooplankton	8(33)	1(6)	1(4)
Other	21(37)	47(55)	33(44)
Mean size of food μg per item ± s.e.	19·6 ± 2(6·5 ± 1)	38·8 ± 7(27·7 ± 4)	59 ± 9(40·4 ± 8)

What happened was that in the presence of green sunfish, the more aggressive of the three species, the other two were forced to take either much smaller or less accessible food items—zooplankton in the case of bluegills, bottom invertebrates in the case of the pumpkinseed. The green sunfish continued to feed on the larger prey among the vegetation which it is well adapted to take with its relatively large mouth and fusiform body. Although such large items are taken by bluegills when they can—energetically they represent more food per unit cost in obtaining it—bluegills are also well fitted for zooplankton feeding. They have long, fine gill rakers, a protrusible mouth and a habit of schooling. In the presence of green sunfish they move out into the open water while the former stays among the vegetation. A comparable shift is made by the pumpkinseed. Its more widely spaced gill rakers do not become clogged as it forages in mud and it moves out over the bottom sediments if it too is prevented from taking the larger weed-living invertebrates by the green sunfish.

In these sorts of ways in freshwater communities many subtle adjustments are being made, any of which may start a chain of further effects; this makes full interpretation of why communities are constituted as they are, very difficult. The vertebrate predators, at the heads of the food webs, clearly have the greatest opportunities of all to cause far reaching changes and adjustments despite their relatively small contributions of biomass and production.

Predation by other vertebrates
It is a pity then that so little is known of the vertebrate predators other than fish. How does the community of S. American rivers respond to changes in the populations of fish-eating reptiles like the caimans and the anaconda? Where do otters, cormorants, ducks, turtles, water snakes, moose, grizzly bears and many

others, all of which take food from freshwaters during at least part of the year, fit in? Thoroughly worked case studies are rare.

The Nile crocodile, *Crocodilus niloticus*, is a very versatile predator, taking foods ranging from large invertebrates to mammals depending on its size and the opportunities offered[74]. Fig. 7.6 shows something of its food relationships in the R. Nile in Uganda. Although traditionally suspected of heavy predation on commercially important fish, crocodiles seem to take less fish in their lifetimes than does the white-breasted cormorant (*Phalacrocorax carbo*) which may form part of the crocodile's prey. When they do take fish, crocodiles may specialize to some extent. For example in Mweru-wa-ntipa, a shallow, swampy lake in Zambia, crocodiles take mainly *Clarias*, a predator on the commercially valuable *Tilapia*.

There is also a probable relation between crocodiles and hippopotami, which feed on land by night and rest in lakes and rivers during the day. Around the Queen Elizabeth National Park in Uganda where the hippopotamus has been protected from hunting its numbers have increased immensely with consequent overgrazing and soil erosion in the foraging area. In the Murchison Falls National Park no such problems have followed the protection of the hippopotamus. Cott[74] notes that, for biogeographical reasons, crocodiles, which prey on infant hippopotami, are absent from the Queen Elizabeth Park, but plentiful near Murchison Falls. Understandably he fears the widespread repercussions which might follow excessive hunting of crocodiles for their valuable belly skins.

A further example of the effects of predation by vertebrates other than fish concerns a small lake in England (Fig. 7.7)[282]. This lake was man-made and lies close to a river richly fertilized with sewage effluent (see Chapter 10). It was originally intended that one basin of the lake (called locally a broad) be isolated completely from the river water (inner broad) and that the other be left open (outer broad). However, leaks in the retaining dams led to ingress of river water into both basins and their water chemistry was similar. Fish had free movement between the outer broad and the river, but were confined in the inner basin. In 1976 (Fig. 7.8) both basins had comparably large phytoplankton crops, turbid water and moderate zooplankton populations. Submerged macrophytes were almost absent. A year later the two basins had become very different, except in their chemistry which remained that of a highly enriched water. The outer basin was much as it had been in 1976, but the inner basin had clear water, sparse phytoplankton, a very large population of both large and small zooplankters, and a substantial development of benthic filamentous algae (*Cladophora*). Twelve species of submerged macrophytes were present, three of them abundant. The benthic invertebrate fauna was also much more diverse. Water from the inner broad when passed through a coarse net and incubated in the laboratory would produce a dense algal population. It seems that grazing by *Daphnia longispina* (O. F. Müll) and *Bosmina longirostris* (O. F. Müll), which were much scarcer in the outer broad, had been able to remove the phytoplankton faster than it was being produced. In most lakes intense zooplankton grazing results

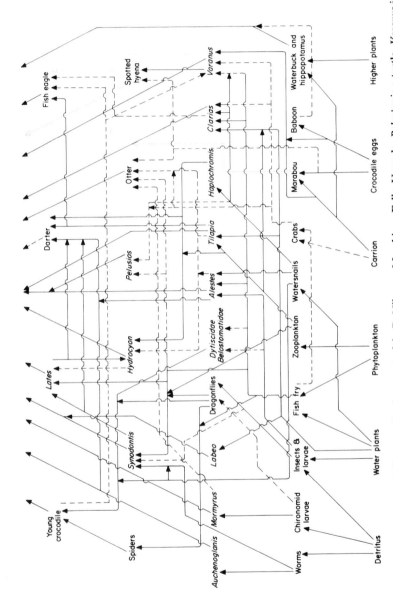

Fig. 7.6. A food web, culminating in the Nile crocodile, from part of the R. Nile near Murchison Falls, Uganda. *Pelusios* is a turtle, *Varanus* is a monitor lizard, other latin names are those of fish species. (Redrawn from Cott[74].)

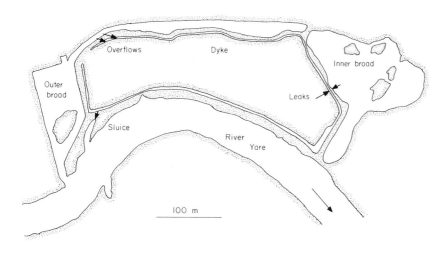

Fig. 7.7. Map of two experimental lakes (broads) at Brundall in Norfolk.

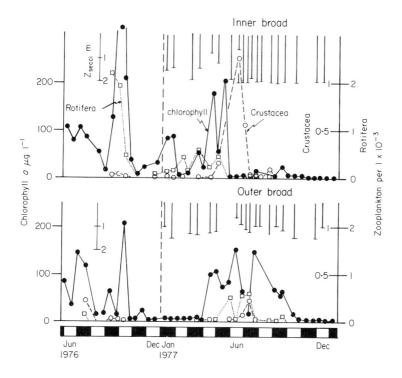

Fig. 7.8. Events at Brundall inner and outer broads in 1976 and 1977. Phytoplankton was measured as chlorophyll a; Z_{secci} is the depth at which a white disc was visible when lowered into the water. (Based on Leah *et al.*[282].)

in development of large and relatively inedible phytoplankters such as the colonial blue-green algae[411]. This did not happen in the inner broad perhaps
because the continued flushing out of the lake by the slight tide of the river did
not allow a long enough period for the relatively slow-growing colonial forms to
develop.

The zooplankton populations in the inner broad in 1977 were exceptionally
large and the mixture of large and small forms suggested that fish predation on
them was minimal(see Chapter 3 & pp. 185–7). Net fishing of the two broads
confirmed that the fish population of the inner broad was negligible in 1977,
though the zooplankton populations in 1976 would indicate normal predation.
The outer broad, with its wide channel to the river had abundant fish in 1977.
The inner broad fish had clearly been greatly reduced in numbers, but no
dramatic fish-kills had been noted and toxic pollutants could be ruled out, for
both inner and outer broads received the same water supply. However, a fish-
eating cormorant had been confined, by a damaged wing, to the inner broad
from late 1976 to autumn 1977. Cormorants take several hundred grams wet
weight of fish per day and it is reasonable to suppose that over the winter
months the bird (plus some visiting cormorants) had been able to reduce
markedly the fish population of this very small body of water. If this is so it
provides a remarkable instance of how a vertebrate predator might considerably
alter the structure of an aquatic ecosystem, though it must be admitted the
circumstances were unusual!

7.4.2 Nutrient importation

Obvious candidates among vertebrates for the importation of nutrients to a
lake are birds which feed by day over a wide land area and roost at night on
lakes where they are relatively safe from their own predators. In some instances
an increase in nutrient loading has followed increases in bird populations due
to human activities. An instance concerning black-headed gulls was mentioned
in Section 1.8. At Wintergreen lake in Michigan migratory Canada geese (*Branta
canadensis canadensis* (L.)) are attracted on their north–south migration in
autumn to the lake. They provide much of the necessary phosphorus and nitrogen to support blue-green algal blooms and other algal populations which may
have given the lake its name[330]. The use of the lake as a bird sanctuary and the
attraction of a pinioned geese flock may have increased the size of the migratory
flock in recent years.

The hippopotamus has a comparable land/water life style. Viner[509] has
estimated that the hippopotamus herd (some 5000 head) which spends the daytime in L. George, Uganda, annually contributes between 15 and 26 tons of
phosphorus and between 76 and 99 tons of nitrogen. Respectively these are
10% and 2% of the total inputs necessary to maintain the lake at its present
fertility.

Two examples of how nutrients may be redistributed by vertebrates in large
waterways have emerged when the vertebrates in question, the caimans (Rep-
tilia) and the sockeye salmon (*Onchorhynchus nerka* Walbaum) have been

reduced in numbers by human activities. In lakes Dalnee and Blizhnee on the Paratunka R., Kamchatka, U.S.S.R., normally 20–26% of the phosphorus loading and exceptionally 35–40% of it has come from the decay of salmon which have died after spawning. The salmon feed and grow in the sea and migrate back up river to spawn. This so weakens them that most subsequently die on the spawning grounds. Commercial fishing of salmon in the Pacific Ocean has reduced the salmon population sufficiently for there to have been a marked decline in the number of spawners, in phosphorus loading on the lakes and in primary, zooplankton and overall fish production in the post-war period of intensified fishing (Table 7.7).

Table 7.7. Effects of migratory sockeye salmon reduction on the ecosystem of L. Dalnee. (From Krokhin[275].)

	1937–1947	1948–1969
Phosphorus load from catchment and direct precipitation (kg lake^{-1}yr^{-1})	374	374
Phosphorus load from death of sockeye salmon (kg lake^{-1}yr^{-1})	132	19·5
Total phosphorus load (kg lake^{-1}yr^{-1})	506	393·5
Effective mean concentration of total phosphorus μg l^{-1}	11·8	9·2
Primary production k cal m^{-2}yr^{-1}	3010	2435
Zooplankton production k cal m^{-2}yr^{-1}	400	280
Total fish production k cal m^{-2}yr^{-1}	67	37

The R. Amazon, draining a vast area where much of the available inorganic nutrient is bound up in the biomass of the surrounding forest, is not a very fertile river, but its very size, like that of the sea, supports numerically large populations of animals though on a unit area basis they are small. Some of the tributaries are even less fertile with solute levels almost as low as the more indifferent grades of laboratory distilled water. During the flood seasons the tributary water is backed by the main flow and the characteristic widenings of the tributary mouths, called river lakes, become moderately deep, if very infertile. Nonetheless it is to the heads of these lakes that many fish move to spawn in the flood season. They feed partly on invertebrates—Ephemeroptera, Trichoptera, Dytiscids—supported ultimately by allochthonous detritus, and partly on animals falling in from the overhanging marginal trees. Phytoplankton and zooplankton crops are low and initially quite insufficient to support the impending hatch of fry.

The fish, however, are followed into the tributary lakes by many other vertebrates—turtles, snakes, marsupial rats and caimans. Caimans—there are four species present, of which the commonest was *Melanosuchus niger* Spix—are crocodilians growing up to 4 m in length with long, toothed jaws well adapted for snapping up fish, their main diet. Where caimans have been removed by intensive hunting for their belly skins[134], fish production has greatly dropped. What seems to happen is that the caimans release nutrients through consumption of adult fish and subsequent excretion into the water. There is evidence, however, that they feed relatively little in the tributary lakes (they may go there perhaps to seek less turbulent water during the main river flood) and that most of the nutrients they excrete come from metabolism of body reserves previously acquired in the main river. These nutrients are apparently sufficient to support increases in the plankton population as the fish fry hatch and effectively help maintain the fish stock used by local subsistence fisheries.

7.4.3 Man

Much of the rest of this book (Chapters 9–12) illustrates how one large vertebrate, man, modifies aquatic ecosystems through predation, in the form of commercial or subsistence fishing, or through changing nutrient loadings or use of waterways. Firstly however, historical perspective needs to be given to this and previous chapters through a consideration of palaeolimnology (see Chapter 8).

7.5 Further reading

The literature on fish is very large indeed. Among others which deal with fish in limnological contexts, such as the Journal of the Fisheries Research Board of Canada, are the specialist journals, Journal of Fish Biology, Transactions of the American Fisheries Society, and Environmental Biology of Fishes. Basic texts on fish biology are Carlander[62] and Lagler *et al.*[277] and more specialist coverage is given in Alexander[2] (functional design) and Brown[50] (physiology). Nelson[381] presents an overall classification of fishes, and McMahon [343] gives a popular account of British fish habitats. A key to British fish is Maitland[322] and tropical freshwater fish receive very good coverage in Fryer & Iles[149], and Lowe-McConnell[300, 301]. Individual volumes on the trout and salmon are Frost & Brown[144] and Hasler[201], Jones[267], and Mills[347] respectively. The literature on reptiles may be approached through Bellairs[28] and Guggisberg[184].

CHAPTER 8
PALAEOLIMNOLOGY, THE HISTORY
AND DEVELOPMENT OF
LAKE ECOSYSTEMS

8.1 Introduction

Lakes are not permanent features of the landscape. The very deepest, such as the Rift Valley Great Lakes of Africa and L. Baikal in the U.S.S.R., may have existed for millions of years and may yet be present some millions of years hence, but the majority of lakes have life expectations of only tens or perhaps hundreds of thousands of years. Since they are at the lowest points of their catchments they become filled inevitably with particles eroded from the land as well as produced and sedimented in the water. The rate at which they are filled in depends on the locality but is often of the order of a few millimetres per year.

In water deeper than a metre or so the deposit laid down is a brownish or greyish lake mud derived from washed in particles, planktonic and littoral detritus and that formed from the partial decomposition of submerged macrophytes. In very shallow water it is often a more structured, fibrous peat derived from emergent macrophytes with their more lignified supporting tissues. Build up of a peat deposit from the edges may result in encroachment of vegetation on the previously open water and eventual conversion of the lake to a swamp which may progressively be replaced by other wetland communities and perhaps woodland. This process is not inevitable, however. In a very large lake, wave action may continually erode the peat deposit of a marginal swamp so that no encroachment occurs. Photographs of the edges of Scottish lochs (lakes) taken over half a century[469] have shown no encroachment, except where the swamp was located on a river delta where allochthonous sediment deposition may have been greater than the rate of erosion by waves. Material eroded from the edges, on the other hand, may be deposited in deeper water as lake mud and contributes to the filling in process in this way.

At first sight lake mud is somewhat unprepossessing stuff—an apparently amorphous, dark coloured, often evil smelling, watery goo. With proper analysis, though, it contains far more information about the conditions in the lake under which it was laid down than we can yet hope to interpret. It contains a record of changes both natural and man-made in the catchment area as well as the lake. It may even harbour a record of global events such as changes in climate, in magnetic field and even the timing of nuclear weapons tests!

Ideally, lake mud has been laid down chronologically from the time of origin of the basin. In most temperate lakes this was 10–15 000 years ago as the polar ice melted back, but in lower latitudes the time-scale is usually longer. In large lakes with steeply descending shores some slumping may upset the chronological sequence, but with experience of a particular lake this can be recognized and interpreted, and from most lakes a continuous sequence of sediment can be obtained using one of a variety of corers designed to take an undisturbed column of the sediment from top to bottom of the deposit.

8.2 Obtaining a core of sediment

The simplest corer comprises a piece of plastic drainpipe, sliced longitudinally in half, then taped together again with waterproof tape. If the water and mud are both shallow, as they might be in lakes formed only a few centuries ago, this can simply be pushed into the deposit until it penetrates the basin material and then pulled up. Success depends on the basin material and sediment being of a consistency such that they stick in the tube when it is pulled up. The overlying water can be siphoned or poured off and the pipe opened to reveal the core.

For deeper water or deeper sediments more sophisticated devices are needed. The Mackereth corer[317] (Fig. 8.1) has proved a valuable instrument in recent years, though several other sorts of corer exist. It comprises a corer tube set as a piston in an outer tube. The outer tube is welded at its base to a large cylinder, about the size of a 50 gallon oil drum, open at the bottom. The equipment is lowered to the sediment and water is pumped from the large cylinder through an outlet, forcing the cylinder into the mud where it acts as a firm base. Compressed air is then injected into the outer tube and forces the inner corer tube into the sediment where it takes a continuous core several metres long. As it reaches its maximum extension the compressed air is diverted into the large cylinder, forcing it and the extended corer tube out of the mud as it is buoyed towards the surface. A valve releases some of the air from the cylinder as it clears the mud so that it travels to the surface steadily, where the apparatus is recovered. Back in the laboratory the core may be pushed from the tube with a piston.

8.3 Dating the sediment

Samples of mud from a core can be dated by a number of methods, mostly using radioactive isotopes. These usually involve measurement of levels of an isotope thought to be produced at a constant rate on the earth and of a derivative formed at a known rate (λ) from its decay. The time (t) during which an amount Ao has decayed to A (Ao–A being measured as the amount of decay product formed or by the concentration of the original isotope left, assuming a constant and known initial concentration) is given by:

$$A = Ao\, e^{-\lambda t}$$

and its derivative, $t = \lambda^{-1} \log e \left(\dfrac{Ao}{A} \right)$

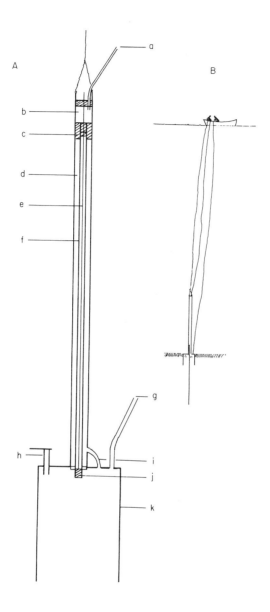

Fig. 8.1. (A) Detail of a Mackereth corer and (B) the corer in operation. The corer is lowered to the sediment and air is sucked out of the anchor chamber (k) through outlet (g). Compressed air is then injected through (a) into chamber (b). This pushes piston (c) down through tube (d) and pushes the corer tube (f) past a fixed piston (j) into the sediment. As (f) fills with sediment, air is displaced from inside it through a fine tube (e). When (c) passes the junction of tube (i) with tube (d), the compressed air is diverted into the anchor chamber (k), forcing it from the bottom of the lake and buoying the whole apparatus back to the surface. A valve (h) releases some of the air in (k) so as to control this ascent. (Based on Mackereth[317].)

Choice of isotope depends on the estimated age of the sediment to be dated. The older the sediment, the longer must the half-life (the time taken for half of a given initial amount to decay) be to ensure that levels of it can still be detected. For sediments up to about 30 000 years old, the ^{14}C method can be used and for those up to about 150 years a method using ^{210}Pb is useful. For sediments, for example in deep tropical lakes, around 10^5–10^6 years old, methods using oxygen isotopes could be valuable.

8.3.1 ^{14}C technique

The ^{14}C method depends on the steady formation of radioactive ^{14}C by bombardment of nitrogen molecules by neutrons at the top of the atmosphere. The neutrons are part of the 'cosmic ray' flux from the sun.

$$^{14}N_7 + {}^1n_0 \leftrightarrows {}^{14}C_6 + {}^1p_1$$
$$\text{(neutron)} \quad \text{(proton)}$$

^{14}C atoms are produced at the rate of about 10^2 cm^{-2} of the earth's surface min^{-1}, giving a ratio of ^{14}C to the stable ^{12}C of about 10^{-12}. It is assumed that carbon isotopes are taken up, ultimately through photosynthesis, into living organisms in this ratio and that ^{14}C then decays with a half-life of 5700 years. The $^{14}C : {}^{12}C$ ratio is measured in material to be dated and compared with the initial ratio as follows:

$$\left(\frac{^{14}C}{^{12}C}\right)_{current} = \left(\frac{^{14}C}{^{12}C}\right)_{initial} e^{-\lambda t}.$$

For ^{14}C it takes, on average, 8200 years for each atom to decay so that:

$$t = 8200 \log_e \left(\frac{10^{-12}}{\left(\frac{^{14}C}{^{12}C}\right)_{current}}\right) = 8200 \log_e \left(\left(\frac{^{12}C}{^{14}C}\right)_{current} . 10^{-12}\right)$$

where t is the age of the material being analysed.

The ^{14}C method is not usable for recent (< 200 yr) sediments, since the widespread burning of fossil fuels with very low $^{14}C/^{12}C$ ratios has upset the previously constant ratio of 10^{-12} in the atmosphere.

8.3.2 ^{210}Pb method

^{210}Pb is present in the atmosphere ultimately as a result of naturally occurring ^{238}uranium in the earth's crust. One of the products of uranium decay is ^{226}radium which decays to the rare gas ^{222}radon. This diffuses into the atmosphere where it decays by a series of short-lived daughters to ^{210}Pb. It is washed out of the atmosphere in rain, where its average concentration is around 2 p Ci l^{-1}, within a few weeks of its formation, reaches the lake sediments and decays with a half-life of 22·26 years to ^{210}Bi. Measurement of the atmospherically derived ^{210}Pb can then give a useful method of dating recent sediments[402]. A correction has to be applied for the ^{210}Pb derived directly from ^{226}Ra decay in the sediments since the initial concentration of this varies from place to place. This is a relatively new method and some controversy exists on how to interpret the results [12,388,398].

8.3.3 ^{137}Cs dating

A third radiometric method used by limnologists is based on a different principle. ^{137}Cs is not a naturally occurring isotope, but was produced and released into the atmosphere by nuclear weapons testing from 1954 onwards, with a peak of such activity between 1959 and 1963. The amounts of ^{137}Cs present in the atmosphere were zero before 1953 and maximal in most areas in 1961/62.

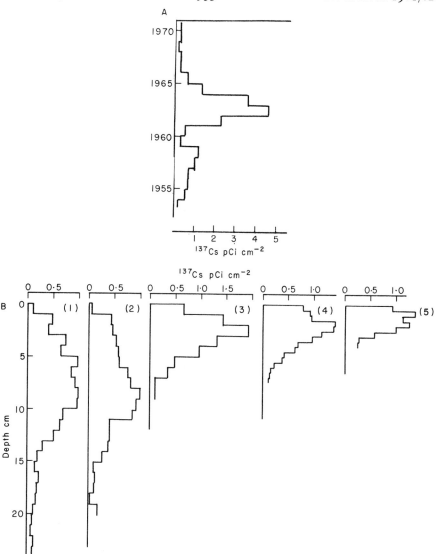

Fig. 8.2. (A) The annual rate of deposition of ^{137}Cs at Windermere, and (B) the ^{137}Cs detected at various depths (in cm) in sediment cores from (1) Esthwaite Water, (2) Blelham tarn, (3) Windermere, (4) Ennerdale Water, (5) Wastwater, all of which are lakes in the English Lake District (see Fig. 8.6). The peak levels correspond to the 1963 stratum of sediment. (Redrawn from Pennington *et al.*[398a] with (2) taken from Pennington *et al.*[398]).

[137]Cs is washed out of the atmosphere by rain and, once in contact with soil or sediment particles, is held very firmly. Its presence in sediment can be detected by measurement of the spectrum of γ rays which it emits and a profile of ^{137}Cs concentration in the sediment gives a 'tail' around 1954 and a peak around 1961/62 (Fig. 8.2). This allows dating of very recent events[398a] at a unique time in palaeolimnology. The method, because of the short half-life of [137]Cs (30 yr) may not be usable at all in about 200 years when most of the isotope will have decayed. That is, of course, unless new supplies of ^{137}Cs have been introduced into the atmosphere. Recently, by comparing past records of the phytoplankton in Rostherne Mere with their remains in the sediments laid down since the early 1960s, Livingstone & Cambray[296] have elegantly demonstrated the accuracy of ^{137}Cs dating.

8.3.4 Non-radiometric methods

Radiometry is, in general, time consuming and therefore expensive. Other methods that are cheaper, if less versatile, are also available. The earth's magnetic field has changed in both its horizontal and vertical components (declination and inclination) in past times and a record kept in London since before 1600 AD shows a change in declination through some 30° since then.

Some minerals, laid down in lake sediments, become magnetized in the direction of the earth's field at the time they are deposited and some of this magnetization, the remanent, persists. The direction of magnetization can be measured in carefully prepared core slices, kept orientated relative to a fixed line on the corer tube, with a sensitive magnetometer and then plotted against depth in the sediment column. An oscillation in recent sediments from east to west and back has been discovered and compares well with the records kept at the London Observatory[79, 320]. For older sediments comparison with ^{14}C dates shows an oscillation in declination with a period of about 2800 yr, with the west and east peaks being contemporary for different lakes (Fig. 8.3). The pattern of remanent magnetization may therefore form a time-scale for other lakes where ^{14}C dating at frequent positions in the core proves too expensive[496].

Under anoxic hypolimnia, in deep lakes, disturbance of the sediment by water currents and burrowing animals may be insufficient to obscure fine variations in deposition between spring and winter and late summer each year. In these cases light and dark layers, or paired varves, may be detectable. The light member of the pair often contains carbonates deposited as a result of intense late summer photosynthesis (see Chapters 3 & 6). The pairs can be shown to be annual by detection of typical spring and late summer sedimented tree pollen[499] in the appropriate members of each pair and dating carried out by counting of the pairs (Fig. 8.4). Such visual varving is unusual, however. Varving detectable by more sophisticated methods may be commoner. X-ray photography of thin longitudinal sections of cores from L. Washington has shown prominent dense bands which are not themselves annual, but their finer striations may be[120]. Similar examination with a stereo-scan electron microscope has shown fine bands of diatom fossils with pairs of layers characterized by spring and autumn species. However, even if these prove to be usable for dating, the method will still be a relatively expensive one for routine use.

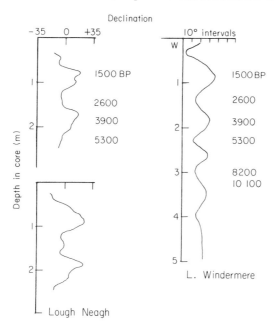

Fig. 8.3. Permanent magnetism, measured as the declination or horizontal component of the magnetization of the sediment in cores from L. Windermere and L. Neagh, Northern Ireland. Two replicate cores are shown for L. Neagh and dates given are based on ^{14}C determinations. (Redrawn from Thompson[496].)

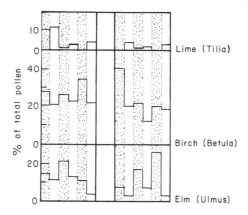

Fig. 8.4. Pollen analyses of two sets each of three paired bands of dark organic and light marl inorganic sediment from McKay Lake, Ontario. The bands are shown to be annual pairs from evidence including the presence of summer-formed lime pollen in the marl bands and spring birch and elm pollen in the dark bands. (Redrawn from Tippett[499].)

8.4 Sources of information in sediments

8.4.1 Inorganic materials

Both inorganic and organic particles are washed into a lake from its catchment and these accumulate in the sediments. Inorganic precipitates produced within the lake are also sedimented and of course the organic matter accumulating from the plankton and the littoral zone also contains elements other than carbon, hydrogen and oxygen. Analysis of sediment, after digestion with strong acids, for particular elements can thus give information on changes and erosion of soils of the catchment, as well as of production within the lake. Some examples of this follow.

In many upland British lakes the catchment area was littered with well fragmented rock when the lake basins were uncovered by the retreating glaciers. This is reflected in relatively high levels of leachable elements like calcium and magnesium and others in the contemporary sediments, which later declined as the rock debris became impoverished[318, 319], and had a marked effect on the productivity of the lakes (see later).

Iodine is a scarce element in soils, but is present, from sea-water spray, in rainfall. It was once thought that the iodine content of sediments might reflect changes in rainfall on the catchment. A consideration of the Vollenweider model (see Chapter 1) would make this unlikely, for increased rainfall would mean not only increased loading of iodine but also an increased flushing rate. The sedimentation rate of iodine in the lake need therefore not increase. Iodine, however, is strongly adsorbed onto clay and humic colloids which are washed into lakes. Its levels therefore seem to reflect the amount of erosion from the soils of the catchment and may indicate soil disturbance through climatic change or human activity, such as deforestation and agriculture[400].

A third example of the use of inorganic analysis of sediments concerns manganese. In deoxygenated hypolimnia manganese II ions, like those of iron II are released from sediments. Manganese II is more soluble than iron II. More of it remains in solution when the lake mixes after stratification and may be washed out of the lake in the autumn and winter periods of high flow. In well aerated hypolimnia manganese is retained in the sediment. Decreases in sedimentary manganese levels may therefore indicate periods of high organic productivity and hypolimnion deoxygenation, or, conceivably, periods when hypolimnion deoxygenation resulted from lowering of the lake level, following climatic changes, and hence reduction in the volume of the hypolimnion relative to a maintenance of the pre-existing fertility.

In highly calcareous catchments, such as those covered with glacial till derived from limestone, lake waters may be saturated with calcium and carbonate ions. The solubility product of calcium carbonate may be exceeded and precipitation of it may follow. On or in the precipitate, phosphate, iron and manganese may be adsorbed or complexed and the water may be extremely infertile as a result[530]. More calcium and carbonate ions enter a lake when such catchments are cultivated and appear to increase the rate at which already scarce nutrients like phosphate are removed from the water. Sediments laid down in various lakes in mid-western America bear evidence of decreased production coupled with an increase in carbonate content of the sediments[528, 533]. Erosion in less calcareous catchments is generally associated with increased, not reduced, fertility.

Much information can be determined from phosphorus determinations on sediments. Because it is relatively insoluble the phosphorus content seems to reflect the rate of loading on the lake, particularly when the rate of sedimentation is taken into account (see section 8.5). It is sometimes possible to detect changes in sewage treatment practice (diversion, expansion of the works) from the phosphorus content of the sediments of lakes into which effluent is discharged [391, 483].

8.4.2 Organic substances

Modern organic chemistry is capable of revealing thousands of separate organic compounds in lake sediments. Many are derivatives of compounds originally deposited which have undergone slow change or diagenesis, and cannot yet be associated with particular groups of organisms. The compounds as yet identified include hydrocarbons, fatty acids, amino acids, sugars, alcohols, ketones, steroids, and plant pigments such as chlorophyll derivatives and carotenoids. In the future it is hoped that the wealth of information revealed by gas chromatography and mass spectrometry on extracts of sediments will be more usefully interpretable than it is at present when perhaps three examples will show its potential value.

The organic matter of sediments comprises partly the refractory materials washed in from the catchment (litter, 'humic acids') and autochthonously produced substances. There has been some question as to which source contributes most. In the English Lake District Mackereth[319] noted that the total sediment depth laid down during the period since the lake basins were carved out by glaciers was broadly similar in lakes such as Wastwater and Ennerdale, which have igneous catchments and are very infertile, and in lakes like Esthwaite and Windermere, which lie in more fertile catchments containing some softer rocks. He concluded therefore that the bulk of the sediment was catchment derived and that organic matter produced in the lake (greatest in Esthwaite, least in Wastwater and Ennerdale) was fully oxidized in the plankton or by the benthos and did not contribute to the permanent sediment. This hypothesis seems to hold for these lakes except in the last few decades when several lakes other than Wastwater and Ennerdale have been fertilized particularly by sewage effluent (see Chapter 10). In these recent sediments organic analysis of carboxylic (fatty) acids has shown a different array of n-alkanoic acids from that in older sediments and in the recent sediments of the unfertilized lakes[76]. Acids with 16, 22, 24 & 26 carbon atoms predominate in the former array, but the C_{16} fraction is absent in the latter. C_{22}–C_{32} acids are abundant in soils but C_{16} acids appear characteristic of algae. The evidence thus indicates a greater preservation of autochthonous material in fertile lakes. Probably this is associated with more severe deoxygenation of the sediment surface in the fertile lakes (and hence limited oxidization of the sedimented compounds) and also with changes in the compounds produced as algae characteristic of fertile waters have grown. Analysis of the carbon skeletons of sedimented hydrocarbons also appears promising, for alkanes between C_{23} and C_{33} appear characteristic of higher plants, C_{31} is predominant in acid peat and C_{27} and C_{29} in base-rich forest soils[77]. Changes in the vegetation of the catchment may thus be detected in sediments and the information used to complement or further interpret studies of the pollen in the sediment (see later).

The greater preservative properties of deoxygenated sediment surfaces are also reflected in the photosynthetic pigment derivatives preserved. In dead cells most pigments are decomposed by oxidation so it is not surprising that a greater pigment content, and a greater diversity of pigments is preserved in sediments laid down under fertile conditions[445]. The contribution of catchment area vegetation to the sediment pigment content is usually small as the litter has been long exposed to the air before it reaches the lake bed. Extraction of chlorophyll derivatives and carotenoids with methanol and acetone and subsequent measurement by spectrophotometry thus helps confirm phases of increased fertility, but if the pigments are separated chromatographically, and separately measured, even more valuable information may be obtained.

The carotenoids present in different groups of algae are often highly specific and have been used to help classify the varied algal phyla. Detection of particular pigments in sediments may then indicate changes in abundance of particular algal groups in the history of a lake. To date, this technique has been applied to the pigments myxoxanthophyll and oscillaxanthin[51] which occur only in the Cyanophyta (blue-green algae). In the sediments of L. Washington (U.S.A.) sediment cores taken in 1967 showed maxima of oscillaxanthin derived from the large populations of *Oscillatoria rubescens* and *O. agardhii* which had appeared in the phytoplankton in the 1950s as the lake became heavily fertilized with sewage effluent[183]. The effluent was diverted completely from the lake by 1967 and progressively the *Oscillatoria* populations decreased. In new cores taken in 1972 it was found that the oscillaxanthin maximum was still present, but now buried under 5·45 cm of new sediments which contained little oscillaxanthin[182]. The pigment thus preserved a record of the blue-green algal growth which characterized the highly fertile phase the lake had passed through. A deeper oscillaxanthin maximum was also detected and seems to record a phase when raw sewage was discharged to the lake for a few years.

8.4.3 Fossils

A wealth of Cyanophytan, protistan and plant remains is preserved in sediments —the heterocysts and sometimes whole filaments of blue-green algae, colonies of the green algae *Pediastrum* and *Botryococcus*, the silica walls of diatoms and the silicified scales and cysts of Chrysophyta, pollen from higher plants in the lake and its catchment area (and perhaps further afield), and lignified cells such as sclereids, fibres and xylem vessels from aquatic plants. From animals there are the tests of certain amoebae, sponge spicules, Bryozoan statoblasts, parts of the exoskeletons of Cladocera and other small Crustacea, mollusc shells, head capsules of chironomid larvae, mite exuviae and even an occasional fish scale.

Because of the closer relationship that algae and plants, rather than animals, have with the physico-chemical environment, it is generally easier to interpret the meaning of fossils of algae, particularly diatoms, and of pollen. The former give information mostly about the lake itself, the latter about its catchment. The occurrence of particular animals is a function not only of the physico-chemical environment, but also to a large extent of competitive and predatory relationships (see Chapters 3 & 7). Nonetheless, animal remains are useful[195] and may even throw light on the past operation of some of the predatory inter-relationships.

Diatom remains

Diatom cell walls are finely patterned, with dots and lines which characterize particular genera and species. They were studied for a long time by naturalist microscopists before they came to the interest of limnologists and consequently a large body of information on the habitats in which particular species occur has accumulated, and considerable reconstruction of past environments is possible from an analysis of the diatoms in sediments. Permanent microscope slides of them can easily be prepared from preparations of diatom walls (frus-tules) made by heating the sediment with an oxidizing agent such as nitric acid and resuspension of the residual diatoms in distilled water. Diatom walls are of two overlapping halves held together by girdling bands and there are four main groups distinguished by their wall patterning (Fig. 8.5).

The Centrales are radially symmetrical and, with some exceptions in the genus *Melosira* are generally planktonic. Other groups are bilaterally sym-metrical and may or may not have a raphe, a longitudinal slit associated with the movement of the cells over a substratum. Those lacking a raphe, the Araph-idineae, may be planktonic or, depending on species, attached to vegetation or rocks or free living on stable sediments. The Monoraphidineae, with a raphe on only one side of the cell, are all species firmly attached to substrata like plants or rocks, while the Biraphidineae, with raphes on both halves of the cell, are usually motile and, again depending on the species may be attached to substrata or may move freely over them, whether rocks, plants or sediment. There are always exceptions to general rules. A Biraphidinean genus, *Nitzschia*, for example, may be planktonic in brackish waters and in some tropical lakes.

With experience, however, an observer can deduce much about the balance of planktonic, epipelic (sediment living) and attached communities and hence about changes for example in water depth and the abundance of weed beds. Certain genera, e.g. *Eunotia* and *Frustulia*, are characteristic of generally infertile water and others, e.g. *Melosira granulata*, of very fertile conditions. There are distinct marine species and distinct freshwater ones and some indicators of brackish conditions. In African lakes, *Navicula elkab* (O. Müll) and *Nitzschia frustulum* Kutz (Grun) characterize saline inland waters[207] and may be used,

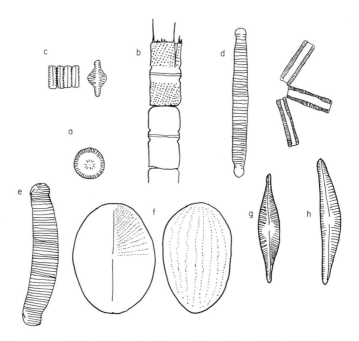

Fig. 8.5. Some representative diatoms. Species are shown after acid cleaning to remove organic matter and reveal the characteristic patterns of their silica walls. (a), (b) Centrales: (a) *Cyclotella*, (b) *Melosira* (seen in side (girdle) view); (c), (d) Araphidineae: (c) *Fragilaria*, (d) *Diatoma*—both valve (top) and girdle views are shown; (e) Raphidioidinae, *Eunotia*; (f) Monoraphidineae, *Cocconeis* seen from each of its two valve surfaces, one of which bears a raphe, the dark line, the other of which does not; (g), (h) Biraphidineae: (g) *Navicula*, (h) *Cymbella*.

in sediment analyses, to recognize drops in lake level and periods of endorheicity (see Chapter 1)[428].

Pollen

The use of pollen preserved in peat and lake sediments has a distinguished history and has formed the basis for recognition that climate has changed greatly over periods of only thousands of years[173, 297, 397]. The walls of many pollen grains are of particularly refractory waxes ('Sporopollenin') which resist the most vigorous oxidizing and corrosive agents. Grains may be concentrated and separated from sediment samples by digestion with hydrofluoric acid. Genera, and often species, are easily recognizable in most groups, though grass pollens for example are not easily separable to species. However, major changes in catchment vegetation can be detected. Agricultural activity is recorded by the appearance of the pollen of characteristic annual herbs which accompany cultivation and evidence of deforestation may be found by a complementary decrease in tree pollen. Such catchment area changes are of immense importance in interpreting events in lakes, as Chapter 1 has shown.

Animal remains

Animal fossils are relatively scarcer than those of plants and may be so fragmented that very expert knowledge may be needed for their identification, let alone interpretation. They may simply be sieved from sediment and concentrated for counting. Some examples will illustrate their potential use.

In lakes in New England and elsewhere, the oldest sediments often contain the remains of a cladoceran zooplankter, *Bosmina coregoni* Baird but this species is replaced in younger sediments by *Bosmina longirostris* (O. F. Müll)[97, 98]. This change appears to be related to progressive filling in of the lakes, decreasing hypolimnion volume, and hence decreasing hypolimnial oxygen concentrations. It seems that *Chaoborus* spp. (see Chapter 6), whose head capsules become more abundant in the sediments at the time of the switch, are favoured by declining oxygen concentrations and selectively feed on the relatively large *Bosmina coregoni* (see Chapter 3). This may upset a competitive relationship between *B. coregoni* and the smaller *B. longirostris* such that the latter become predominant.

A second example concerns certain Cladocera which have been detected abundantly in very early sediments of Esthwaite in the English Lake District [181]. These species, although widespread, are found frequently in sub-Arctic lakes, and include *Chydorus sphaericus* (O. F. Müll), *Acroperus harpae* Baird, and *Alona affinis* Leydig. They occurred in Esthwaite only during a phase characterized by pollen data as the Allerød amelioration and seem to indicate a slight warming, to sub-polar conditions, from intensely cold conditions both previously and subsequently.

8.5 General problems of interpretation of evidence from sediment cores

Vallentyne[506] has pointed out that 'no anatomist or physiologist in his right mind would ever base a study of the life history of an organism on the analysis of its accumulated faeces. This is, however, precisely the position of a palaeo-limnologist with respect to the developmental history of a lake. Sediments are lacustrine faeces, the residue remaining after lake metabolism'. Clearly, therefore, interpretation of lake development is fraught with difficulties and it is wise to examine some of them before considering interpretations from particular lakes.

Sediment is not necessarily deposited uniformly. Higher rates of deposition may be measured near inflows and in the centres of basins which are conical in shape. Water movements and gravity tend to move fine sediment from the edges to the deepest parts in a process now called sediment focusing. Cores from several places may thus be necessary to obtain a full picture, though even in quite large lakes it is remarkable how similar the picture obtained from different parts of the lake bed is, at least qualitatively. Depending on the lake, microfossils may be deposited unevenly and may or may not be redistributed by water movements. Davis[94] found that relatively heavy oak pollen grains were evenly distributed over the sediment of a Michigan lake, yet lighter ragweed pollen settled slowly and was carried to the windward edges before reaching the sediment. On the other hand she found that the surface few millimetres of sediment were resuspended during mixing periods and that the material could be recycled upwards into the water perhaps four times before becoming part of the permanent sediment. This process tended to even out irregularities in previous sedimentation, both in space and in season.

Benthic animals also may progressively mix the surface sediments as they form, thus evening out annual variations into a sort of moving average and smoothing long-term trends. Davis[95] found that tubificid (Oligochaeta) worms, whose feeding was centred at 3–4 cm depth, displaced sediments upwards and displaced larger particles, e.g. of pollen grains, less than smaller ones. The worms fed with their mouths downwards and their anuses upwards.

Much evidence comes from diatoms and in general these are preserved well. There is some selectivity, however. In deep lakes only a fraction of the diatom silica polymerized during wall formation reaches the permanent sediment[394] and dissolution may occur in the surface sediments of shallow lakes[17]. Long, thin diatoms, e.g. *Synedra*, *Asterionella*, tend to break easily by abrasion with inorganic particles or in invertebrate guts, and may then more easily be redissolved. Some very thin-walled genera, e.g. *Rhizosolenia* are rarely preserved at all. Centric and littoral diatoms seem to persist best, compared with planktonic Araphidineae, though intact specimens, even of these, are to be found in very old lake muds.

In interpretation, much depends on knowledge of the current autecology of particular species, yet unfortunately this is usually poorly known. Much more confidence can be placed on the detection of whole arrays of species commonly associated with particular habitats than on the occurrence of particular species. For example, though it is not a diatom but a desmid (Chlorophyta), *Pleurotaenium trabecula* is normally found in infertile waters of low pH and low total solids content. It also occurs, however, on the sediment of reed swamps around highly fertile, alkaline lakes with high total dissolved solids content. This is because high dissolved CO_2 levels are to be found in both habitats[364]. In the former this is because low pH favours the persistence of free CO_2 (see Chapter 3) and in the latter because high bacterial decomposition rates produce abundant CO_2. The occurrence of *P. trabecula* alone, if it were preserved in sediments, could not diagnose a particular habitat, but the occurrence of a variety of genera of desmids would certainly indicate low pH and soft water conditions.

A further problem of reliance on single 'indicator' species of diatoms is that the diagnostic patterning of the wall may be changed as dissolved silicate levels fall[27]. One 'species' may thus change into another! Sufficient laboratory work has not been done yet to determine how widespread is this phenomenon.

Perhaps the greatest problem, and possibly the most misleading feature of sediment core interpretation, is the way in which the results are expressed. Early workers often calculated the percentage representation of a particular species or group of fossils in their total count from a sediment sample of undetermined weight. This gave relative changes in frequency. Thus a decrease in frequency of a species A, relative to that of a species B would be interpreted as a decline in the incidence of A and an increase in B. The true situation might have been an increase or decrease in both species, with B increasing at a greater rate or decreasing at a lesser rate than A.

An improvement has been the expression of results on the basis of per unit weight of sediment, since this gives some indication of absolute changes if the sedimentation rate has been constant. However where sedimentation rates have changed greatly an absolute increase in a species may have been accompanied by a decrease in its numbers per unit weight of sediment if the rate of sedimentation of inorganic or organic matter has increased at an even greater rate. The only reliable way of expressing results of counts of fossils or of chemical analyses is thus in terms of amount laid down per unit area of lake bed per year and means that comprehensive dating of the core must be carried out. This is expensive and has not been done for most cores. All of the above limitations should be borne in mind in the following accounts of results from particular lakes.

8.6 Blea Tarn, English Lake District

Hills of hard, Borrowdale volcanic rocks up to 600 m high surround the small, upland lake called Blea Tarn in the English Lake District (Fig. 8.6). It is 3·4 ha in area, 8 m in maximum depth and is surrounded by tundra-like moorland of grasses and sedges and by *Sphagnum* bogs. There are some plantations of conifers but the only human disturbance of the catchment is a small sheep farm, and its effect on the lake is likely to be slight. Habitation of the area in the past is also known to have been only sporadic so Blea Tarn provides an example of a lake and its catchment largely under control of physico-chemical events.

A core, taken from the lake[204] was a little over 350 cm long (Table 8.1) and was dated from changes in the pollen content. Distinct phases of vegetation change, registered by pollen, have been recorded in north-western Europe and dated originally by radiocarbon measurement. These phases are numbered I–VIIb (or VIII in more southerly areas) and fall into two groups, the late-glacial (late Weichselian) phases I–III of tundra vegetation, and the post-glacial (Flandrian) IV–VIIIb in which various sorts of woodland developed in north-western Europe as the climate warmed.

Blea Tarn was formed in an ice-carved depression in the debris eroded from the hills by the glaciers and had water in it by about 14 000 BP (before present). For 4000–5000 years its biological production was low—relatively few diatom frustules and little organic matter are present in the clay and silt late-glacial sediments. Around 11 900 BP there was a brief period of warming, the Allerød interphase (named after a site in Scandinavia where it was first detected in deposits) in which the sedge, grass and dwarf willow tundra was briefly diversified

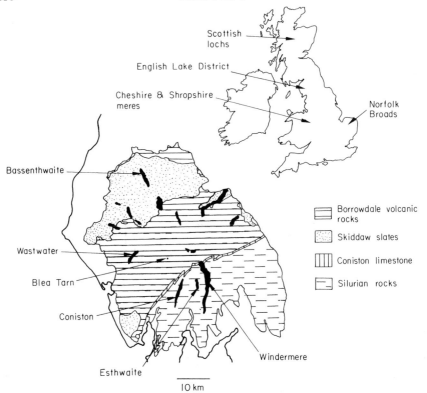

Fig. 8.6. Geographical location and geology of the English Lake District. The Silurian rocks of the southern part of the Lake District are softer and give more subdued mountainscapes and ionically richer waters than those of the Skiddaw and Borrowdale series. (Modified from Macan[310].)

with birch trees and juniper bushes. The lake sediments were then slightly richer in organic matter and diatoms; the diatoms present in the late-glacial period included few phytoplankters and were of a mixed collection, some characteristic of slightly alkaline water.

Final melting of the ice a little more than 10 000 years ago in this area led to marked changes in the general area and lake. Hazel (*Corylus avellana*) colonized, followed by birch (*Betula*), oak (*Quercus*), elm (*Ulmus*) and, in the wetter phase of VIIa (called the Atlantic period), alder (*Alnus glutinosa*). The lake sediments became less inorganic and ultimately were of largely organic, brown lake muds. The diatom fossils indicate a post-glacial phase of high fertility, with a well-developed plankton. Species such as *Asterionella formosa* Hass. & *Rhopalodia gibba* (Ehr.) O. Müll, and the genera *Fragilaria* & *Melosira* were common and indicate an increase in nutrient loading. This may be attributed to the presence of much-pulverized (and hence easily leached) rock left by glacial action and an increase in run off, as temperatures increased, to transport the nutrients to the lake. The organic matter of the sediment was probably catchment-derived from the deepening soils of the woodlands near the lake. As phases

Table 8.1. Palaeolimnology of Blea Tarn, Langdale[204].

Depth (cm)	Zone	Date BP	Sediments	Pollen	Sedimentation rate (mm yr^{-1})	Diatoms and interpretation
0–147	VIIB	Sub-boreal 4900	Brown mud Detritus gyttja	Ash, elm declines	0·32	Very infertile conditions
147–227	VIIA	Atlantic 7400		Oak, elm birch, alder (Forest period)		*Eunotia* (infertile water)
227–327	VI / V / IV	Boreal 10 200 POST-GLACIAL	Clay gyttja	Hazel peak		Acidophils increase; *Asterionella, Fragilaria, Melosira*; Fertile, alkaline indicators increase
327–343	III	LATE-GLACIAL	Pink clay	Artemisia Empetrum Lycopodium	0·36	Diatoms scarce, cold? tarn completely frozen over
343–349·5	II/III	10 700		Empetrum & Artemisia, Juniper, Birch		
349·5–353·5	II	Allerød	Dark mostly organic silt			*Campylodiscus noricus* (alkaliphilic)
353·5–357	I	11 900	Unvarved pink clay	Salix grasses, sedges	0·07	*Pinnularia, Cymbella, Fragilaria*
Below 357		Pre- 14 900	Varved clay	no pollen		No plankters

IV and V progressed from around 10 500 BP to 7400 BP the fertile lake diatom flora did not persist. Progressively it was replaced by one of less fertile, more acid waters characterized by *Eunotia* and *Pinnularia*, which has persisted to the present day with little change in 7000 years.

The decrease in fertility of the lake is attributed to exhaustion of the readily-available nutrients in the glacial debris and subsequent dependence on the slow weathering of fresh basal rock at the base of the soil profiles. This is an extremely slow process and studies on the composition of present day waters of upland lake district tarns show that it differs little from that of rainfall[316].

8.7 Esthwaite

A glacier flowing down a small river valley carved out the basin of Esthwaite (Fig. 8.6) and dammed the lake with a moraine at the foot of the valley. It is a larger lake (100 ha) than Blea Tarn and lies among more subdued scenery of softer Silurian rocks which are more readily weathered than the Borrowdale volcanic rocks around Blea Tarn. For all but the last few thousand years, however, Esthwaite had a similar history to that of the smaller lake, and rather more information is available on the chemistry and animal remains of the sediment as well as on diatoms and pollen.

The late-glacial phases (I–III) (Table 8.2) supported tundra around the lake and clays were deposited in the presumably rather cloudy water as the fine debris of the glacial till was eroded. The warming in phase II was marked by the appearance of Cladocera in the water, particularly of species of Chydoridae, a group usually associated with the shallow littoral zone, but diatoms were not well preserved, perhaps partly because of abrasion by the silt and clay. The levels of sodium, potassium, calcium and magnesium in the sediments were around 10, 30, 5–10 and less than 10 mg g^{-1} dry wt. respectively.

With the start of the post-glacial period the surge in production noted for Blea Tarn was recorded in the sediments. The leaching of broken rock fragments led to increases particularly of calcium (22 mg g^{-1} dry wt.) and phosphorus in the sediments. The latter increased from about 3·5 to 5 mg g^{-1} dry wt. as phase VI was succeeded by the early part of phase VIIa. Diatoms (*Melosira arenaria* Moore, *Epithemia*, *Fragilaria*) of a fertile phase and planktonic Cladocera numbers increased markedly. Analysis of the organic component of sediments laid down in phase VIIa has shown a ratio of n-alkanoic to branched and cyclic monocarboxylic acids similar to that found currently in the surface sediments of fertile lakes.

As phase VIIb was entered about 5000 years ago, however, the regression to less fertile, more acidic conditions found in Blea Tarn took place. Calcium levels fell to about 5 mg g^{-1} dry wt. and phosphorus levels to about 2 mg g^{-1} dry wt. *Cyclotella*, *Tabellaria*, *Eunotia*, *Anomoeoneis*, *Gomphonema*, and *Cymbella* in the diatom flora collectively indicate a decline in fertility. Increases in sodium and potassium levels were recorded in sediments of zones VIIa and VIIb perhaps

Table 8.2. Palaeolimnology of Esthwaite. (Compiled from Cranwell[75], Goulden[181], Mackereth[318], Pennington et al.[398a], Round[436].)

Depth (cm)	Sedimentation rate (mm yr⁻¹)	Date BP	Sediment/ Pollen	Diatoms	Animals	Organics	mg g⁻¹ Ca	P
0-150	Current 5·6-11	VIIb (upper)	Plantain	*Asterionella, Melosira,* Fertile phase	Chironomids increase from 100 cm up, also *Bosmina longirostris*.	0-5 cm 82% n-alkanoic 18% branched cyclic acids	4-5	2mg g⁻¹ falls with dilution by increasing sediment
150-300	0·55	VIIb	Oak, birch alder	*Cyclotella, Gomphonema, Cymbella*				
		5100 BP		*Eunotia, Anomoeoneis.* Acidic indicators, infertile	*Bosmina coregoni* predominant			
300-410	0·46	VIIa	Alder increase Brown mud from here up	*Melosira, Stephanodiscus* Fertile phase		336-382 cm 82% n-alkanoic 13% br. cyclic acids		3·5 mg g⁻¹
410-437	0·19	8900 BP	Banded clay/ mud	*Melosira arenaria, Epithemia, Fragilaria* Fertile phase				
437-460		V	Pine, Hazel					
460-478	0·16	IV	Grey clay	Poor flora				
478-498		10 300 BP III			Littoral arctic forms		22	
498-515		II	Grey silt/clay Salix, sedges				5-10	
515		I	Laminated clay					

reflecting increased rainfall. Much of the sodium, in particular, of English Lake District waters is provided from droplets picked up by the prevailing winds from the sea only a few tens of km to the west.

In Esthwaite, in contrast to Blea Tarn, the unproductive phase has not continued to the present day. During the latter part of phase VIIb, or the last 3000 years there has again been an increase in fertility. At first this was moderate but very recently, in the 19th century, it has been marked. *Asterionella* and *Melosira* have reappeared in the diatom flora, and chironomids became abundant about 1600 years ago. At about the same time the cladoceran *Bosmina longirostris* appeared and partly replaced the previously recorded *B. coregoni*. Sedimentation rates have increased in the 19th century to 5·6–11 mm yr^{-1} from a mean post-glacial rate of only 0·48 mm yr^{-1}.

The reason for these changes is to be found in human activity in the catchment area[310]. Pollen of a common agricultural weed, the plantain, *Plantago lanceolata* is found in sediments younger than about 3000 years, and around 3700 BP neolithic peoples are known, from archaeological evidence, to have moved into the Esthwaite catchment area, though not to any extent into that of Blea Tarn. Forest clearance for fuel and agriculture leads to release of nutrients previously stored in the woodland biomass, to greater soil erosion and perhaps to greater leachability of nutrients from the excreta of domestic stock. The clearances were intensified by a new wave of colonization around 900 BP by Norsemen and Norse-Irish, perhaps retreating from political disturbances in the Isle of Man. Many place names in the English Lake District are of Norse derivation.

During the 19th century the English Lake District became popularized particularly by the poetry of Southey and Wordsworth, who both lived there, and since the building of a railway, completed in 1847, the populations of both residents and tourists have increased. In turn this has led to improved mains sanitation and it is the sewage effluent from the popular tourist village of Hawkshead, on the inflow river to Esthwaite, which has led to the latest phase of fertilization, or eutrophication of the lake in the last thirty years or so.

8.8 Pickerel Lake

The prairies of N.E. Dakota, U.S.A., form a great contrast to the uplands of the English Lake District, though their lakes also were formed by ice and are no more than 10 000 or 11 000 years old. A large ice block, calved from the Continental glacier as it melted back, was buried amid the tons of rubble of limestone and shale washed out from under the glacier. As this block itself finally melted it left in its place a basin up to 25 m deep which now holds Pickerel Lake. The lake sediments now occupy about 8 m of the basin and provide a record of lake development[205] in an area which is now generally drier than north-western Europe and which lies in the grassland rather than woodland biome.

Four phases (Table 8.3) in the surrounding vegetation have been deduced by pollen analysis. After the retreat of the ice, spruce (*Picea*) and tamarack (*Larix*) forest had developed by 10 500 BP, and by 9400 BP this had diversified

Table 8.3. Palaeolimnology of Pickerel Lake, Dakota[205].

Depth in core (cm)	¹⁴C dates	Sedimentation rate (mm yr⁻¹)	Pollen	Diatoms	Interpretation
0				Fewer, varied, alkalibiontic, increase in halophiles. Some Melosira granulata	Some enrichment (increase in Fragilaria crotonensis and in Ambrosia pollen). Agricultural fertilization and disturbance
100		1·05	Prairie grasses. Some increase in woodland		
200	2700 BP				
300		0·91		Many, no halophobes, % planktonic decreases	Transition
400	4200 BP		Blue stem grass prairie. Ambrosia Artemisia, Composites, Chenopods. Some wood. Seeds of wet-mud plants	Many, alkaliphilous, no halophobes, more plankton Melosira granulata, Stephanodiscus. Some sandy layers	Eutrophic (M. granulata) Transition to open prairie (summer droughts), fluctuation in lake level. Increased erosion (sand layers). Evaporative concentration—some brackish taxa
600	9400 BP		Mixed deciduous woodland. Quercus Ulmus, Betula Alnus, Abies Acer, Fraxinus	Fewer. Fragilaria. Benthic. (Molluscs appear). Acidophilous species decrease Many, benthic, alkaliphilous Fragilaria, Epithemia. Some halophilic	Change to more alkaline water
700		3·82			
800	10 500 BP		Forest. Picea, Larix	Acid/neutral, Acidophilous, halophobe. Eunotia	Spruce forest, acid soils, low nutrient loading
825	Glacial till				

to a forest of oak, elm, birch, alder, fir (*Abies*), sycamore (*Acer*) and ash (*Fraxinus*), probably with some grassy openings in it. The climate then became progressively drier for the forest was replaced by prairie grassland, with grasses including *Andropogon* (blue-stem), Compositae (*Ambrosia* (ragweed) and *Artemisia*) and annual herbs of the family Chenopodiaecae. This vegetation has persisted since 9400 BP, though since 4200 BP there has been some patchy deciduous woodland development.

The drying of the climate, and its effect on the catchment vegetation is reflected also in events in Pickerel Lake. In the early forest phases the effect of the raw conifer forest humus was to neutralize the alkaline ground water of the calcareous glacial deposits, for the diatom flora was then a mixed one. There were some indicators of neutral to acid waters, such as *Eunotia* & *Tabellaria*, which do not appear in the younger sediments, while genera now found in fertile, alkaline waters (*Fragilaria, Epithemia, Mastigloia*) were also present. Diatoms of strongly alkaline habitats were absent and the mainly benthic flora, coupled with the remains of aquatic plants (*Potamogeton pectinatus* L., *Najas flexilis* (Willd.)Roslk.& Schmidt,*Najas marina* L.)testifies to poor plankton development.

As the deciduous woodland replaced the conifer forest and then itself was succeeded by grassland, the diatom flora underwent a transition and molluscs appeared in quantity in the lake. Diatom indicators of neutral to acid water disappeared and the flora was generally sparser. Then, as the prairie developed, a major increase in diatoms took place. There was a high production of phytoplankters of fertile lakes—*Melosira ambigua* (Grun). O. Müll., *M. granulata* (Ehr.) Ralfs, *Fragilaria crotonensis* Kitton, & *Stephanodiscus niagarae*—and some species characteristic of brackish waters were found. The sort of climate that favours prairie has summer droughts, giving increased mineralization of nutrients, great fluctuations in water level and in ionic content of the water. The fluctuations in levels are indicated by layers of sand, eroded from the marginal beaches, in the sediment profile and by the presence of seeds of annual higher plants characteristic of exposed wet mud. Evaporative concentration of ions in summer was demonstrated by brackish diatom taxa and the increased nutrient loading by the characteristic plankton species.

Again this burst of production was not sustained, for after 4200 BP some woodland development may have stabilized the soil and decreased the erosion rate. Presumably this was associated with a less extreme, somewhat damper climate, and was linked with a mixed diatom flora less dominated by indicators of high fertility. Towards the top of the sediment profile a further enrichment, though not as great as that between 9400 and 4200 BP has been noted. *Melosira granulata* & *Fragilaria crotonensis* have again increased and coupled with an increase in ragweed pollen suggests that this is associated with agricultural activity and fertilization of the land during the past two centuries.

8.9 Lago di Monterosi

Not all lakes are as young as those formed by ice action, but few of the older ones have been studied by palaeolimnologists. The Lago di Monterosi in the Campagna, 40 km from Rome, in Italy, is one. It lies in an old volcanic crater which is almost circular and about 600 m in diameter. At present the lake is a moderately fertile one with extensive beds of aquatic macrophytes (*Nymphaea*, *Myriophyllum*, *Ceratophyllum*) and a plankton not untypical of soft lowland waters, but it has changed greatly during its existence[240].

The basin was formed by a volcanic explosion a little more than 26 000 years ago, when northern Europe was still heavily glaciated. At that time the Roman Campagna was probably a cold, dry steppe with large mammals like the mammoth grazing over it. Sediments from the lake in its first three thousand years (period A, Table 8.4) indicate an alkaline, rather fertile lake with remains

Table 8.4. Palaeolimnology of the Lago di Monterosi. (After Hutchinson *et al.*[240].)

	Years BP	
	0	–As in C_2.
C_2		Productivity less than during C_1, but more than in B. Diatoms, some blue-green algae and chironomids, but *Sphagnum* increased. More intensive farming (pigs, grain) and soil erosion.
	1200	–Decline in productivity and eutrophy, decrease in Chironomids. Fewer blue-green algae but eutrophic diatoms present.
C_1	AD 10 BC 340	In this period, dramatic changes. Increase in Ca^{2+}, rate of sedimentation, Chironomids and Chydorids. Marked increase in eutrophy. Blue-green algae very common.
	2800	–Surroundings more wooded (*Abies* and *Quercus*). Less *Sphagnum*. Probably a little less acid. Historically, human activity in the basin probably began increasing. Chenopodiaceae pollen.
B	5000	–Lake shallow, but still with low sedimentation rate. Increase in acidophil diatoms and *Sphagnum*. Water plant flora decreased. Chrysophycean cysts indicate oligotrophic conditions, Chironomids and Cladocerans very rare.
	13 000	–Sedimentation rate low and little organic matter deposited. Probably local plant cover well developed, with grasses. More water-plant pollen—lake shallower.
	23 000	–Early lake dilute, slightly alkaline (due to Na and K bicarbonates). Moderate sedimentation rate. Few chironomids, diatoms of wide tolerance. Sponges and pollen of *Myriophyllum*, Nympheaceae and *Potamogeton*. Surroundings cold, dry, steppe-like with *Artemisia*.
A	26 000 BP	–Origin of lake by volcanic explosion.

of *Gloeotrichia* (a blue-green alga), *Chaoborus* and *Glyptotendipes* (Chironomidae) and of aquatic plants of fertile habitats. Sponge spicules, also of species from such habitats, remain from what was probably a richly developed invertebrate fauna in the weed-beds. Such conditions might be expected. The climate was cold—the Wurm II glacial maximum in the north had still to be approached —but the crater rim and surrounding permeable catchment of fresh ash, lapilli and scoria provided fresh rock surfaces for leaching of nutrients.

This early productive phase, seen also in other lakes discussed above, did not last beyond 23 000 BP, again presumably as the fresh deposits became leached out and the run-off waters less rich in dissolved ions. There followed a very long period (B) of reduced production and low sedimentation rates (0·0044–0.0075 cm yr⁻¹) which lasted for 20 000 years with little change. *Sphagnum*, a plant of acid to neutral infertile waters, colonized. The sponges of alkaline waters declined and chironomids became scarce. Woodland of oak, fir and hazel developed in the surrounding area, and stabilization of the soils by the forest probably also reduced nutrient loading on the lake. Some minor changes could be ascribed to changing water levels related to climatic fluctuations.

Then, dramatically, the lake became again very fertile. There were maxima of blue-green algae and some green algae (*Pediastrum*). *Melosira granulata*, characteristic of very fertile lake plankton, appeared and there was a rich cladoceran and chironomid fauna; calcium was deposited heavily in the sediments and the sedimentation rate increased nearly sixfold to at least 0·038cm yr⁻¹. *Sphagnum* disappeared. The date for this transition given by ¹⁴C analysis is 2220 ± 120 years BP and after corrections have been applied for fluctuations in the rate of formation of ¹⁴C and for variability in the method, the date most likely falls between 150 B.C. and 285 B.C.

No evidence of natural events could be found to explain the changes which occurred but there was a major interference by man in the catchment at the time. The Roman consul Lucius Cassius Longinius had a road built around 171 B.C. to give rapid passage between Rome, eastern Tuscany and the upper Arno valley, where unrest, which was to culminate in the second Punic war, was about to break out. The ancient Via Cassia runs around the southern edge of the catchment of the Lago di Monterosi, and the disturbance of road building in what is quite a small catchment (little larger than the lake itself) seems to be the only sensible explanation of changes in the lake. The roadworks seem to have disturbed the vegetation cover and altered the drainage pattern slightly so that water previously draining away from the lake burst through as springs (the Fontana di Papa Leone) near its edge. This water percolates through calcareous deposits and explains the increased calcium deposition in the lake during the period in question. Soil erosion caused by removal of vegetation explains the increase in sedimentation rate.

The immediate effects of the road building were short and the lake settled back into a less productive phase soon afterwards. It has never quite returned to its state in period B, however, because of the permanent alteration in the hydrology caused by the springs, and because, in the post-Imperial period agriculture and deforestation caused further changes and disturbance. *Sphagnum* did reappear in the fifteenth century and the numbers of chironomids and crops of blue-green algae have greatly declined. There was a nadir in the fortunes of the Monterosi area in the fifteenth century and these seem related to a period of decreased production sandwiched between the post-roadbuilding period, when pollen of nettles (*Urtica*) and Chenopodiaceae indicate cultivation and stock rearing, and the last two hundred years of farming prosperity.

8.10 Filling in of shallow lakes

Sediment is deposited in all lakes and even in slow-flowing rivers, so that theoretically all lakes should eventually be made so shallow that they are converted to areas of reed swamp, with a river flowing through them, if they were fed originally by surface water. In a very deep lake, however, at sedimentation rates, where man has not disturbed the catchment, of the order of a millimetre or so per year, this process may take a very long time. L. Tanganyika in Africa, for example, has a maximum depth of 1500 m and a current sedimentation rate of around 0·5 mm yr^{-1}. This should give it a future span of about 3 million years —enough for climatic and geological events to intervene and change the surrounding area markedly before such a state is reached.

Lakes of only a few metres depth, however, may fill in not only by sedimentation, but also by peat accumulation at the margins and encroachment of swamp from the edges towards the middle. An example will illustrate this process.

8.11 Tarn Moss, Malham

Malham Tarn, in Yorkshire, England, is a small lake situated in glacial drift and fed by nutrient-rich spring water. In the past it has been larger, but encroachment of aquatic plants has now completely filled about half of its original area[408]. About 8 m of peat have been laid down in parts where encroachment has been complete (Fig. 8.7). The earliest deposit, from the late glacial phase (see above), is boulder clay deposited as outwash from a glacier, and immediately overlying this are several thin layers of inorganic materials—clays with stones up to 1 cm diameter. These were laid down in the period about 12–13 000 BP,

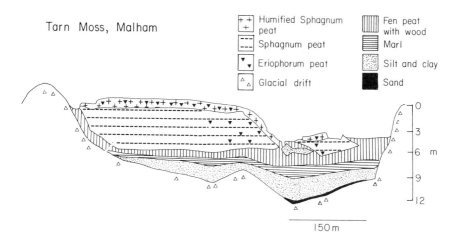

Fig. 8.7. Profile, reconstructed from peat borings, of part of Tarn Moss, which occupies part of the basin of Malham Tarn, a small lake in Yorkshire, England. (Based on Piggott & Pigott[408], and also available in Pigott & Pigott[407a].)

just after the lake formed, and represent the washing in of loose material from the surrounding, probably not well vegetated catchment. Pollen analyses show that a tundra vegetation then predominated (pollen phases I–III). The lake water was probably turbid with suspended clay, and light penetration limited. It is not surprising that aquatic plant remains are not plentiful then. The water cleared as the clay was deposited and the sediments become silty and calcareous (94–95 % $CaCO_3$), and the oospores of *Chara* (stoneworts) were abundant. Shell fragments of at least six snail species also testify that animal and plant communities appropriate to highly calcareous water (from the inflow springs) were present. The *Chara* probably formed most of the sediment with the marl deposited on its surface. In the remaining lake similar conditions still prevail. However, in the shallower water emergent plants colonized the calcareous sediment and laid down peat. Grass and sedge remains have been found but the peat is relatively well decomposed and is amorphous and black in parts. Such peats are characteristic of nutrient rich conditions and are formed by a diverse flora known collectively as fen in the U.K., alkaline bog in the U.S.A., and as minerotrophic mire elsewhere. Fen conditions persisted for several thousand years while concomitantly the vegetation of the catchment had changed from tundra to birch forest to hazel scrub. The water table in fen is approximately at the surface of the peat. Temporarily dryer phases allowed some drying out of the peat surface and invasion by birch trees in parts. The remains of these are to be found in brush wood peat within the fen peat. Fen vegetation, laying down fen peat, is still present around the tarn edges where nutrient-rich water still has access. However, some few thousands of years ago in the middle of the fen mat the peat built up to such an extent that even after compression by new peat laid on top, it was above the level of the groundwater table. At this point rainwater became the main source of water to the peat surface and, by leaching, caused the surface to become acid. Conditions were thus created for bog plants, which, by definition, occupy acid peat soils, to colonize and *Sphagnum* species took over from the fen plants. Many *Sphagnum* species are able to maintain the interstitial water, in the mat of moss, acid by ion exchange. They adsorb cations such as Mg^{2+} and Ca^{2+} and release H^+ ions. The *Sphagnum* peat, which is easily recognizable, sometimes even to species level, from the distinctive structure of *Sphagnum* leaves, is about 3–4 m thick in the centre of the encroachment area of Malham Tarn. As it grew thicker it dried out and heather (*Calluna vulgaris*), whose woody remains and leathery flowers are preserved, colonized. In recent centuries a general drying out has allowed cotton grass (*Eriophorum vaginatum*) to invade and cotton grass peat forms the most recent peat layer.

All small temperate lakes where vegetation is encroaching have not behaved exactly as Malham Tarn has (Fig. 8.8). If local climate is dry, capillary action may provide sufficient salts to the fen peat surface to prevent acidification and bog may never invade. Alder and willow woodland (carr) may invade instead, as happens in the drier areas of eastern England and western Wisconsin. If the groundwater is acid, bog may invade immediately without the intervention of a fen stage, as happens around many American and Scandinavian lakes set in

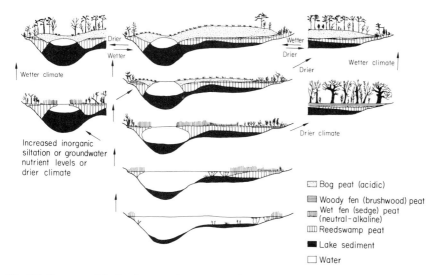

Fig. 8.8. Some possible developmental sequences of vegetation in small lake basins. The central column shows a sequence from open water through reed swamp and fen to domed bog, which might occur in a moderately wet climate and result in floating bog in the deeper parts of the basin. To each side are variations dependent on long-term climatic changes, reflected in woodland development or on erosional changes in the catchment, which might increase the ionic content of the groundwater sufficiently to retard or prevent the vegetation surface becoming sufficiently leached for acid enough conditions to develop and favour bog formation. (Based on Sinker[464].)

sandy glacial drift. Either bog or fen stages may dry out sufficiently to allow colonization of hardwood forest, or in very wet climates the bog may form the climax vegetation as happens in the maritime seaboards in Maine, British Columbia and Sweden. Indeed the bog may grow so well that it forms a floating mat or quaking bog (Fig. 8.8).

8.12 Consensus

Shallow lakes generally have higher total phosphorus levels in the water than deep ones. They also give an appearance of high fertility because of their weed-beds and the rich invertebrate fauna these often support. The higher fertility of shallow lakes is, of course, a feature mostly of the nature of their catchments which have fertile, easily eroded soils or human activities, which *inter alia* have favoured also rapid filling in. The catchments are also generally of subdued relief and the lake basins formed in them were consequently never very deep.

The apparent correlation between deepness and infertility, shallowness and fertility, has led to an association with the process of filling in of one of steadily increasing fertility, or eutrophication. This has been embodied in a long-held idea that lakes therefore become naturally more fertile with time. The most important result of palaeolimnological studies has been to counter this idea.

Sediments, themselves the agents of filling in, contain the evidence (Blea Tarn, Esthwaite, Lago di Monterosi) that over long periods lakes may become less fertile as they become more shallow. This is a much more reasonable general expectation than one of increasing fertility. The upheavals which make lake basins might be expected initially to expose readily weatherable minerals, but these are in finite supply and must eventually become leached out. Some parallel can be found in the conversion from fen to bog (see Section 8.11) and indeed the words eutrophic and oligotrophic were originally coined to describe this process.

On the other hand, just as 'natural' eutrophication is certainly not inevitable, neither is 'natural' oligotrophication. Pickerel Lake became naturally more fertile as climatic conditions changed to favour prairie rather than woodland in the surrounding catchment. The important point, however, is that this change in lake fertility was not an internal change in an undisturbed catchment. It was a change consequent on major disturbance of the catchment by climate. All of which underlies the general principle that what happens in a lake is determined by what happens in its catchment. Often related to climate, the direction of

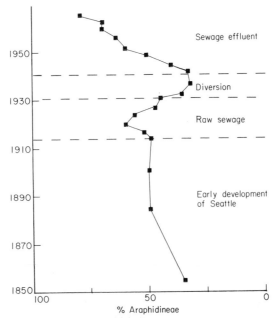

Fig. 8.9. Recent history of sewage disposal to L. Washington, Washington, U.S.A., reflected in its recent sediments. Seattle lies on the shores of the lake and was founded in the mid 19th century. Fertility of the lake is reflected in the percentage of Araphidinean fossils in the diatom assemblage of the sediments. The greater the fertility, the higher this proportion. Raw sewage was discharged to the lake in the early 20th century, but this was diverted to the sea around 1930. Later, treated sewage effluent was discharged to the lake, creating progressively more fertile conditions throughout the post-war years until the late 1960's (not shown on the graph) when the effluent was diverted and the lake became again much less fertile (see Chapter 10). (Redrawn from Stockner & Benson[483].)

change may reverse, perhaps several times in the development of an ancient lake such as the 4·5–5·0 million years old L. Biwa in Japan[223].

The major 'happening' in many lake catchments is now the activity of man. Occasionally this may lead to less fertile conditions in the lake (see Section 8.4.1) but usually it has led to increased fertility or 'artificial' eutrophication. Often this has led to undesirable consequences but is not irreversible (see Chapter 10). Man-induced eutrophication is of ancient origin. The Neolithic peoples began it in Esthwaite, the Romans in the Lago di Monterosi. The tradition has been enthusiastically continued since for increasing human populations have not only continued to disturb catchment areas through cultivation, deforestation and addition of agricultural fertilizer, but have also developed a short circuit in the earth's phosphorus cycle.

Human sewage used to be discharged to the land by means ranging from the most primitive and direct, to the septic tank, from which an effluent slowly percolated to the soil and sub-soil. The phosphorus compounds richly present in the sewage were readily precipitated and retained by the soil particles. The need for stricter public health measures as populations have increased has meant that larger volumes of sewage have had to be dealt with and the final product of modern sewage treatment is an effluent, rich in phosphate, which is most easily disposed of to the waterways. This subject is dealt with at length in Chapter 10, but for the moment Fig. 8.9 shows the precise record to be found in its sediments of the changing practises of sewage disposal around L. Washington, U.S.A., since the city of Seattle was established on its banks.

8.13 Further reading

Further literature relevant to this chapter falls into two groups—one concerned with identification of microfossils and techniques for preparation of them for analysis, and one concerned with problems of interpretation of sediment cores. In the former group are all the various keys and manuals listed for other chapters and it is unnecessary to repeat them here. The philosophy of interpretation has been covered, largely from the point of view of geologists, though this is still relevant, in Cushing & Wright[86] and Imbrie & Newell[247]. Frey[142] and Winter & Wright[541] include an all round coverage of work on lake sediments, the former in the form of symposium, and Birks & West[36], Godwin[173], Moore & Webb[351] and Pennington[397] provide sources particularly in the field of pollen analysis. Horie[223] has edited a series of collections of papers on the ancient L. Biwa in Japan and these volumes are available from the L. Biwa Research Project, Otsu, Japan. Moore & Bellamy[350] is a general account of peat and peatlands and Godwin[174] provides a fascinating account of the changes of an area of fenland in eastern England, drawn mainly from personal experience.

CHAPTER 9

FISHERIES AND FISH PRODUCTION

9.1 Introduction

The supply of fish, whether for recreation or food, is what many laymen immediately think of when aquatic habitats are mentioned. Fish production is also of great interest to limnologists and encompasses two main lines of scientific work. Firstly, there is the problem of how great is the absolute productivity of the fish community in a waterway, and its division among all the species present, and how it is related to other components of the ecosystem. Largely because of the difficulties in reliable sampling of highly mobile populations there are relatively few comprehensive data on these aspects. Secondly, there is the mass of data on fish yield, the harvest collected by man in the operation of fisheries. Yield is often wrongly called production by fisheries scientists, but in a fishery which is being maintained indefinitely it is always less than the true production. This is because not all species of fish are sought in a fishery—some are inedible, and it may be undesirable or uneconomical to try to catch the smaller ones. In a general way, however, the yield should at least reflect the production. A high yield can be derived only from a high production and a low production will limit yield.

9.2 Relationship of production to overall fertility of the ecosystem

It would be surprising if the production of fish did not bear a close relationship to the fertility of a waterway as expressed, for example, by its total phosphorus content, or some parameter related to this. Oglesby[385] has summarized data on annual fish yield (Y) as g dry wt. m^{-2} yr^{-1}, and mean summer phytoplankton chlorophyll concentration (C) in μg l^{-1} for the epilimnia of stratified lakes, or the entire water column of ones mixed to the bottom in summer, in the following regression for 19 lakes, mostly in the northern hemisphere:

$$\log_{10}Y = 1 \cdot 17 \log_{10}C - 1 \cdot 92, (r^2 = 0 \cdot 84).$$

As might be expected, a similarly significant regression was obtained with phytoplankton primary productivity (P in g C m^{-2} yr^{-1}) of the lakes:

$$\log_{10}Y = 2 \cdot 00 \log_{10}P - 6, (n = 15, r^2 = 0 \cdot 74).$$

Without doubt, these regressions ultimately reflect a causative relation, however indirect. The variability in the regression is quite large ($16 - 26\%$ of the variation is unexplained) and might be accounted for in several ways: fish yield is not the same as production, and allochthonous organic matter may significantly alter the potential food supply of the fish whose prey may not depend only on organic material synthesized in the lake.

There are other ramifications, however. Fish production depends on successful breeding (or in fisheries terms, recruitment) and this may demand a specific physical structure of the habitat—sandy beaches or weed-beds for instance. The spatial habitat was shown by Werner et al.[524] to be the major determinant of how the fish community was made up in two Michigan lakes. Small fish species seemed confined to dense areas of weed where they were less vulnerable to predation. Some general measure of the structure of the habitat is given by mean depth of the lake. The shallower this is, the more likely there is to be a greater relative availability of structural 'features' compared with that of open water, and hence the greater the proportion of spawning area is likely to be. Most fish spawn in shallow water.

Recognition of this is given in the morphoedaphic index of Ryder et al.[440], which is also reasonably correlated with fish yield (see Chapter 2). The index is calculated as the total dissolved solids content of the surface water divided by the mean depth. Total dissolved solids should generally be related to the supply of potentially limiting nutrients like phosphorus and nitrogen compounds, but may be significantly increased in brackish or peaty lakes where waters are stained with brown organic substances. These may have no positive effect on algal or aquatic plant production and may even depress it. Use of conductivity measurements (the ability of a water to conduct an electric current, hence a measure of its ionic content) instead of total dissolved solids obviates problems with organic compounds but is still misleading in one direction in saline waters and in the other in waters artificially enriched with nitrogen and phosphorus from human activities (see Chapter 10). Use of total phosphorus measurements would be preferable but such data are fewer, and total phosphorus is not easily measured, particularly away from well-equipped laboratories.

The TDS, conductivity or total phosphorus content of a water already includes in itself an allowance for lake depth, as demonstrated by the Vollenweider relationship in Chapters 1 and 2. The morphoedaphic index thus contains an extra weighting for depth and in theory should give a closer relationship with fish production than a simple regression with chlorophyll, phytoplankton productivity, or total phosphorus concentration. Oglesby[385] found a slightly less good relationship using the index with TDS or conductivity, but a more critical comparison using total phosphorus might show otherwise.

For the moment the extra importance of the depth relationship can only be guessed at, but a number of studies suggest that any relationship attempting to predict fish yield from physical and chemical parameters may need to be much more complex than those simply relating water chemistry and production at lower trophic levels. Fig. 9.1 shows the relationship between annual yield of fish and mean lake depth in a series of African and Canadian lakes. A hyperbolic function is found which becomes linear when both co-ordinates are expressed

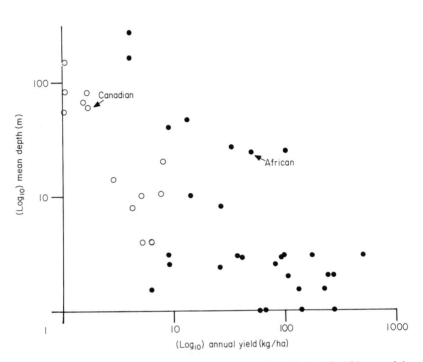

Fig. 9.1. Relationship between depth and fish yield in series of (tropical) African and (temperate) Canadian lakes. (Plotted from data in Fryer & Iles[149] and Rawson[420].)

Table 9.1. Fish yields in fresh waters. Yields are given in g (fresh wt.) m⁻²yr⁻¹. Dry weights are about 25% of these.

Locality	Latitude	Yield (*if production)	Reference
Bere Stream	50° 45′N	19·8–44·0*	Mann[327]
Devil's Water	50° 45′N	30*	Mann[327]
Docken's Water	50° 45′N	14*	Mann[327]
R. Thames	51° 30′N	43*	Mann[326]
Shelligan Burn	56° 20′N	12·4–25*	Egglishaw[123]
English Lake District and			
Pennine Rivers	55°–55°N	6–10*	Le Cren[285]
L. Naroch, U.S.S.R.	55° 25′N	6·1	Winberg *et al.*[540]
L. Batorin	55° 25′N	11·8	Winberg *et al.*[540]
Loch Leven	56° 10′N	7·3–71·3	Thorpe[497]
L. Sabanilla, Cuba	23°N	22*	Holčik[216]
L. Luisa, Cuba	23′N	27·6*	Holčik[216]
Gompak R., Malaysia	3°N	90	Bishop[37]
L. George, Uganda	0°	90–150	Dunn[112]
African lakes other than L. George	5°N–15°S	0·4–50	Lowe-McConnell[300]

logarithmically. However, the scales are quite different for the two series. For a given depth the African lakes have much greater yield than the Canadian ones —by at least an order of magnitude. Perhaps this is fortuitous and reflects more intensive fisheries in Africa, where fish is a major subsistence protein food; or perhaps it reflects the greater numbers of herbivorous fish in tropical lakes; or perhaps again there is some influence of the more muted seasonality where growth of underwater plant cover is concerned in tropical waters. Table 9.1 shows that tropical lakes on average might yield more fish than temperate ones (as expected from the models discussed in Chapter 1) but that a considerable overlap occurs and a tenfold difference is certainly not found. Perhaps the whole confusing situation stems from attempts to analyse data which are not yet sufficiently sophisticated.

9.2.1 Measurement of fish production

The measurement of fish production is not an easy business, mainly because of the problem of sampling a most unevenly distributed population. The sample taken is likely to constitute a significant part of the population in a river section or small water body and therefore most of it must be returned alive after measurement. This is particularly important when the fish are about to spawn.

Fish production may be described in the same terms as those for other animals:

Ingestion − Egestion = Assimilation = Respiration
+ Production of new tissue and gametes + Excretion.

Production is estimated as the increase in weight per unit time, due allowance being given to the fact that losses to predators and disease must be included, as must be the large proportion of the body weight that the gametes comprise, particularly in females.

The first problem is to estimate the absolute population size per unit area of waterway. It may sometimes be possible to sample the entire population in a pond by draining out most of the water and concentrating the fish into a small volume where they may be scooped out. This, however, is destructive of habitat and is unsuitable for most investigations. The entire population may also be poisoned by rotenone (an extract of the roots of a higher plant, *Derris* spp. Leguminosae) or other toxins. The dead fish float to the surface or are netted from the bottom. Clearly this can only be the terminal sampling of an investigation.

Electrofishing, in which an electric field created in the water stuns the fish, may be used to recover almost the entire population in a small, shallow area such as a stream stretch. The method is potentially dangerous to the operators as a quite powerful alternating current or direct current is created between two electrodes, or a single electrode and the ground. Too strong a current will kill the fish. With AC fields the fish are stunned and must quickly be collected in nets as they rise to the surface. With DC fields they move towards one electrode and must be removed from the water before they are killed by touching it. AC operation is particularly useful in turbid water where the fish cannot easily be seen.

The most widely used method of sampling fish is to use a net or trap which removes a sample of the population. These fish are marked or tagged in some way to make them identifiable then released again. After a period to allow them (it is hoped) to mix randomly with the rest of the population, this is again sampled and the number of marked and unmarked fish counted. The ratio of recaptured marked fish (n_{rm}) to the total number recovered in the second sampling (N^1) is supposed to be equal to the ratio of the number of fish originally marked (n_m) to the total population (N):

$$N = \frac{n_m \cdot N^1}{n_{rm}}$$

N may be placed on an areal basis from the area of the lake. In a big lake the population may move only over a restricted area which must be determined by marking fish and then sampling over a wide area to discover the extent to which the marked fish have moved. Practical problems involved are in ensuring that the samples are not selective of particular fish sizes, and that the marking technique does not alter fish behaviour or increase the chance of death. These ideals are probably never attained.

Where fish are moving up or down river on migration it is relatively easy to catch an unbiased sample by traps which span the entire width of the river. Where lakes are concerned seine nets are useful. These involve the laying out of a small meshed net in a wide arc with the ends of the arc at the shoreline of a suitable shelving beach. The top of the net has floats to keep it at the surface and the bottom of the net is weighted to hold it at the bottom. After setting, the net is hauled in from the shore. Seines may also be set in the open water, but must have a rope threaded into the bottom so that as the net is drawn in, it is closed at the bottom into a 'purse'. This requires a mechanical winch on the boat used. Fish escape around seine nets but this may be minimized by an elongated sock of netting bulging out at the centre and called a cod end. The fish tend to move into this as the net is hauled in.

Trawling is also a relatively non-selective method for bottom-living fish. The trawl is a bag of netting, kept open either by its resistance to moving through the water or by a wooden beam across its entrance, and pulled by a sufficiently powerful boat. Trawling, however, at least temporarily destroys the bottom habitat and is impractical in weedy areas. It may also radically interfere with the breeding of fishes spawning in nests on the bottom.

Gill nets are highly selective. The net is floated passively in the water at a desired depth and fish moving against it may move through it and pass out the other side, or they may be too big to penetrate it at all, or they may move part through it until the widest part of their body becomes jammed and their gill covers catch on the netting as they try to move back. Such nets, dependent on their mesh size and to some extent the slackness with which the net is set, are very selective for particular size ranges. A set (or fleet) of different mesh sizes may be used for experimental fishing, but most fish are damaged on recovery. Marking and tagging take many forms. Some of them do not discriminate one marked fish from the next and include notches or holes cut in the fins—these may grow out after a time but are usually recognizable for several years—or subcutaneous injection of dyes, though these fade progressively. Tags are small plastic or metal plates, which may be individually numbered, and which are surgically inserted into the skin or muscles, through the opercula, fins, or lower jaw. The tags are sometimes lost during the fishes' life and almost certainly the shock of their insertion and the slight impediment they convey increase the chances of death of the tagged fish by predation or disease. Fin clipping in which a small notch is cut from one or more of the fins, is probably the best compromise method. A certain degree of individual recognition, of samples clipped at different times, for example, can be attained, with minimal interference with the fish.

9.2.2 Growth measurement

By use of a marked sample the fortunes of a fish population can be followed throughout life. The individuals of most temperate fish species spawn over a limited period of the year and the newly spawned generation or cohort can be treated as a unit recognizable in successive years because of the annual or other periodic rings laid down in the skin scales or in the otoliths (the ear bones). When a population is sampled it is held in an aerated tank while each individual, or a random sample if there are very many, is weighed and its length measured. A length/weight graph may be used to avoid the need to weigh as well as measure in future samplings. The fish are usually lightly anaesthetized to calm them during these measurements. Scales may be removed from a part of the body, often high on the back just below the dorsal fin, where their rings are known to be clear. If this is done carefully with blunt forceps the fish should suffer little damage, though its chances of becoming infected by fungi, protozoa or bacteria subsequently may be increased. Removal of otoliths requires dissection and is fatal.

The scales may be used simply to age the population and identify separately the different cohorts (often called year classes) or they may be used on an old fish to estimate the growth in previous years, for the distance between the rings is believed to be proportional to the growth in the year the ring was laid down. Rings, however, may be reabsorbed during periods of starvation.

For a given cohort an Allen curve (see Chapter 4, Fig. 4.5) may be plotted relating the mean weight of fish to the numbers remaining in the cohort. At first there are very many small fry, but as individual weights increase the numbers decline. The area under the curve as it approaches the abscissa when the last survivor dies gives the total production, and production in a given year can be found by determining the area under the curve between points representing the times in question. The curve is roughly hyperbolic, but has irregularities—numbers decline in winter in temperate lakes though the increase in weight of the survivors may be negligible. It may even be negative as fat stores laid down the previous summer are used up. A similar decrease in weight occurs during spawning when the weight converted to gametes can be determined from the area under the curve as it reverses direction during the spawning period. By adding up the production of each of the several year classes in a given year, and by doing this for each species, a measure of the total fish production may be obtained.

Use of scale ring measurements allows determination of the changes in growth of individual fish from year to year, and may be used to relate this to the vicissitudes of the environment. It cannot be used to calculate total production, however, unless population numbers are known in each past year. Since so little extra effort is needed to weigh the fish once captured, the 'Allen curve' method must usually be preferred, for production studies, to scale measurements which are more tedious to make.

9.3 Inland fisheries

Bald statistics can be very misleading—only a few per cent of the world's protein food comes from freshwater fisheries, but most of the world's protein is produced as meat, fish and eggs and consumed by the few hundred million people of the western industrialized world. The bulk of the protein that is available to most of the world's population is in the form of fish. In some coastal areas the bulk is of marine fish, but deep in the continents of Africa, South America and Asia freshwater fish are critically important. In Africa and South America most of the fish are caught from wild populations; in Asia culture in ponds has been important for perhaps 4000 years and pond culture is also skilfully managed in Eastern Europe. From L. Victoria in 1970 as much as 100 000 tonnes (fresh weight) were harvested, compared with only 25 000 tonnes of marine fish from the nearby East African seaboard; though catch statistics are difficult to collect from primitive areas, perhaps as much as 1 400 000 tonnes yr^{-1} of fish are taken from African waters, 40% of it from rivers and flood plains[300]. In comparison, commercial fisheries in inland lakes in the Western World, for example in the St. Laurence Great Lakes of the U.S.A. and Canada, are of minor significance. Commercial fisheries in the west for prized fish such as the Atlantic salmon and the eel serve a luxury market, though they are of great importance to local economics. Fisheries in Europe and North America are most importantly recreational fisheries and in economic terms may be very big business overall. In 1972 it was estimated[379] that perhaps £160 million were spent on angling in England and Wales by almost 3 million fishermen. The total population was about 50 million, so angling is probably the major British participant sport. The costs of renting fishing rights and of buying land at the edge of good fishing waters is also very high—for salmon fishing the cost was as high as £750 ($1500) per fish caught per year even in 1971.

Except where the fisheries constitute a small part of the ecosystem, as in the case of many remote tribes in the tropics, fishing must be actively managed if a reasonable harvest is to be sustained from year to year, for fishing represents an increase in mortality rates for populations where this had previously balanced their birth and growth rates. It is important that the fishing mortality should only replace natural mortality or, if it increases total mortality, it should do so only to the extent that there can be compensatory increase in growth and recruitment.

9.3.1 Primitive fisheries

Subsistence fisheries are to be found everywhere in the tropics and were also common in temperate regions before the industrialization of modern agriculture increased supplies of land-produced protein. Indeed the methods used have a world-wide similarity, being variously based on the rod and line, spear or lance, basket trap or some form of movable fence or primitive net. Nets are, by and large, a relatively modern development, for natural fibres rot easily and although cotton nets were introduced to Africa by the Arabs in the 19th century, it is the lightness and durability of man-made fibres that has given most impetus for conversion of primitive fisheries to commercially based ones. Early synthetic nets were (and still are) made in parts of Africa from the rayon picked from the casing of worn out car tyres!

Primitive tribes have possibly been well integrated with their ecosystem for thousands of years and a thorough quantitative ecological study of these rela-tionships for a fishing tribe would be very interesting. Thesiger's study[495] of the Arab tribes of the Tigris–Euphrates marshes hints at close relationships with the use of marsh plants, fish and the domesticated water-buffalo, and Rzóska[441] summarizes the yearly migration of Nilotic tribes of the upper Nile Swamps as they move their cattle in rhythm with the seasonal flooding of the riverine grass-lands. Some clue to the rich detail that might be unearthed comes from an account of Africa in the 1930's by Worthington & Worthington[544]. It serves also to describe some 'primitive' fishing methods which belie a rather close knowledge of the principles of freshwater ecology.

On L. Albert, the Banyoro tribe collect grass or brushwood and tie it into bundles which they lower to the lake-bed in 6–10 m of water attached to a line buoyed by pieces of ambatch wood. Ambatch (*Aeschynomene profundis*) is a leguminous swamp tree with a light corky wood used also for making canoes. Overnight the bundles become colonized by small fish, mainly *Haplochromis* species, presumably taking cover from predators. The bundles are hauled up and the *Haplochromis* baited alive on small hooks on lines with which the larger tiger fish (*Hydrocynus vittatus*) are caught. In turn, the tiger fish are used to bait large barbed hooks to catch nile perch, the ultimate quarry, for its flesh is prized and the fish are large.

Baskets, woven from papyrus or other reed-like stems or pliant tree branches are widely used. Women of the Jaluo on N.E. L. Victoria move into shallow water in groups of seven or eight, each with a small basket on her head and a much larger, wide-mouthed basket to hand. The women converge in a circle and simultaneously sweep the large baskets through the encircled water, scooping out the fish and depositing them in the head baskets. Pelicans feeding in the same area apparently employ similar methods—driving fish into a small area which they surround before scooping with their mouths.

The Jaluo also use baskets together with a several hundred metre long fence of woven papyrus stems which they haul in like a beach seine net. As the fence comes close to the shore a large number of baskets constructed with a non-return internal passage are put in the water. Fish concentrated by the fence move into them, and may then be readily removed to land. Such fences and baskets are also widely used to catch river fish which are forced to move into the baskets as they swim up or down river.

Knowledge of the habits of river fish is used also by the Baganda at the Ripon falls (Uganda) where barbel (*Barbus* spp.) move upstream. At eddies and hollows in the rapids, where exhausted fish congregate to rest, groups of men periodically spear the fish or catch them by dragging a hook through the water. This and the technique of poisoning fish with extracts of various plants, e.g. some *Euphorbia* spp., constitute perhaps the least sophisticated of fishing methods.

Fish caught are cooked and eaten fresh, or, more often, dried in the sun after removal of their guts and scales. Sometimes they are smoked and the product is edible for a few days before it rots. Dried fish are used also for barter and may be distributed in the hinterland by foot or bicycle, where larger transport cannot penetrate. The system is efficient on this scale in providing people with protein over a small area. More widespread use of the fish, and the supply of larger and more distant markets has meant much greater organization of the fisheries. Not only should they be managed, but large catches may mean large nets, mechanical aids, larger boats, docking facilities and cold storage. Economic considerations then become as important as ecological ones in running the fisheries. The tropical fisheries may be entering this stage, as trawling is contemplated to exploit fishes not presently touched by the usual gill net commercial fisheries of African lakes, but for the moment, the L. Victoria *Tilapia* fishery discussed below is representative.

9.3.2 Commercial fisheries

Fisheries do not differ from any other commercial venture in that the aim of their management must be to maximize production, the yield of harvestable fish. The particular problem of fishing, however, is to sustain the maximum yield from year to year by harvesting no more than the equivalent of the annual increment of fishes becoming large enough to be worth catching plus the annual growth of those already fishable. It is relatively easy for fishermen when not subject to control to remove more than this annual recruitment and growth, in which case the fishery will eventually become extinct.

Fig. 9.2 shows the increase in biomass of a newly-hatched cohort of fishes. The rate of increase is at first high but declines as the maximum potential biomass, represented by the curve's asymptote, is reached. Simultaneously some members of the cohort die and the rate of mortality is at first high, but declines

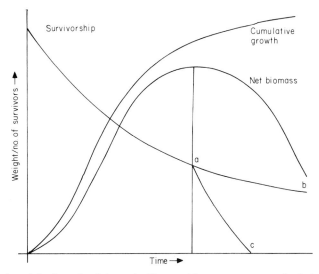

Fig. 9.2. Ideal exploitation of a fish stock. The net biomass represents the balance between cumulative growth and death, and at some time reaches a peak. Provided spawning has by then occurred the peak represents the time from which most efficient removal of fish can take place. The natural survivorship curve, ab, is replaced by the steeper fishing mortality curve, ac, as fishing removes biomass before it can be further diverted to disease organisms, parasites or other predators. (Based on Fryer & Iles[149].)

as the fishes become older, bigger and less vulnerable to predators. The net effect of growth and death is to create the biomass curve; growth exceeds death in the early part of the cohort's existence, but eventually mortality predominates. There is a point (a) at which the biomass of the cohort is maximal. The aim of a fishery is to catch that portion of the biomass which, after the maximum has been attained, would be lost to natural mortality. The more rapidly the fish can be removed after the maximum the better. The intention is to replace the natural survival curve ab with the fishing mortality curve ac, the area between the curves representing the yield. It is important that fishing does not remove fishes of an age at which growth exceeds natural mortality, for this will reduce the potential yield and may also remove fish that have not spawned. This may reduce the potential recruitment to the next season's fishery and is called overfishing. It is also important that fishing mortality should only ever replace natural mortality, and never exceed it. Fish that can safely be removed are thus mature and should not be allowed to grow to their potential maximum size—by which time most of the biomass of the cohort will have been lost by natural mortality.

The year to year changes in fish population can thus be represented by[439]:

$$P_2 = P_1 + (R + G) - (F + M)$$

where P_1 and P_2 are the fish stocks (biomass) in two successive years, R is the annual recruitment of mature fishes to the fishery, G is the growth made by those already fishable but not yet removed, F is the annual mortality due to fishing and M the natural annual mortality. In a well-run fishery, ideally $P_1 = P_2$ and $M = 0$, but this is never achieved, for natural environmental fluctuations will alter R and G in ways that a fisheries manager cannot predict in time for him to regulate the amount of fishing (F). P should, however, fluctuate around a mean without showing a tendency to increase (underfishing) or decrease (overfishing). The year to year changes in a fishery can be monitored and fishing methods regulated through devices such as controls on net mesh size, number of nets allowed, and season of fishing, to keep F at the desired level.

This is done by thorough sets of equations, or models, such as that outlined in the next section. Much sophistication has been achieved in the development of such models for marine fisheries[83, 84, 185] where very good data can be obtained from fish landings at a very few ports. Most commercial freshwater fisheries, however, are subtropical or tropical and involve large numbers of fishermen operating individually or in small groups around the indented margins of large lakes. Catch statistics are then not easy to obtain. A fishery for *Sarotherodon* (*Tilapia*) *esculentus* (Graham) in L. Victoria, however, is notable in that landings are made over a restricted area where fishery scientists managed to account for changes in the population[158, 159]. The fishery has now been over-exploited[145] as a result of the ignoring of scientific predictions for political and commercial reasons, and thus provides a doubly interesting example.

9.3.3 The North Buvuma Island fishery

The first essential of fishery management is that the fished population should be recognizable and discrete. In a small lake the entire population of a given species may provide recruits, but in a large lake there may be several different stocks or sub-populations, separated by geographical barriers, such as deep water, for inshore species, or stretches of swamp. Such sub-populations may be distinguished on the basis of slightly differing blood proteins, detected serologically, by minor morphological differences, or if neither of these exist, by careful observation of the movements of tagged fish. The absolute size of such a 'unit stock' may be determined by the mark, release, recapture method outlined in Section 9.2.1 but an absolute measure of population size, as opposed to a relative one, is not usually necessary for fisheries management. The catch obtained (numbers or weight landed) per unit fishing effort (hours of fishing, numbers of nets set per night, total length of nets set, or numbers of trawls made), called the stock density, is such a relative measure. Proof that catch per unit effort is a measure of population size is necessary, for almost all fishery management is based on this assumption.

Let N_x be the number of fishes in a cohort at the beginning of a season's fishing and N_{x+1} those remaining to join the next season's stock.

Now,

$$N_{x+1} = N_x e^{-Z}$$

where Z is the coefficient (a fraction) of total mortality per season, the sum of the coefficients of mortality due to fishing (F) and to natural causes (M). An exponential mortality seems reasonably to fit practical experience.

The number of fish dying per year is:

$$N_x - N_{x+1} = N_x(1 - e^{-Z}).$$

The catch or yield, Y, to the fishery from the cohort is:

$$Y = F/Z . N_x (1 - e^{-Z})$$

and if several cohorts, numbering s, are exploited in the fishery, the total yield Y' is:

$$Y' = F/Z \, N_x (1 - e^{-Zs}).$$

The catch per unit effort is proportional to Y'/F, for the fishing mortality, F, reflects the amount of fishing done.

$$Y'/F = \frac{N_x}{Z} (1 - e^{-Zs}).$$

This relates the catch per unit effort to numbers of fishes in the fishable population and may be elaborated as follows.

The fishing mortality, F, is a function of the fishing intensity, f, that is the effort, e, expended per unit area, A, and the index of catchability, q. q is a measure of how vulnerable the fish are to the particular gear used.

$$F = fq = \frac{e}{A} . q$$

and

$$q = \frac{FA}{e}$$

Since

$$F . N_x \propto Y' \text{ and } F \propto \frac{Y'}{N_x}$$

then

$$q = \frac{Y'}{N_x} . \frac{A}{e} = \frac{Y'}{e} . \frac{A}{N_x} = \frac{d}{D}$$

where d is a function of the catch per unit effort and D is the absolute population density of the fishes N_x/A. Since

$$d \propto Dq$$

and $d = Y'/e$, and since q is a constant for the gear used, catch per unit effort is therefore proportional to D.

The aim of management is to predict future populations of the fish and the effects on them of different fishing intensities. This is best done separately for each cohort, and depends on recognition of separate cohorts. In temperate regions annual scale or other rings enable this, and length of fish is usually well correlated with age. For tropical fishes like the Buvuma *Sarotherodon* the scale rings may be irregular or laid down at six-monthly spawning intervals. For practical management of the *Sarotherodon* fishery it was most convenient to consider the fish in successive length classes rather than age classes.

Consider a cohort of fish of a given length range, e.g. 23·0–23·9 cm, designated as b. N_b is the number of such fish and they die at the rate M_b from natural causes and F_b from the fishery. It takes a time t_b for a fish to grow through the length range b, so that the number of fish surviving into the next length group, (b + 1) is:

$$N_{b+1} = N_b e^{-(F_b + M_b)t_b}$$

This represents an instantaneous abundance, for F and M are instantaneous mortality rates. A measure of the average abundance $\overline{N}_b$ is given by integration of the above equations for all lengths in the range considered:

$$\overline{N}_b = \frac{N_b e}{F_b + M_b} (1 - e^{-(F_b + M_b)t_b})$$

The problem now becomes one of estimating t, M, and F so that N_b can be predicted for successive length groups until the cohort is completely removed. N_b is given by catch per unit effort after a correction has been applied for any differences in selectivity of different sorts of fishing gear used.

Estimation of t_b, F_b and M_b for the Buvuma Sarotherodon *factory*
t_b can be calculated from measurement of the rate of growth of the fish. Fish grow proportionately more slowly the bigger they become and this growth is often mathematically expressed by the von Bertalannfy equation:

$$L_{b+1} = L_\infty - (L_\infty - L_b)e^{-Kt_b}$$

where L is length, and L_∞ the asymptotic maximum length attained by the fish as its growth rate drops in old age. K is a constant defining the rate of deceleration of growth. Fish in the length groups b, (b + 1) etc., must be aged and L_∞ established from a plot of L against age. K can be calculated from rearrangement of the equation, expressed graphically and an estimate of t_b can then be obtained.

Fishing mortality is related to fishing effort and a graph can be plotted of total mortality for a length or age class against effort (Fig. 9.3) from several years data. Total mortality is given by:

$$N_{b+1} = N_b e^{-Z}$$

where Z is the coefficient of total mortality and N_{b+1} and N_b are the numbers (given by catch per unit effort) of the cohort at successive times. The intercept on the total mortality axis gives the mortality in the absence of fishing, M, and the curve gives the fishing mortality, F, for any given degree of effort.

Fishing mortality can also be inferred by tagging fish and finding the proportion of tags recovered by the fishery. The fishing mortality in the *Sarotherodon esculentus* fishery can be regulated very closely by the use of different sizes of mesh in the gill nets that are used. For example $4\frac{1}{2}$ in (11·4 cm) nylon gill nets capture fish of mean length 29·3 cm with a standard deviation of 1·48 cm.

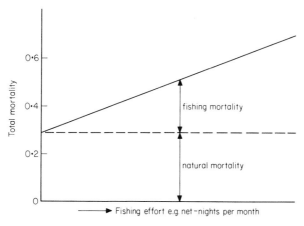

Fig. 9.3. The determination of natural mortality from the relationship between total mortality and fishing effort.

The original equation for N_b, above, can now be solved for all length groups (approximately age cohorts) in the fishery and the following parameters calculated for future periods of the fishery: biomass, the product of abundance and average weight of the length group as a total for all length groups; numerical catch, the average abundance multiplied by the relevant fishing mortality in each length group; catch in weight, the product of numerical catch and mean weight as a total for each length group.

Table 9.2 gives values for M_b and F_b for the L. Victoria *Sarotherodon esculentus* fishery. It can be seen that 5 in mesh (14·2 cm) does not catch fish

Table 9.2. Mortality and net selectivity in the Buvuma Island *Sarotherodon* fishery. (From Garrod[158].)

Length (cm)	Total mortality	Natural mortality	Fishing mortality	% of catch in a particular length category for nets of particular mesh size		
	per year	per year	per year			
(b)	(Z_b)	(M_b)	(F_b)	4 in	4·5 in	5 in
23·0–23·9	0·275	0·01	0·265	9·9		
24·0–24·9	0·307	0·01	0·297	38·6		
25·0–25·9	0·325	0·01	0·315	82·5	4·4	
26·0–26·9	0·372	0·01	0·362	98·9	19·1	
27·0–27·9	0·396	0·016	0·380	65·5	51·6	1·0
28·0–28·9	0·465	0·024	0·441	24·0	89·5	11·0
29·0–29·9	0·517	0·122	0·395	4·9	98·0	32·0
30·0–30·9	0·625	0·266	0·359	0·5	82·6	73·5
31·0–31·9	0·761	0·536	0·225		46·9	99·0
32·0–32·9	0·929	0·790	0·139		25·5	91·5
33·0–33·9	1·314	1·018	0·296		10·2	63·5
34·0–34·9	1·996	1·124	0·872		4·1	34·0
35·0–35·9	4·954	1·138	3·816			17·5

below 26·9 cm length but is very effective in killing the 31·0–31·9 cm category, whereas 4 in mesh (11·4 cm) kills fish optimally at length 26·0–26·9 cm and not at all above 31 cm.

Fig. 9.4 shows the effects of various levels of fishing effort (which is proportional to fishing mortality) on the steady state biomass of the stock, the weight of catch, and the catch per unit effort. The effects are more extreme at the smaller mesh sizes because they crop fish whose growth rate is still higher than the natural mortality.

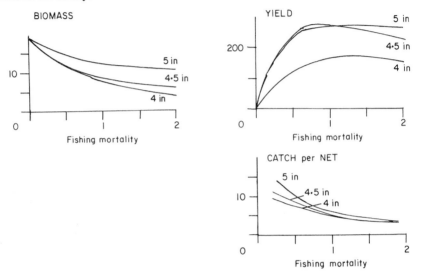

Fig. 9.4. Predictions of the fishery model for the Buvuma Island *Sarotherodon* fishery for different levels of fishing mortality (related to fishing effort) and different mesh sizes of gill net. (After Garrod[158].)

There would thus be advantages in using only 5 in mesh nets for these keep the population biomass high, thus favouring the maintenance of spawning and recruitment and keeping the fishing mortality high, between 0·5 and 1·0. Decreasing the mesh size further does not increase the catch, but decreases the stock slightly and the catch per unit effort markedly. This means an uneconomic use of labour and gear. For the fishery in question, F = 1·0 is equivalent to the setting of 46 000 net-nights per month or about 1500 nets per night. Each net is a standard length.

For various reasons the fishery has not been so rationally exploited as it might have been with the help of Garrod's model. Mesh sizes much smaller than 5 in have been used and events have taken the following course (Fig. 9.5).

Between 1953 and 1956 the fishery was fairly stable, with F = 0·3 and 153 000 net-nights fished per month. The model shows that a catch per net of 1·44 fish was obtained, which was marginally profitable for the fishermen. It was not the maximum yield that could be sustained biologically, but the amount of fishing needed to attain this is often uneconomical. In 1955 and 1956, for unknown reasons (perhaps poor recruitment) the catch per net fell. Legislation in force forbade nets of less than 5 in mesh and had been in force since the 1930's. Illegal $4\frac{1}{2}$ in nets were used and the yield temporarily went up because a section of the population previously unexploited was now being fished. The minimum 5 in net legislation was repealed.

Under the new conditions the model would predict a catch of 1·8 fish per net-night and catches of around 2·0 per net-night actually recorded provide part of the validation of the model. The increased catch provided incentive to

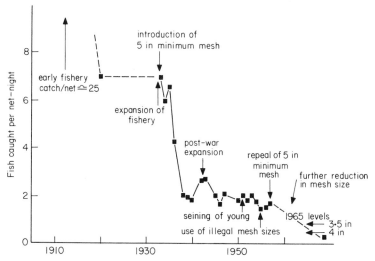

Fig. 9.5. History of the *Tilapia* (*Sarotherodon*) *esculenta* fishery in the Kavirondo Gulf, L. Victoria. (Redrawn from Fryer[145].)

the fishery, which led to more nets and a renewed decrease in catch per net to 1·36. This was similar to that previously, but with a 25% reduction in biomass of the stock. It is believed that this reduced the rate of recruitment to marginal levels.

In recent years both 4 in and $3\frac{1}{2}$ in nets have been used! These certainly remove some fish before they have begun to breed and must mean that the yield of the fishery can now only be temporarily sustained at a very low level. The catch per net was 0·35 in 1968. More recent statistics are not available[145].

9.3.4 Recreational fisheries, hatcheries and pond culture

In Great Britain freshwater angling is a major sport. In North America and mainland Europe it is equally important recreationally, while the fish caught are more usually eaten rather than retained in a keep net and thrown back at the end of the day. The British taste runs mostly to the Salmonid fish, the Cyprinids and other 'coarse' fish being considered to have a muddy flavour. Partly this is true; for some species it is prejudice! Nonetheless sport fishing probably has a negligible effect on the species sought compared with environmental fluctuations and other human activities of man which pollute or seriously modify the habitat (see Chapters 10 & 12). Recreational fisheries fall into two categories—game and coarse. Game fish (salmon, trout) are actively fished. The bait is often a replica in feathers and wire of the adult of an aquatic insect which is cast into the water at a spot where the fisherman's experience tells him the fish may lurk ready to strike at his artificial fly. Game fishing for salmon is a sport of the upland regions largely, but brown trout fishing is countrywide in Britain. Anglers of coarse fish (unfortunately called coarse fishermen) are lurking predators rather than active hunters. Their bait is usually real food—worms, or maggots, or bread and flour pastes, and they may throw extra bait into the water to attract fish to their vicinity. Surveys have shown that two-thirds of British fishermen (about 3 million people in all) seek coarse fish and prize a dense population of well-grown, hungry fish of a variety of species[284, 286]. Crowded and hungry fish, however, are rarely well grown. There would appear[379] to be a greater demand for pike, tench, carp and barbel than the available supply.

Management of recreational fisheries is largely one of habitat maintenance—restriction of pollutants, diversion of conflicting interests such as water skiing, conservation of weedy spawning areas and provision of suitable bank edge habitat for the fishermen themselves. There are 'close' seasons, during spawning, when fishing is not permitted and licensing systems give some check on numbers of fishermen, though their main function is to finance restocking and fisherman-amenities. There is relatively little that can be done to regulate natural populations when they are decimated by outbreaks of disease or decreased recruitment, except in overall habitat management. Stocking an established waterway with fry is generally self-defeating unless a major kill of preexisting fish by an isolated pollution incident has occurred, or when a new lake, formed by flooding of a quarry or gravel pit for example, has become available.

There has been a tendency, however, in recent years in Great Britain to remove the natural fish fauna of small privately owned lakes by poisoning with rotenone, and to replace it with monospecific populations of brown trout or rainbow trout (*Salmo gairdneri*)[323, 324]. The latter is not endemic to Britain but grows well in waters which would not otherwise easily support game fish. It does not usually breed where it is stocked, though there are some isolated breeding populations, and must usually be annually restocked from hatcheries and fish farms. A self-sustaining coarse fishery costing little to operate is thus replaced by a game fishery at the cost of restocking on the one hand and loss of a diverse fish (and also amphibian) fauna on the other. Such schemes are usually run by fishing clubs.

Hatcheries and fish farms have also been established to provide table fish, again usually trout, in Britain though coarse fish like carp, prized some centuries ago in England, are skilfully cultivated in Eastern and Central Europe. The management of such farms must be carefully planned particularly for salmonid fish, whose tolerances to low oxygen levels (see Chapter 7) are not great. Overfeeding and the accumulation of uneaten food and faeces may lead to disastrous deoxygenation at night, and the relatively dense monospecific fish populations like all domesticated animals, may suffer epidemics of protozoan, helminth, bacterial, viral and fungal parasites and diseases. Adult fish are usually carefully stripped of their milt and ova just before they are ready to spawn and fertilization and egg rearing is carefully controlled in hatcheries to give as high a survival of young fish as possible.

It is in Asia, however, that fish culture is most widely practised. The fish cultured are coarse fish and management varies from the most carefully controlled to an almost incidental facet of village life. The ponds are usually fertilized with organic matter. Ideally this should be sterilized, but sometimes it comprises human excreta and permits maintenance of various parasite cycles (see Chapter 11) linking humans and fish. Such fish culture, however, may produce very high yields and in China an empirical understanding of the aquatic ecosystem has led to stocking of fertilized ponds with several species together. The white bighead (*Hypophthalamichthys molitrix* (Val.)), the striped bighead (*Aristichthys nobilis* Richardson), the grass-carp and the Chinese black roach (*Mylopharyngodon piceus* Richardson) are cyprinid fish which respectively eat phytoplankton, zooplankton, aquatic macrophytes and benthic invertebrates. There can be little more efficient conversion of food to fish flesh than in such a pond, and such polyculture may point one way to maximizing the supply of protein in densely populated areas.

Fisheries are but one way in which fresh waters are affected by man. The next three chapters discuss others.

9.4 Further reading

A standard handbook of methodology for the determination of fish growth and production is Bagenal[16] which also lists reference works for fish identification world-wide; and a wide survey of the ecological basis of fish production is Gerking[166]. Mills[347] is a comprehensive study of salmon and trout management, and Behrendt[26] a perhaps less detached but highly interesting practical account of private fishery management. Cushing[84] is a relatively advanced treatment of the dynamics of marine fisheries and Cushing[85] a simpler précis. The management of inland fisheries in Africa is discussed by Fryer & Iles[149] in some detail, and the assessment of fishery potential by Henderson et al.[208]. Relevant journals are the *Transactions of the American Fisheries Society, the Journal of Fisheries Research Board of Canada, the Journal of Fisheries Management, Journal of Fish Biology* and *Journal of Fish Diseases.*

CHAPTER 10
THE USES OF WATERWAYS;
POLLUTION, WASTE DISPOSAL AND
WATER SUPPLY

10.1 Introduction

Water was crucial in the evolution of living organisms; their biochemical systems are all adapted for function in an aqueous medium. Our developed societies have also followed this dependence in parallel ways. We each use on average more than 5 m^3 of water per day, or more than sixty times the volume of our own bodies. The figure includes use of water for drinking, cooking, laundry, bathing, domestic and industrial waste disposal, the cooling of electricity generating plants at power stations, and, in dry areas, irrigation. Only about a tenth of the total is used for purely domestic use, but even this amount (500 l) contrasts strongly with less than 10 l per day used per person in most primitive human societies and probably only about 1–2 l used by African bushmen.

Some of the water we use comes relatively uncontaminated from upland lakes and rivers, but in the lowlands, which are always the more densely populated, water must be used and reused many times as it passes ultimately from atmosphere to land to sea in the turn of the hydrological cycle. It is not that there is an overall shortage of fresh water in the world but most of it is contained in lakes and rivers remote from the centres of population and transport of it from them is expensive or impracticable. Much effort therefore goes into treating effluent or 'used' water before discharging it back to lakes or rivers and in controlling its quality. Ecological principles underly all these processes, and the word 'pollution' has been used to cover all those changes to water which reduce the range of uses to which previously the water could have been put. A polluted water is a less versatile medium than an unpolluted one. If we are to use water, as we must, to retain our current, comfortable standard of living, then some pollution is inevitable. Technology is available to remove any pollutant at its source, but may be very expensive. The level of pollution to be tolerated depends on the future use to which the water must be put and is determined by a balance between this, economic feasibility and sometimes aesthetics.

Though not actually being processed through houses, factories and farms, water may still be used by societies. Hydroelectric power, commercial fisheries, drinking water storage, navigation and recreation all demand water of greater or lesser quality, and such needs must also be taken into account in determining acceptable pollution levels. Particularly in managing fisheries and drinking water supplies an understanding of limnology is essential, and the decisions that must be made are difficult. The English Midlands, for example, with their high populations, important industries and abundance of potentially polluting effluents are also one of the strongholds of recreational fishing, and although some rivers there are grossly and unacceptably polluted, many have been managed in order to accommodate both buoyant industry and vigorous fish.

10.2 Pollutants

The most common pollutants of water are high levels of organic matter, plant nutrients, suspended mineral particles, deoxygenating substances and heat, and even small quantities of poisons like heavy metals, pesticides, some organic chemicals, acids and radioactive substances. Few are entirely new to living organisms, for, with the exceptions of some synthetic chemicals, natural discharges of all of them occur somewhere in the world, and specialized communities of living organisms have evolved to tolerate them. The communities which develop are generally not diverse, for the habitats created are, by the standards of most waters, 'extreme'. Conditions created by pollution from human activities tend to be more extreme, but only in degree. Thus the discharge of raw sewage, or blood from a slaughterhouse, to a river creates conditions paralleled by the dunging of a water hole by large mammals in the dry savannah regions, or the washing of an animal corpse into a stream or small river, or the conditions in the hypolimnion of an extremely fertile lake. Leaching of agricultural fertilizers or discharge of plant nutrients in the effluent from sewage treatment works are extensions of the conditions in naturally fertile catchments, and the suspension of fine particles washed from china clay or other mineral workings or construction sites is a normal feature of the streams draining mountain glaciers. The hot springs which bubble out of the rocks in volcanic areas are often both very hot, sometimes very acid, and loaded with a range of, to most organisms, toxic heavy metals in high concentrations. Heavy metals are common also in streams draining ore bodies as well as mineral mine workings, which may also be very acid.

In all such natural habitats a living community is present. This does not mean that grossly polluted habitats should be acceptable—many are not at all pleasant places to be in—but it does mean that suitable organisms can be harnessed to treat some pollutants biologically, as happens in sewage treatment, or that specifically adapted organisms may be used as indicators of pollution where chemical analysis is inadequate to detect small concentrations or intermittent discharges. Particular attention will be given here to pollution by heavy metals, organic matter, heat and plant nutrients, and its control.

10.2.1 Heavy metal pollution
Heavy metals are those, conventionally, with a specific gravity of five or more, and of the forty or so such elements not normally present only as radioactive isotopes, manganese, iron, copper, zinc, molybdenum, cadmium, mercury and lead have attracted most notice. Some of them are required as trace elements by

living organisms, though others (Cd, Hg, Pb) do not appear to be. At more than 'normal' levels even the trace elements are frequently toxic though the mode of action is often obscure. Selective inhibition of particular crucial enzymes seems one likely reason. Normal levels of these elements in unpolluted waters are around $1 \mu g \, l^{-1}$, with zinc concentrations around $10 \mu g \, l^{-1}$; effluents from metal industries and from the exposure of metal sulphides to air and water in mine waste dumps may increase these concentrations by several orders of magnitude. The bacteria and blue-green algae which colonize hot, volcanic springs may tolerate extremely high levels of these metals, and also of fluoride and arsenic[65], and in 'normal' streams, polluted by heavy metals, highly resistant strains of algae may be selected for[537]. Dense populations of green algae (Chlorophyta) may frequently be seen in the water of recirculating fountains playing over the bronze accoutrements to the statues of long dead worthies in European cities! Fauna fare less well, however, and in Welsh mountain streams polluted by the waste from old lead mines, both zinc and lead levels may lead to a depauperate invertebrate fauna of insect larvae, *Tanypus nebulosus* and *Simulium latipes*, and some flatworms. Crustaceans and oligochaetes seem to be particularly susceptible, and fish are absent from such streams[265]. Salmonid fish tend to die at lower concentrations than coarse fish[266].

Death of specific organisms is not the only consequence of heavy metal pollution. There may be reduced growth rates of those that survive and accumulation of the metals in their bodies by factors of many thousands over the concentrations found in the environment. There may be therefore, effects on amphibious but unresistant predators, such as water birds.

It would be convenient to be able to establish the levels at which each heavy metal causes physiological or behavioural changes or death in each species likely to be subjected to heavy metal pollution. Sophisticated apparatus has been designed, in which fish or other organisms can be exposed to known levels of metal ions in recirculating laboratory streams, and the LC_{50} or (lethal) concentration at which 50% of a test population die in a set time can be established. Unfortunately the LC_{50} varies not only with species, but with life cycle stage, concentrations in the water of other heavy metals, oxygen, bicarbonate, organic matter and temperature. Many heavy metals are more soluble, for instance, at low pH, and are precipitated as presumably less active carbonates and hydroxides at pHs above 7. There are, in turn, purely analytical problems of determining the concentration of active ion in an effluent. In particulate or colloidal form or chelated with organic compounds, it may be detected by the analytical method but not by the organism.

In practice, therefore, the amount of heavy metal which a particular factory might be allowed to discharge has to be decided in Great Britain by the pollution control officer of one of the Regional Water Authorities (the bodies responsible at present for pollution control in British waterways), on the basis of experience, judgement and inspired guesswork. This amount, called the 'consent', takes into account not only the concentration but also the total volume permitted to be released to a river. The consent must also allow for the periods when natural flow in the river will be lowest (and concentration of pollutant therefore highest), for other effluents upstream and downstream, and in deference to political realities, the economic state of the industry and area. Too stringent a standard may cost the industry so much in special extracting plant to remove the pollutant that it becomes unprofitable, and an appeals procedure discourages a pollution officer from allowing too great a safety margin[502]. In the final analysis he hopes to maintain at least some fish populations and to minimize obvious visual effects of the pollutant; the Water Authority has powers to prosecute (after warnings) if the industry consistently exceeds the consent once the period in which it may appeal is over. Other countries, of course, have different control systems. The European Economic Community, for example, would prefer absolute standards for pollutant levels, but the British system contains a high degree of flexibility.

10.2.2 Heat pollution

Most temperate waters do not exceed about 20°C in summer; sub-tropical and tropical rivers may reach 30–35°C or occasionally 40°C. Heat pollution occurs when these temperatures are increased by more than a few degrees. Usually this is because water is withdrawn from a river to cool generating plants at power stations and is discharged up to 15°C warmer. In some areas the water must be cooled in cooling towers or lagoons before discharge to the river. Such plant may be very costly and unaesthetic though it is theoretically possible to return the water at its original temperature. In Britain most very large power stations are sited on estuaries since no British river is sufficiently large to cool even a moderately big power station in its freshwater reaches. Small stations are sited on rivers, however, and the large rivers and lakes of the continental land masses can be used for the largest stations and undoubtedly will be further used in the future.

Natural hot springs, and some rivers draining areas where such springs are common, allow some basis for prediction of the effects of heat pollution, as long as it is remembered that naturally hot waters are chemically rather extreme, whereas thermally polluted waters, except for a change in oxygen saturation (see Chapter 2) are usually chemically 'normal'. The percolating water which becomes superheated by hot rocks close to the surface in volcanic regions such as Yellowstone (U.S.A.), Iceland, Japan and New Zealand often has a high pH of 8·5–10, or sometimes a very low pH of less than 3, high levels of silicate ($> 10^2$ mg Si l^{-1}), hydrogen sulphide (10^0–10^1 mg l^{-1}), fluoride (10^0–10^1 mg l^{-1}), arsenic (up to 2 mg l^{-1}), sodium and phosphate[64].

Some bacteria grow in boiling water[45, 47]. They are heterotrophic on the small amounts of organic matter picked up by the spring water and may be detected by suspension of sterile slides, on which the organisms grow, in the spring. Photosynthetic organisms do not grow above 73–74°C, however, though chemotrophs may flourish up to 90°C. From 74 to 60°C in springs at least in N. America and in Japan, a few species of the blue-green algal genus *Synechococcus* grow, and the diversity of blue-green algae increases to include filamentous genera like *Mastigocladus* and *Phormidium* at the lower end of the range, together with thick underlying mats of photoheterotrophic flexibacteria of the genus *Chloroflexus*. The striking orange undermat of *Chloroflexus* with its overlying blue-green algal layer maintains its abundance largely because no animals which might graze it can exist above 45–50°C. The biota above 50°C is entirely prokaryotic in alkaline springs, though in acid ones a peculiar eukaryote, *Cyanadium caldarium*, may persist up to 55 or 60°C[106]. *Cyanidium* is a flagellate, has phycobiliproteins and a pH optimum of between 2 and 3.

Some protozoans might survive at 60°C but multi-celled animals, almost all of them Ostracods (a group of small crustaceans), at the higher levels tolerated, do not persist above 50°C. They are joined, around 45°C, by some rotifers, water mites, more ostracods, nematodes, gastropods, and insects, particularly larvae of ephydrid flies. At around 40°C a few amphibians and fish may be found, though with only one or two species in any given thermal stream. As the water cools further, the species lists lengthen to the high diversities of 'normal' waters[65, 171].

Water and mud samples from cool habitats, kept in the laboratory at 35–38°C become more and more dominated by blue-green algae and at 42°C few eukaryotes persist. Experimental slides suspended in heated, flowing water had diatom populations at 20–30°C, Chlorophyta at 30–35°C and blue-green algae at 35+°C in one study[395]. Clearly, therefore, both field and laboratory evidence indicate that very serious changes might be expected if waters are heated to around 40°C, and such levels might easily be reached downstream of power stations in warm temperate and tropical regions where summer water temperatures may naturally approach 35°C.

Below such levels predictions are less easy to make. Temperature has a major effect on growth rates of poikilothermic animals and relatively little heating may change balances of species in invertebrate and fish communities which cannot presently be predicted. Lethal temperatures for freshwater fish have been determined by laboratory experimentation—around 30°C for Atlantic salmon, sea and brown trout, 33·5°C for roach, 35°C or over for tench, carp and stickleback—but these can only be rough guides to how a community will react to warmed effluent which does not raise the temperature to the lethal level. A few degrees temperature rise may alter growth and feeding rates, perhaps increasing the former, but ultimately decreasing the rate of survival. It may inhibit some digestive enzymes or rephase a spawning cycle such that the fry no longer hatch at the time that a crucial food organism is present in abundance. Incidence of fungal and bacterial diseases of fish also increases in warmer waters as does susceptibility to other pollutants.

The temperature structure of a water body near a discharge of hot water is also complex and fish may easily avoid the hotter and lethal regions so that the subtler effects of temperature become much more important in predicting the effects of heat pollution, though they are far more difficult to determine than the temperature lethal to an organism.

Fish may be attracted in winter to warm effluents, and carp culture has been practised in power station effluents in the U.S.S.R. There may be much future in such fish culture where the needs are for rapidly growing fish which are cropped young, but it does not follow that such high growth rates will be advantageous in a natural, balanced, aquatic community. Enough is not yet known, to be sure, but since integration of life cycles, feeding relations and production has become closely fitted to the natural fluctuations of the environment through natural selection, and since temperature is related to so many other features of the aquatic environment, it seems likely that even a moderate rise in mean water temperature could have serious long-term effects.

10.2.3 Pollution by organic matter

Input of organic matter is a normal feature of aquatic ecosystems. Indeed the detritivores of streams (see Chapter 4) and the sediment communities of slow-flowing rivers and lakes (see Chapter 5) depend upon it for most of their energy. The organic matter is relatively rapidly converted to inorganic substances or left in a refractory state. The main difference between natural organic input and pollution by organic matter is that the former tends to be in large packets, like leaves, with a low surface to volume ratio, or is relatively refractory when finely divided, while the latter is usually soluble or finely divided and very labile. The bacteria which immediately colonize it need much oxygen to decompose it and organically-polluted water rapidly becomes deoxygenated. In rivers, loss of most invertebrates and fish follows and the pollution community comprises a mass of filamentous bacteria (including 'sewage fungus', *Sphaerotilus natans* and others), colourless flagellates, grazing ciliate protozoons and anaerobic chemoautotrophic bacteria like *Beggiatoa*. Progressively, as the organic matter is decomposed, this community may be replaced by one in which filamentous algae like *Cladophora* predominate. These are stimulated by the copious release of ammonium and phosphate from the decomposition, and support numerous chironomid larvae, which may cause a nuisance when they emerge. If the stream bottom is muddy, numerically rich communities of oligochaete worms and chironomids may develop. These further process the organic matter indirectly by consuming the abundant bacteria, and as oxygen levels begin to rise again, *Asellus* and other moderately tolerant invertebrates become abundant. Eventually, as the water again becomes fully oxygenated the 'clean water fauna', with a high species diversity, is able to return[241].

Organic pollution used to be far more common in Great Britain than it is now, because for reasons originally of public health, the most usual source, domestic sewage, is treated to remove much of its labile organic content before discharge to a river. Some tidal rivers still receive raw sewage, and organic discharges from, for example, the food industry and intensive stock rearing units do occur, but by and large, organic pollution in the developed world is well understood and adequately controlled. It is controlled through use of the same biological processes that break down organic matter in a river, but they are concentrated into the small area of a sewage treatment works rather than along several kilometres of waterway. The deoxygenating ability, or oxygen demand of a sewage effluent, the product of a sewage treatment works, may be crudely assessed by measuring the rate at which oxygen disappears from solution in a sealed bottle kept for 5 days in darkness at 20°C. Dilution of rich effluents is necessary for the test, otherwise complete deoxygenation would occur in hours leading to an underestimate of the total demand. The biological oxygen demand (BOD) of crude sewage is around 600 mg O_2 l^{-1} 5 day^{-1}, and of unpolluted river water less than 5 mg O_2 l^{-1} 5 day^{-1}. Good sewage treatment reduces the BOD of the discharged effluent to, at most, 30 mg O_2 l^{-1} 5 day^{-1} and usually less than this.

10.3 Sewage treatmemt

The raw material pumped to a sewage treatment works is mostly water, with 1–2% of particulate and colloidal organic matter and a host of dissolved compounds. Domestic waste is only one component, and the bulk of this is bathwater and kitchen water, rather than faeces and urine. In urban areas there is drainage water from the streets and some factory effluents, though these are strictly controlled by a consent system so that metal and other poisons do not inhibit the organisms harnessed at the works to remove organic matter. The street drainage includes grit and stones and somehow large objects like dead cats find their way into the sewers. These must be removed by sedimentation and a stout grille respectively. It may also be necessary to remove fibrous 'rag' material which may interfere with pumps at the works. The variety of objects received is great and engineers profess puzzlement at some of its origins.

After this screening the sewage has few particles larger than a few millimetres —passage through the sewerage pipes has generally broken up larger lumps— but a macerator now completes the process. At this stage the 'foul water' is greyish brown and turbid and has a BOD of 600 mg O_2 l^{-1} 5 day^{-1} still. It is led into large tanks, the primary sedimentation tanks where particulate organic matter settles out as a sludge, and the BOD of the overlying water decreases by 30–40%. During the few days of this settlement there is some decrease in numbers of pathogenic bacteria and viruses, as these are digested by protozoa or fail to balance, by growth, the natural mortality. The primary sludge is pumped to digesters, while the supernatant water enters the secondary stage.

The sludge digesters are enclosed tanks in which the sludge is heated to 30–35°C by methane produced by bacterial fermentation of itself. The methane is recycled to burners and may be used to generate electricity to power the works. Fermentation is carried out mainly by *Methanomonas* spp., which, being heterotrophic but anaerobic, cannot completely break down the organic matter to carbon dioxide and water. Such bacteria are found in swamp and lake muds, where the methane they produce has been called marsh gas, and were so abundant in the grossly polluted canals of seventeenth and eighteenth century England that the clouds of methane they produced sometimes could be set on fire, to the detriment of canalside buildings. After several weeks' digestion the residual sludge is considerably less obnoxious and should have an almost pleasant, earthy smell. Its disposal is something of a problem, solved by compressing it to remove water or composting it with urban refuse on rubbish tips, giving or selling it to local farmers for fertilizer if it does not contain large quantities of heavy metals, or dumping it, in the cases of very large works near the coast, from specially designed ships out at sea.

Meanwhile the supernatant water from primary sedimentation is usually treated in one of two ways—the trickling filter or the activated sludge processes. The former is the older method and removes a slightly lower proportion of the residual BOD than the latter, which, however, requires more control and attention and is therefore slightly more expensive. Both depend initially on the same principle—breakdown of organic matter by a similar community of bacteria and protozoa as that which responds to gross organic pollution of a river.

Trickling filters are typically circular beds of rock fragments each a few cm in diameter. The beds are up to 50 m diameter and about 2–2·5 m deep. Usually they are set on a valley slope below the primary sedimentation tanks so that water moves under gravity through pipes to the centre of the bed and out along four arms extending radially from the centre. Gushing out through holes in the arms the water moves the arms by jet propulsion so that the bed is evenly sprayed. As a new bed is brought into use a film of bacteria and grazing protozoa develops on the rock fragments and converts the organic load into bacterial and protozoon cells and CO_2. The process may be aided by aerating the water before it enters the filter. Growth of the bacteria would soon clog, waterlog, and deoxygenate the filter were it not for grazers on the bacterial and protozoon film. These are a few species of oligochaetes and fly larvae, whose respiration removes yet more organic matter as CO_2, and a smaller proportion on emergence of the flies, sometimes to cause a nuisance in neighbouring housing areas.

The water trickles through the bed and out into channels. At this stage its BOD has been reduced to about 60 mg O_2 l^{-1} 5 day^{-1}, and it comprises a rich inorganic solution containing phosphate, ammonium and nitrate ions, among others, with sloughed off bacteria, faeces of invertebrates, and residual refractory organic matter. The BOD is reduced to 20 mg O_2 l^{-1} 5 day^{-1} by settling of these solids in tanks to form a secondary sludge. This may ultimately be dried and disposed of to farmers or dumps, or digested with primary sludge.

The activated sludge process is carried out in large tanks. Primary effluent is led into the tanks, seeded with a floc of particles of bacteria and protozoa, largely ciliates, kept back from the previous batch, and vigorously aerated and agitated. This promotes rapid growth of the floc, or activated sludge and some rotifers and nematodes may graze it. The plant is usually run on a carefully regulated and monitored continuous flow system to maintain a high ratio of floc to incoming organic matter. The effluent is taken to ponds where the floc is settled then pumped to sludge digesters, while the supernatant may be the final effluent or may be further 'polished' by physical filtration or percolation through adsorptive activated carbon columns.

Sewage works often operate both secondary processes and may recirculate effluent through more than one trickling filter; many engineering refinements may be incorporated—Sweden, for example, has a completely underground activated sludge plant which creates no aesthetic problems. There are also other methods of sewage disposal. In areas of low population, where it may be uneconomical to build a treatment works, each house may have an underground septic tank into which the domestic sewage flows. The sewage is decomposed mostly anaerobically and effluent percolates into the surrounding soil, while a remaining sludge must be pumped out into a tanker at infrequent intervals and transported to a treatment works for digestion. Mini-sewage works have been developed, comprising tanks of glass fibre reinforced plastics only a metre or so high which allow sewage from a small group of houses or a farm to pass over a large internal spiral surface on which the bacteria and protozoa grow. Maintenance is easier than that of septic tanks, whose effluent pipes may clog, flooding the tank and the garden under which it may be set! In tropical regions the sewage may simply be led into shallow lagoons where it is bacterially oxidized, aided by the photosynthetic oxygen production of dense populations of phytoplankton (usually Chlorophyta including *Chlorella*, *Scenedesmus*, and *Chlamydomonas*) which develop in the nutrient-rich medium.

From the point of view of BOD and turbidity the standards of 20 mg O_2 l^{-1} 5 days^{-1} and 30 mg l^{-1} suspended solids, set by a Royal Commission in the U.K. in 1913 for final effluents are easily achieved. If they are not, the fault lies usually in the management of the works, pumping machinery breakdown, or overloading of the works during heavy rainstorms when some diluted raw sewage has to be released to a river. New sewerage schemes usually have separate pipe systems for street drainage and foul water to obviate this problem. Even good effluents, however, may contain pathogenic gut bacteria and viruses such as those of poliomyelitis and infectious hepatitis, though these should be greatly diluted by the river flow and undetectable a little way downstream.

Through its potential pathogens and also the high levels of phosphate (around 10 mg PO_4–P l^{-1}) and other plant nutrients contained in it, sewage effluent is not harmless. In parts of Europe, effluent may be the raw material for the domestic drinking water supply within a week of discharge in summer, and the high phosphate loading, which results from discharge to slow-flowing rivers or lakes, increases phytoplankton production sufficiently to cause serious problems. This enrichment, or eutrophication (see Chapters 1 & 2), has become very common as increasing human populations have been served, in the developed world, by mains sewerage. It has a number of consequences which will be examined in Chapter 12 but is not an insuperable problem. Firstly, however, it is necessary to look at the related problem of storage and provision of drinking water.

10.4 The supply of drinking water

Ideally, drinking water is taken from upland lakes or storage reservoirs where a rocky catchment provides a low nutrient loading. The lakes consequently have small phytoplankton populations, well-oxygenated hypolimnia and no farm or human effluents to introduce potentially pathogenic bacteria. Filtration costs are low and only minimal amounts of chlorine or hypochlorite are needed to sterilize the water. A safe and palatable product is readily produced.

Though there are urban areas near enough to such supplies, or rich enough to pipe them over quite large distances, the bulk of urban populations live in lowland regions where river catchments are naturally fertile. Such catchments are also agriculturally valuable and eutrophication by fertilizer run-off and leaching occurs. The concentration of human population, evaporation from the land in summer and hence low summer river flows mean that much of the local surface water is recycled frequently through both the sewage treatment works, becoming considerably more fertile in the process, and the water purification works. Some areas are fortunate enough to be able to draw on naturally rock-filtered underground water supplies, but the raw material for most lowland drinking-water is a richly fertile phytoplankton growth medium potentially also containing pathogenic microorganisms. When stored in a reservoir, such water supports dense populations of phytoplankton and if the reservoir is deep enough the entire hypolimnion may become completely deoxygenated in summer. This means that filtration and sterilization costs are high and that a high proportion of the reservoir volume may be unusable because of build-up of toxic H_2S in the hypolimnion (see Chapter 5) in summer, when water is generally in short supply. Large spring diatom populations and particularly the blooms of blue-green algae (see later) developing in summer may give unpleasant tastes to the water, which it is usually necessary to remove with activated charcoal, entailing further expense.

These problems may be tackled at source by removing as much as possible of the phosphate from the effluent at the sewage works (a process known as tertiary treatment in North America, where it is common, and phosphate stripping in Great Britain, where it is virtually unused) or by attempting to manage the reservoir in such a way as to minimize growth particularly of the summer algal populations. Treatment of cause is probably ultimately less costly than treatment of symptoms, particularly when other consequences of eutrophication are considered (see Chapter 12).

10.4.1 Phosphate removal

It is not difficult to remove between 70 and 95 % of the phosphorus from sewage effluent. Most of it is in the form of soluble reactive phosphate and can be chemically precipitated as iron, aluminium or calcium compounds. The former two are most efficient and a decision on which to use will depend largely on cost and local availability. It is greater than proportionately more expensive to remove 90 % of the SRP than to remove 70 % and careful nutrient budgeting of a waterway may indicate that the lower efficiency may be acceptable. Much depends on the uses to which the water will be put and guidelines have been worked out for Canadian lakes, which may be generally applicable (Table 10.1). Use of models like Vollenweider's (see Chapters 1 & 2) allows calculation of the permissible loading of phosphorus that will not give more than a specified

Table 10.1. Maximum permissible average summer concentrations of chlorophyll a (μg l^{-1}) and total phosphorus (μg P l^{-1}) in Southern Ontario lakes for different categories of use. (From Dillon & Rigler[101].) Phosphorus and chlorophyll levels were related in these lakes by the significant regression:

$$\log_{10}[P] = \frac{\log_{10}[\text{chlorophyll } a] + 1.14}{1.45}$$

U.K. lakes falling into the various levels are also given.

Level of water classification	Use and appearance of water	Depth at which Secchi disc visible (m)	Chlorophyll a level	Phosphorus level	U.K. lakes falling into each level
1	Swimming in clear water, highly oxygenated hypolimnia suitable for game fishing	5	2	10	Scottish highland lochs, upland Welsh tarns, westernmost English Lake District lakes, Wastwater and Ennerdale and mountain tarns
2	Water recreation, preservation of game fishery not imperative	2–5	5	19	Some English Lake District lakes, e.g. Windermere
3	Aesthetic appearance (very clear water) for swimming not required. Emphasis on fairly productive coarse fisheries (bass, wall-eye, pickerel, pike, maskinonge, bluegill, yellow perch). Hypolimnetic oxygen depletion common and possibility of fish kills in winter by deoxygenation under ice	1–2	10	30	Esthwaite, some Shropshire Meres, Upton Broad
4	Suitable for coarse fishing, but with increased danger of fish kills in winter; hypolimnetic oxygen depletion very severe	less than 1·5	25	56	Some Shropshire Meres, Martham Broad
5	Not applicable or included in the Ontario classification. Fish kills infrequent only because most British lakes do not freeze over for long in winter	possibly much less than 1	greater than 25	greater than 60	Most Norfolk Broads (see Chapter 12), the London drinking water reservoirs and some Shropshire Meres

phosphorus level. Planning of sewage treatment can then be undertaken appropriately. Canada and some states of the U.S.A. now routinely install phosphate stripping into new and renovated treatment works, and the practice is increasing in Scandinavia. The United Kingdom presently lags behind in this respect.

About half of the phosphate in sewage effluent comes ultimately from digested food, but the remainder in Britain is supplied by domestic phosphate detergents, use of which has increased greatly since the 1950's. The phosphate in detergents is used as a 'builder', as opposed to the 'surfactant' which dislodges greasy dirt from laundry. The builder prevents redeposition of the dirt as precipitates with calcium ions in hard waters and appears essential to the high standard of laundry now expected. Phosphates are ideal builders[113]. They are not toxic even to infants consuming them out of curiosity, nor do they corrode the metals of washing machines. From the viewpoint of water quality and reducing the costs of phosphate stripping, it is desirable to find alternative builders which are equally safe but also environmentally harmless. Some of the substitutes tried, the rather alkaline silicates, borates, and carbonates, probably do not satisfy the former criterion, but there has been much interest in the use of nitrilotriacetic acid (NTA). NTA is a chelating agent which forms complexes with metal ions. It is a more efficient builder in keeping calcium ions in solution than is sodium tripolyphosphate, the commonly used phosphate builder, and was incorporated in the late 1960's into North American and Swedish detergents, after extensive screening of its properties and human and environmental safety. It is widely used now in Canada and in Sweden, but not in the U.S.A. In late 1970, injections of NTA into rats were found to increase the toxicity of, and incidence of birth defects caused by, mercury and cadmium compounds. The experiments were not designed as well as they might have been, and the relationship of intraperitoneal injections of quite high levels of NTA to the considerably lower levels likely to be absorbed through effluents recycled as drinking water is obscure to say the least. On publication of these tests, however, NTA use was discontinued in the U.S.A. and, for psychological rather than rational reasons, NTA may never be used there again. Citrates, however, have now been shown to be effective builders, with no human or environmental drawbacks. Legislation has now been drawn up governing States which border the St. Lawrence Great Lakes, restricting the phosphate content of detergents to 0·5%, while in Canada there is a restriction of 2·2% (Greater Lakes Water Quality Agreement 1972, updated 1978). Previous phosphate levels in detergents exceeded 50%.

In crowded countries, the management of lakes has to accept the loading of phosphorus provided by an existing population. Where populations are lower, and active development is taking place, matters can be more ordered. In N. America and Scandinavia many people would like to build holiday cottages in lake districts which are presently undeveloped. Such building means greater export of phosphorus to the local waterways, but can be regulated since the phosphorus loadings from different land-use-types, per individual of population, and for different means of sewage disposal (septic tanks, mains) can be calculated.

Dillon & Rigler[102] have developed a series of equations based on refinements of the Vollenweider model which allow planners to calculate the effects
of different degrees of development without even leaving their offices! The
equations presently apply to areas of Canada, but a similar approach could be
made anywhere.

10.4.2 Reservoir management; coping with the symptoms

A number of factors have contributed to the situation in Britain where nutrient
loading on drinking water reservoirs is high and where immediate hope of
reducing the loading is low. These include the conservatism of the water industry
which has not yet attempted a cost/benefit study of phosphate stripping, equal
reticence on the part of the Government to stimulate research into phosphate
substitutes in detergents, and the size of the problem in a crowded and also
heavily-farmed country. Collingwood[72] believes that the problem of agricultural run-off is sufficient to maintain algal crops high enough (40–50 μg l^{-1}
chlorophyll a in summer) to keep the cost of water treatment at very high levels.

Emphasis has thus been placed on attempting to minimize phytoplankton
growth in reservoirs. Obvious ways involve the use of algicides, less obvious
ones, parasites, grazers and vigorous mixing of the water column. Interesting
suggestions, unlikely to be used, include the shading of the reservoir with floating
opaque table tennis balls (these would tend to pile up on a lee shore in any
wind!) and the use of explosives. The latter, and the less drastic mixing of the
water column follow from some of the properties of the blue-green algae, discussed later.

Algicide use is widespread, costly and somewhat uninspired. Copper sulphate
has frequently been used. It kills blue-green algae but these are often soon
replaced by copper-tolerant Chlorophyta in equal abundance. Other algicides,
mostly organic, are also available. A number of parasites—bacteria, viruses and
chytrids (see Chapter 3)—have been isolated or identified for some common
phytoplankters, but either cannot yet be cultured at all, or not on the scale
needed to seed a large reservoir, at reasonable cost. Husbanding of the zooplankton is also attractive, but not yet practically possible. In the case of parasite
control the evolution of resistant strains would be anticipated and heavy zooplankton grazing often causes development of the larger, gelatinous and un-
palatable green and blue-green algae (see Chapter 3), which may be a worse
nuisance in water filtration than the smaller forms removed by grazing.

Some success has come, however, from manipulation of the water column
by aeration and mixing. Part of the reason for this success depends on the
physiology of some of the blue-green algae which cause problem 'blooms' and
it is necessary first to understand how these form.

10.4.3 Formation of blue-green algal blooms

Blooms are scum-like aggregations of some genera of Cyanophyta (particularly
Aphanizomenon, Anabaena, Microcystis and *Oscillatoria*) which form under
limnological conditions favouring high fertility at the water surface. The word

'bloom', derived from its original German origin 'Wasserblüte' should be confined to this use and not applied to any increasing population of algae as it now commonly is.

The genera concerned (Fig. 10.1) are able to regulate their buoyancy physiologically. Their cells contain gas vesicles which are bundles of adjacent prismatic bodies, with protein walls, permeable to atmospheric gases but not to water, which contain a gas mixture in equilibrium with that dissolved in the surrounding water. Some are weaker than others and can be collapsed by application of external pressure (about 7 atmospheres) to the water in which the cells are suspended. If the vesicles are removed from the cells (this can be done by ultra centrifugation following mechanical breakage of the cell wall and membrane) rather more pressure is needed to collapse them (about 11 atmospheres). This is because in the cell they are already subjected to considerable internal turgor pressure. The recent literature on gas vesicles may be approached through Walsby[516].

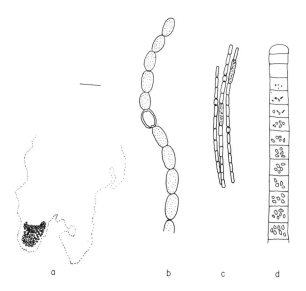

Fig. 10.1 Bloom-forming algae. (a) *Microcystis aeruginosa*, (b) *Anabaena flos-aquae*; (c) *Aphanizomenon flos-aquae*, and (d) *Oscillatoria rubescens*. Scale bar represents 10 μm.

Since they may occupy up to 30% of the cell volume, gas vesicles can make the cells less dense than water, so that they float upwards. This relative movement, like sinking (see Chapter 3), has advantages in promoting nutrient uptake. However, as the cells rise into regions of higher light intensity in the water column, their photosynthetic fixation of carbon dioxide increases and the concentration of soluble organic substances in the cell rises. This is accentuated if available phosphorus and nitrogen supplies are low and limiting growth rates, as frequently they are, even in fertile lakes in summer, when blue-green algae are at their greatest abundance. As the internal concentration of soluble organic substances rises, so also does the internal turgor pressure and some of the vesicles, the weaker ones, collapse. The cell may then become denser than water and starts to sink. This mechanism prevents the cells moving into the highest light intensities at the water surface which, in summer, may be lethal for these organisms[422].

As the cells sink into less well-illuminated parts of the water column, conditions are such that the rate of gas vesicle formation is greater than that of overall cell synthesis and division. The concentration of gas vesicles in the cell therefore increases, the cells become positively buoyant again, and start to float upwards. This cycle keeps them in water strata most suitable for their growth.

If the water is extremely fertile the phytoplankton crop as a whole may, in summer, build up to such a concentration that the euphotic zone becomes very shallow and much less deep than the epilimnion. The cells then may spend much of their time at very low light intensities and the differential rates of gas vesicle synthesis and cell production lead to formation of cells so buoyant that they rise to the surface very rapidly on calm days. The mechanism by which increased turgor pressure bursts the weaker vesicles is unable to operate, for photosynthesis is inhibited at the high extreme surface light intensity and the cells are trapped at the surface forming the characteristic paint-like scum. This may cause problems of taste and odour as it decays or if it is later wind-rowed at the lake edge.

10.4.4 Turbulent mixing in reservoir management

Vigorous mixing, either by introducing incoming, relatively warm, river water through jets in the base of the reservoir, by the use of pumps floating on rafts, or by vigorous bubbling of compressed air have been found sometimes to reduce the growth of blue-green algae though not necessarily of diatoms, which normally occur during periods when lakes and reservoirs are naturally well mixed. Some account of these may be found in Ridley[429], Ridley et al.[430] and Steel[474].

The mechanism is obscure but may lie in preventing blue-green algae from regulating their position in layers of the water column optimal for their growth. A useful feature of artificial mixing, even if it does not diminish algal growth, is prevention of hypolimnion formation and deoxygenation. This increases the volume of water available for treatment and supply and also creates a greater volume of fish habitat than would otherwise be present.

10.4.5 Treatment of water for domestic supply

Water for drinking must be safe and palatable. The former refers to its content of dissolved substances likely to be harmful (phenols, arsenic, heavy metals, selenium, cyanide, nitrate and others), which must meet standards decided nationally, but often adapted from World Health Organization recommendations, and also to its freedom from the agents of infectious disease. Typhoid, paratyphoid, cholera, amoebic dysentery, salmonellosis, infectious hepatitis and poliomyelitis are usually transmitted in water, and the low incidence of them in

N. America, Britain and most of Europe is a tribute to the standards maintained by the water industry. Pathogenic microorganisms do not readily survive for long outside their hosts and several days storage in a reservoir reduces markedly any contamination from sewage effluent. A normal gut bacterium, *Escherichia coli* Type II, is common in the effluent and more resistant to the external environment than the pathogens. Drinking-water standards are based on detection of fewer than miniscule numbers of this bacterium in the piped supply.

To achieve palatability it is usual either to filter the reservoir water rapidly under pressure or slowly under gravity through sand. In the latter case, a community of algae, bacteria and protozoa helps to bind together the suspended particles in the water, but in rapid filtration coagulation with aluminium salts is first necessary. Rotating drums of very fine mesh steel fabric, continuously back-washed to prevent clogging, may be used as a first step in filtration. Subsequent sterilization may be by chlorine, hypochlorite, ozone, ultraviolet radiation or electrolysis with silver and carbon electrodes. The latter releases minute quantities of silver which is the active agent.

Water-borne disease problems are present also in the Tropics where they are more varied and extensive. The next chapter examines some of them, and the biological problems involved in storing water in tropical countries.

10.5 Further reading

Pollution is now a subject on which there is an enormous amount of literature, much of it describing case histories, or detection methods. Hynes[241] is still an excellent account and may be amplified with reference to Esch & McFarlane[129], Krenkel & Parker[273] and Parker & Krenkel[393] (thermal pollution); Muirhead-Thompson[373] (pesticides); Hart & Samuel[196], Klein[272], Oglesby *et al.*[386] and Whitton[536] (general), and American Public Health Association[10] (analytical methods).

Eutrophication is now also well-covered in the literature and key volumes include Likens[293], National Academy of Sciences[378], Vallentyne[507] and Vollenweider[512]. Vallentyne[507] is particularly valuable for the clarity of its writing. Hutchinson[238] is a comprehensive review and Porter[412] an interesting inter-disciplinary study of part of the eutrophication problem. A current important European problem is the rising level of nitrate, particularly in groundwater. Burden[54] and Taylor[494a] discuss this. USEPA[505] is a survey of methods for the restoration of lakes and water quality where they have been altered particularly by eutrophication.

Specialist pollution journals include *Water Research, Journal of the Water Pollution Control Federation* and *Proceedings of the Society for Water Treatment and Examination.*

CHAPTER 11
HYDROBIOLOGICAL PROBLEMS IN
THE TROPICS—MAN-MADE LAKES
AND WATER-BORNE DISEASES

11.1 Introduction

New dimensions are added to the problems of water management in the Tropics and sub-Tropics. The rivers flowing through teeming Asian cities are no doubt as organically polluted as any ever have been in the world, but much of the Tropical area has a low population density and great rivers flow through vast areas peopled at most with small villages. On the one hand these rivers, the Niger, Congo, Zambezi, Nile, Orinoco, Amazon, Ganges and Brahmaputra, for example, are a resource which might be tapped for hydroelectric power, or irrigation, and which provides vital fisheries for peoples generally short of first-class protein. Yet on the other hand the rivers, their tributaries and the irrigation channels led from them, harbour the invertebrate and fish vectors of some widespread and debilitating diseases, where life cycles turn unhindered in the absence of the relatively strict public health measures of the Western World.

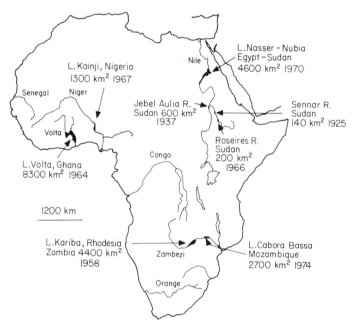

Fig. 11.1. Some major man-made lakes in Africa, with their areas and dates at which the lakes began to form. (Based on Lowe-McConnell[299].)

The damming of some of these rivers, particularly in Africa for reasons of electricity generation and irrigation, has created huge lakes, hundreds of kilometres wide or long. With them have come problems not fully anticipated in their planning and there is still controversy over the ultimate balance of costs and benefits. Nonetheless such lakes (Fig. 11.1) as Kariba, on the R. Zambezi, Volta on the R. Volta, Kainji on the R. Niger, and Nasser-Nubia on the Nile have been fascinating to limnologists, and studies of them have allowed prediction of possible problems in future man-made lakes, and increased understanding of fundamental limnology.

11.2 Man-made lakes—the early stages of filling

A generalized picture[338] has been derived from studies on the major new African Lakes, complemented by the many Russian studies of barrages placed on rivers in the U.S.S.R., though details vary according to local conditions. As the water rises behind a newly-closed dam it tends to be turbid from suspended sediment in the river and from erosion of the newly-flooded soils at the water's edge. Production is then low, for the turbidity prevents much phytoplankton growth. Gradually, however, as the new lake becomes larger the volume to circumference ratio increases and the products of erosion are much diluted. Progressively, also, the incoming river silt is deposited in deltas at the river and stream mouths as the current flow is decreased by the water mass. In L. Kariba there was no early turbid phase for river silt is deposited in vast swamps just upriver from the lake basin, but in other lakes it has been prominent.

Flooding of the old river valley vegetation kills it and it starts to rot underwater. This has two consequences. Firstly, there may be deoxygenation of the bottom water with production of H_2S. In L. Volta the whole of the water was deoxygenated for a time as the flooded softwood forests began to decay. Secondly, the decomposition and the waterlogging of previously terrestrial soils releases nutrients such that the conductivity of the water may rise, and there is a considerable internal loading of nitrogen and phosphorus compounds. The rise in L. Kariba was from 26 mg total dissolved solids in the river to 67 mg in the early lake. This in turn stimulates increased growth of phytoplankton, aquatic macrophytes and fish, though in the latter case only partly for this reason. L Kainji which has a very short water retention period and hence is frequently flushed out consequently did not undergo these stages, nor did L. Nasser-Nubia, where 80% of the basin was of rock and desert sand.

In L. Kariba, however, the most spectacular example of this burst of production was the rapid colonization, in the first few years after the dam was closed in late 1958, of a floating fern, *Salvinia molesta* D. S. Mitchell (Fig. 11.2), then called *S. auriculata* Aubl. *S. molesta* is a robust plant, up to 30 cm long with overlapping leaves densely covered with firm hairs. These prevent wetting and waterlogging of the plant, which also has air spaces which give it buoyancy. *Salvinia* spp. normally can reproduce sexually—sporocarps are produced on trailing

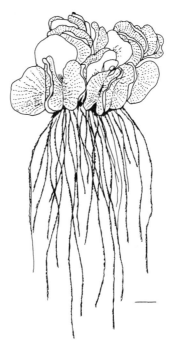

Fig. 11.2. *Salvinia molesta* Mitchell. Scale bar represents 1 cm. (Redrawn from Cook *et al.*[73].)

underwater stems—but *S. molesta* vigorously produces stolons and can rapidly build up a dense mat on the water surface, which may itself support a 'sudd' vegetation of other species, like *Scirpus cubensis* rooted in it. Subsequent investigation has shown that *S. molesta* sporocarps are mostly empty and sterile, and that in the formation of gametes, meiosis is so irregular that the plant may be completely sterile. It seems to be a pentaploid hybrid of two related South American species, *S. biloba* Radii and *S. auriculata*, perhaps of horticultural origin[348].

S. molesta was very much a problem plant on the new L. Kariba in the early 1960's for it impeded navigation and the use of fishing nets, and created anaerobic conditions in the shallow margins of the lake which were otherwise likely to be most productive of fish for the shoreline villages. *Salvinia* spp. do not form monospecific stands in South America, where they are endemic, and it is a mystery how the hybrid species has reached East and Central Africa and also Ceylon and Indonesia, where it does form such stands. Probably it was introduced by botanists, aquarists or water gardeners, as were other noxious water plants like the water hyacinth, *Eichhornia crassipes*. Removed from contact with their endemic competitors and grazers, such plants have frequently become problems demanding expensive control. *Salvinia* was known in the R. Zambezi at Kazungula above the Victoria Falls in 1949 but the river flow presumably prevented build up of large populations until the lake started to form downstream. By 1962 *Salvinia* covered a quarter of the lake, 10^3 km².

Around 1962 considerable concern was felt about the *Salvinia* problem in L. Kariba, for it seemed that it might permanently hinder navigation and fishing on the lake, and might eventually also cause problems in the turbines of the power station built underground next to the dam. At the Gebel Auliya dam on the R. Nile, *Eichhornia crassipes* (Fig. 11.3), a very bulky, beautifully-flowered, floating angiosperm has continually to be controlled to prevent its spread into irrigation ditches below the dam. It would rapidly block the water flow in them. Masses of the S. American *Eichhornia*, originally introduced as a business gift to Louisiana, now clog canals in most of the southern States of America and cost millions of dollars per year in control by herbicides and other means, to which an entire scientific journal is devoted!

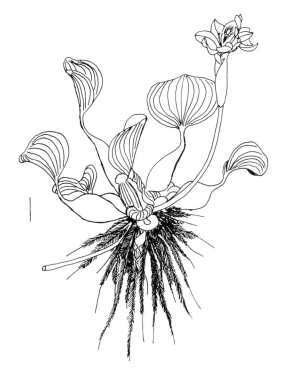

Fig. 11.3. *Eichhornia crassipes* (Mart.) Solms-Laub, Water hyacinth. Scale bar represents 3 cm. (Redrawn from Cook *et al.*[73].)

Fortunately the growth of *Salvinia* in L. Kariba declined in the mid-1960's as the supply of nutrients, provided by the rotting of flooded vegetation, was flushed out of the lake. The water replacement time of Kariba is about 4 years. There is evidence that nitrogen may be slightly more scarce than phosphorus in some African waters and *Salvinia* growth now seems limited by the availability of nitrate and covers less than 10 % of the lake surface, mainly in sheltered inlets. It seems that wave action, in the middle of the lake, would in any case have prevented its covering the whole surface.

Fears that serious floating weed problems would beset other new lakes have not been realized, though early periods of high biological production have been characteristic of most. The reason for the Kariba problem lay in the presence of an exotic species able to exploit the new conditions in the absence of any competitors.

As the internal nutrient loading declined in the new lakes, the hypolimnia progressively became more oxygenated (Fig. 11.4); the lakes have settled to a less fertile state, and the initially highly productive fisheries have also declined.

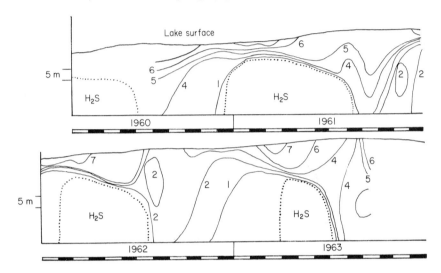

Fig. 11.4. Depth time diagram of oxygen concentration (mg l^{-1}) for L. Kariba during four years as the lake first filled completely (1958–1962) and just afterwards. Dotted line encloses depths and dates during which H$_2$S was formed. It was not formed after 1967. (Based on Harding [192a].)

11.3 Fisheries in new tropical lakes

The fishes of tropical rivers usually have catholic tastes in food as they are faced with greatly changing conditions during the year. In dry seasons they are confined to a narrow river bed, where predation on the crowded populations by carnivorous fish and reptiles may be high and other food may be scarce, while in the wet season, when the rivers flood over a much wider area, they move out into inundated swamps and grassland (see Chapter 7). There an abundance of detritivorous invertebrates grows among the flooded vegetation, which also provides spawning sites and cover. Rise in river level may be the 'trigger' which induces spawning in many of these fish, in contrast to those of temperate rivers which respond usually to changes in temperature or photoperiod, or both.

The stages of filling of a man-made lake create an extended period of rising river level for the original river fish fauna, and not surprisingly most of the river species flourish initially. As the lake level stabilizes some of them disappear from the main waterbody but may persist near river mouths if they need flowing water for spawning. In L. Volta the Mormyrids (elephant snout fish), which were important river species, are examples[403]. In contrast, species which feed on submerged macrophytes, detritus and periphyton, such as *Sarotherodon galilaeus* (Artedi), *S. zillii* (Gervais) and *S. niloticus* (L.), have much increased their proportion of the total fish population in the new L. Volta. Some niches in the lake may be unoccupied because food sources, scarce in the original river, may have no appropriate species to exploit them. No predator on open water zooplankton was present in L. Kariba until two such fish species *Limnothrissa miodon* (Blgr.) and *Stolothrissa tanganyicae* Regan were introduced from L. Tanganyika in 1965. This might be seen as an unwise move in view of the initial problems caused by the unintended introduction of *Salvinia molesta* but just as the latter now forms no problem and increases the diversity of habitat in the lake, the fish introductions seem not to have been retrograde in this case.

As well as the provision of suitable spawning habitat during the initial filling stage, a second factor has contributed to the early high fish production of new tropical lakes. This is the high production of invertebrates in the littoral zone where many of the original trees may remain, dead but standing, for several years. The submerged branches became covered with abundant periphyton, stimulated by the high nutrient levels, which in turn supported grazing invertebrates. Notable among these in L. Kariba and L. Volta was a mayfly, *Povilla adusta* Navas. Its larvae burrowed into the bark and rotting wood of the softwood trees flooded in the southern part of L. Volta, and took advantage of holes bored by beetles (*Xyloborus torquatus* Eichh.) in the hardwood trees around L. Kariba. The beetles attacked the drowned trees when they were exposed to the air during the annual drawdown (see below) of the lake level once the basin was full. *Povilla* normally is found in dead papyrus stems and rotting submerged branches in African lakes.

About a fifth of the future basin of L. Kariba was cleared mechanically of trees and bushes before inundation, and the debris burned. This was to create areas free of snags for the nets of a future fishing industry. At L. Volta such clearance was not done for economic reasons and because it was thought that the softwood vegetation would soon rot down. Fishermen in L. Kariba, however, have found it most profitable to fish, with appropriate small scale methods, in the areas where submerged trees still remain because fish production is apparently higher there. The submerged trees are an ephemeral habitat, however, which must eventually disappear under the action of wood borers, bacteria and water movement. The production they foster represents one of the last phases of the initial productivity surge of the new lakes, which gives way, after a few years, to a less productive, relatively stable ecosystem.

11.4 The stabilized phase of new lakes

The experiments unwittingly carried out in the creation of a new lake were of great importance in the development of understanding of limnology. According to long-held ideas on lake fertility, the initially fertile water should have remained highly productive and gradually increased in fertility, as nutrients from the catchment area were presumed to accumulate in the basin. The fact that these lakes became less fertile as an initial supply of nutrients was washed out or fixed in sediments is a major piece of evidence that lakes do not maintain their fertility unless an external loading of nutrients is continually applied. The sequence of early turbid phase, subsequent productive phase and then a decline in fertility is also mirrored in the development of natural lakes, e.g. those of the English Lake District, in the absence of human intervention (see Chapter 8).

For the few limnologists working on newly forming lakes in the Tropics, the almost weekly dramatic changes must have been most exciting. Soon after the

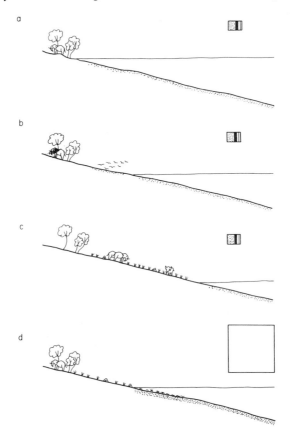

Fig. 11.5. Cycle of events during the annual drawdown and flooding at the edge of L. Kariba. (a) Represents the lake at maximum level with a moderate chironomid fauna in the surface sediments (vertical lines, *Tanytarsus* sp.; stippling, *Cryptochironomus linderni*; white, *Chironomus transvaalensis*; black, other chironomid species). As the lake level falls, stranded benthic invertebrates are fed upon by birds (b) and vegetation grows on the drying mud (c). Large mammals move in to graze, depositing dung over the shore. When (d) the lake refills the decomposing vegetation and dung provide a large food source for large populations of *Chironomus transvaalensis* which then develop. (Redrawn from McLachlan[339].)

closure of the Kariba dam, for example, hundreds of square kilometres of partly submerged woodland were festooned with communal spiders' webs, catching clouds of insects emerging from the expanding lake[18]. The stable phase has been equally interesting[338].

At the beginning of the rainy season it is normal practice for the lake levels to be lowered by several metres (drawdown) to accommodate the impending flood from upstream. This exposes, on shallowly shelving shores, wide expanses of lake bottom. On these in L. Kariba large populations of benthic inverte-brates, particularly of six species of chironomid larvae, are stranded and are immediately eaten by predaceous flies (*Lispe nuba* Widemann) and wading birds. A lush grass grows on the exposed mud flats as the wet season begins and this attracts herds of elephant, buffalo, impala and zebra whose grazing results in deposits of much dung (Fig. 11.5). As the water levels rise again and the large mammals retreat from the flats, the dung and remaining rotting grass provide a rich supply of nutrients for the overlying water and localized heavy phyto-plankton populations develop. They also provide much labile organic matter for the newly recolonizing benthic invertebrates, which initially are dominated by only one species, *Chironomus transvaalensis* Kieffer. As this species processes the initial heavy load of organic matter, under presumably quite extreme deoxygenation at the mud surface, it is superseded by the more diverse chironomid fauna which was present at the start of the cycle. *C. transvaalensis* was abundant in the early filling phase of L. Volta under rather similar conditions of abundant labile organic matter. In a sense then, the annual drying and flood-ing cycle at the margins of a new lake still reflects some of the major changes which went on as the lake formed.

11.5 Effects downstream of the new lake

Once a new lake has filled, the average river flow below the dam may only be a little less (because of evaporation over the expanded water surface of the lake) than it was before. If water is removed from the lake for irrigation, the flow may be less at some times but greater at others, in the dry season when the irrigation network drains to the river. The seasonal pattern of flow is, in most cases, con-siderably changed and this may cause great changes in the lower river ecosystem.

Firstly, the silt load carried by the river is deposited in the lake itself, and the turbidity of the outflowing water is much decreased. In the R. Nile this silt pro-vided an annual fertilization of delta lands and also a detrital food source which supported a valuable inshore sardine fishery in the Mediterranean Sea off the Nile delta. The fishery has declined greatly since the closure of the Aswan High Dam which impounds L. Nasser-Nubia, and agriculture on the delta now requires increased use of artificial fertilizers[442]. In the R. Niger, below the Kainji dam, fisheries seem to have been at least temporarily affected because traditional fishing methods depended on the fish being unable to see the nets used in the turbid waters, and the regulated river flow no longer drains some of the down-river swamps at times essential for rice cultivation in them[248].

Below the Volta dam a valuable shell fishery could have been completely eliminated if the dam had been located only 10–15 km further downstream[280]. The clam, *Egeria radiata*, is collected from submerged sand banks free of weed in the river and sun-dried for sale. The adult clams thrive in fresh water, but the spawn and juveniles (veligers) require water with about $1°/_{oo}$ salinity. Spawning occurs in the dry season when, before impoundment of the river, sufficient salt water moved in from the coast to give a salinity of $1°/_{oo}$ about 30 km from the river mouth. During filling of L. Volta the river flow was reduced and the $1°/_{oo}$ level moved upriver to about 50 km from the sea. This caused some problems in that the collecting industry was displaced from its usual markets but these were not serious. With impoundment the dry season flows have in-creased so that the $1°/_{oo}$ level is now only 10 km from the coast, and had the dam been placed much lower downstream would have been displaced into an area of the estuary unlikely to have provided suitable habitat for the clam.

Fish migration is another feature of the original river system which may be interfered with by a dam, and has been most studied in temperate rivers. Anadro-mous fish, like the salmon, must be 'helped' over the dam by a pass or lift. This may comprise a cascade of water moving through a stairway of pools through which the fish can jump in easily attainable stages. Alternatively the fish may be directed by underwater concrete flanges into a pool at the foot of the dam which is incorporated into a lift for ascent of the dam wall. More problems are en-countered in safely directing juvenile fish downstream past the dam and many may be killed or damaged by the power station turbines.

11.6 New tropical lakes and human populations

Modern limnological studies are incomplete if they ignore the effects of man on lakes and rivers. The creation of new lakes in the Tropics has led to some new problems and intensified some pre-existing ones. The new ones are largely out-side the scope of the book for they are sociological and psychological. River valleys in the tropics are relatively densely populated and the new lakes have displaced as many as 50 000 people (Kariba), 42 000 (Kainji), 80 000 (Volta) and 120 000 (Nasser–Nubia). Although the moving of the villages was planned well in advance, the modification of a culture closely geared to the seasonal flooding and shrinking of the river and the fishing and farming opportunities presented by this was no easy matter. During the period after the move there was evidence of increased mortality, not all of it attributable to infectious disease. It is, how-ever, with water-borne diseases in the tropics that limnological study can aid public health measures. Water-borne diseases were, of course, features of the rivers before inundation, but the increased shoreline of the lakes and particu-larly the networks of irrigation channels which are associated with some lakes, e.g. L. Nasser-Nubia, and with the many small lakes formed by damming for irrigation or fish culture have exacerbated locally what is a widespread and serious problem.

11.7 Trematode diseases

Three groups of water-borne diseases other than the bacterial and viral ones mentioned in Chapter 10, are important to man—the trematode (flatworm) diseases carried by freshwater snails, the nematodes or filariae (round worms) transmitted by arthropods, particularly insects, and the viral and protozoon diseases carried by mosquitoes.

The best known trematode disease is schistosomiasis or bilharzia. World-wide it infects between 150 and 200 million people at any one time. The causative animal is one of three *Schistosoma* species, *S. mansoni*, *S. japonicum* and *S. haematobium*, which exist in the human body as pairs comprising a cylindrical female, around which is folded a slightly larger male. The pair move around the blood stream and produce eggs which are shed in the faeces (*S. mansoni*) or urine, but the adult population itself does not multiply in the body. It may increase only by repeated infection from contaminated water until sufficient pairs are present (up to 400) for serious symptoms—cystitis or colitis, blockage of blood vessels, sometimes brain infection—seriously to affect the victim. It is a debilitating rather than a killing disease.

Eggs released by urination or defaecation into a lake margin or slow-flowing river hatch out into miracidia, about 0·1 mm long, which can live in the water for 24 hours. If the habitat is suitable they may come into contact with a gastropod snail of the appropriate genus which they quickly penetrate. *S. mansoni* is carried by snails of the genus *Biomphalaria*, *S. japonicum* by *Oncomelana* and *S. haematobium* by *Bulinus*. Such snails are herbivores feeding largely on periphyton or macrophytes and are favoured by dense weed-beds, such as have formed in the western margins of L. Volta (largely of *Ceratophyllum demersum*). Previously, schistosomiasis was uncommon in the faster-flowing reaches of the R. Volta where current flow prevented much weed growth. At present some 69% of children on the shores of L. Kariba have *S. haematobium* and 16% *S. mansoni*, and the effect of the lake formed behind the first Aswan dam, built in the 1930's on the Nile was to increase the *S. mansoni* infection rate from 2 to 75%.

In the snail the miracidia divide and metamorphose into thousands of tailed cercariae, each about 0·4 mm long, which are released in clouds into the water. There they are infective for up to 8 hours and the life cycle is completed when they contact the bare skin of a person, who might at the time be washing, fishing or excreting into the water. The cercariae secrete a substance which enables their rapid penetration (3–5 min) often through a hair follicle, after which they develop into adults.

The disease is rife among riverine and lake peoples. *S. japonicum* is found in Asia, particularly in flooded rice paddies, and *S. mansoni* and *S. haematobium* favour gently running water, as in irrigation channels and still water respectively. Bilharzia can be treated by drugs or surgery for blood vessel blockages, but could be prevented by proper sanitation if irrigation schemes were not increasing in number faster than public health measures. In the short term, molluscicides have been used and much research is now devoted to controlling the host snail populations by fish or by competition with introduced snails, e.g. *Marisa cornuarietis*, which are unable to carry the metacercariae.

Other snail-borne trematodes are less widespread. In S. Korea and China, the practice of eating raw freshwater crabs and crayfish (*Potamon*, *Sesarma*, *Eriocheir* and *Astacus*) steeped in rice wine leads to spread of a lung fluke, *Paragonimus westermani*, carried by these as well as by snails. In Nigeria this disease also infects crab-eating mongooses. A parasite of the liver, *Opisthoreis sinensis*, which, as well as snails of the genera *Bulinus* and *Parafossarulus*, infects freshwater food fish is transmitted by the practices of eating raw fish and freshwater shrimps (apparently thought to be a cure for nose bleeding in parts of the East), and using human faeces to fertilize fish ponds.

Aquatic plants are involved in the transmission of the tropical Busk's fluke (*Fasciolopsis buski*) and the world-wide liver fluke (*Fasciola hepatica*). Busk's fluke infects and obstructs the bowels of up to 85% of some populations in China, Assam and India and is passed via faeces to freshwater snails (*Planorbis* spp.). The cercariae, when they emerge, encyst on the roots or fruits of aquatic plants like *Trapa natans*, the water chestnut, which are eaten, particularly by children. Domestic pigs are an alternative host and in Thailand the penning of pigs over channels connected with rice fields promotes both the growth of rice and the transmission of Busk's fluke. Liver flukes are similarly transmitted in developed countries by faecal contamination by sheep or cattle of water, *Lymnaea* sp. (the snail host) and edible water cress (*Rorippa nasturtium-aquatica*), which harbours the cercariae.

11.8 Filarial diseases

Few people can fail to have been horrifically fascinated by photographs of elephantiasis, a gross swelling of the lower limbs and in males sometimes the scrotum. Elephantiasis is a symptom of several diseases (filariasis) caused by nematode worms, and transmitted by mosquitoes. The greater part of the mosquito life cycle is spent as an aquatic larva and most means of control of the disease must be aimed at this stage, though individuals can be treated with modern drugs. Mosquitoes are such important vectors of disease[460] that a generalized account of their life history is necessary. Details, varying from species to species, may be crucial to control of a particular vector. Eggs are laid singly or in rafts of up to several hundred on suitable water surfaces. The precise nature of this water surface—whether it is organically polluted or not, or overhung with vegetation or among emergent or floating macrophytes—often determines whether oviposition occurs. The eggs are denser than water but are coated with a waxy layer which allows them, if floating, to be supported in the surface tension film.

After a few days, larvae hatch from the eggs. The larvae are air breathers and may attach to the surface film of the water by a pair of open breathing tubes (*Culex*) or lie along the surface so that oxygen can diffuse in through the body spiracles. Some inject their breathing siphon, below the water surface, into the root or stem air spaces of floating or submerged macrophytes (*Mansonia*). The larvae generally feed by filtering particles from the water with brushes on their mouthparts and undergo several moults before transforming into rather distinctive shaped pupae. These too may have breathing tubes, but instead of extending, if present, from the eighth abdominal segment they project from spiracles on the thorax. Pupae do not feed, and move rapidly from the surface if disturbed. Within a few days the adult emerges from them. Males do not take blood meals but may feed on plant nectar. Females may also feed on nectar but usually require at least one blood meal before they are able to lay fertile eggs. Their mouthparts are highly specialized ducts carrying salivary secretions containing anticoagulants, piercing stylets, and a tube for sucking blood. The whole is housed in a protective sheath[263].

In equatorial Africa, the Pacific islands, the Far East, the West Indies and S. America, Bancroftian filariasis is caused by the nematode *Wuchereria bancrofti*. There are about 250 million current cases, with the heaviest infection in India, and the parasite may be carried by several genera of mosquitoes—*Anopheles*, *Aedes*, *Culex*, *Mansonia* and *Psorophora*.

The adult nematode lives in dilated lymph vessels in the human host and blockage of these causes the swelling called elephantiasis. The adults produce many juvenile forms called microfilariae between about 2200 hr and 0200 hr, except in Fiji, Samoa and Tahiti, where the mosquitoes bite during daytime and the microfilariae are produced accordingly. The mosquito ingests some microfilariae with its blood meal and these form sausage-shaped larvae in the muscles. Eventually the larvae migrate to the head of the mosquito and into the labella or proboscis with which the mosquito penetrates skin to suck its blood meal. Returned to a human host the larvae mature to form reproductive adults in about three months.

Mansonia, the mosquito vector of another filariasis, *Brugia malayi*, can be controlled by weed clearance since its eggs are laid on the undersurfaces of floating water plant leaves, e.g. *Eichhornia crassipes* and *Pistia stratiotes*, and its larvae can only breath by tapping the air spaces of the plant roots. In general, Bancroftian filariasis is a disease of urban areas since its mosquito vectors can breed in puddles and drains, temporary bodies of water, but *Mansonia* need more permanent weed-beds in larger bodies of water, generally in more rural areas, and *Brugia* is more common in these.

Three other nematode diseases, the first much-feared, are carried by arthropods with aquatic stages—onchocerciasis (river blindness), loa loa and guinea worm. Onchocerciasis is carried by blackflies, or buffalo gnats (*Simulium* spp.), and is a disease associated with fast-flowing rivers where *Simulium* thrives (see Chapter 4). Its incidence may decline over a river stretch permanently flooded by a dam, but may be rife around the dam wall, where cascading water provides ideal conditions for *Simulium*, although there it can be controlled by use of insecticides added to the emerging water. There are some 20 to 50 million cases in Africa and perhaps 200 000 in South and Central America; vast areas bordering rocky streams are uninhabitable because of the disease whose vector has a flight range of more than 20 miles. Microfilariae are released into the blood of infected victims from adults living in nodules in the skin. When adult *Simulium* bites it scarifies a small area with its short proboscis and microfilariae are attracted to the site by a salivary secretion and are sucked up with the blood meal and transferred to the next individual bitten. The microfilariae cause the worst symptoms of dermatitis, skin thickening, giving an aged appearance, dwarfism (by invasion of the pituitary gland in children), and fluffy opacities of the eye, leading to glaucoma, cataracts and eventual blindness.

Another filarial disease, loa, loa, infects 75–90% of forest dwellers in Zaïre and 25% in Nigeria. It is carried by the softly softly fly, *Chrysops* sp. (Tabanidae), which prefers to bite just below the knee, and breeds in densely-shaded, forest streams floored with leaves and mud. The nematode may cause encephalitis if the microfilariae infect the spinal fluid, but can be treated by drugs or by surgery. The adult worm, which can be 70 mm long, sometimes migrates across the eye at a rate of 1 cm min^{-1}, long enough for it to be removed, under local anaesthetic, with a bayonet-edged needle passed under it.

Another nematode worm that can be removed mechanically as it passes close to the skin surface is the Guinea worm, *Dracunculus medinensis*. It can be wound out on a thin piece of wood over 3–4 weeks (it is 5–8 cm long) and may be the 'fiery serpent' of the Israelites, which Moses recommended should be set upon a pole! The adults live in blisters under the skin and release larvae when the blister softens and bursts in contact with water. The intermediate host is the zooplankter *Cyclops* which is ingested in drinking water. The disease is mainly one of relatively arid regions where drinking water supplies are scarce and intensively used, such as in Mauritania, N. Ghana, Chad and S. Iran. It causes arthritis and secondary infection, e.g. by tetanus, which may penetrate along the track of the adult worm moving beneath the skin. Large quantities (up to 500 ml) of pus may be released if a female bursts deep in the tissues. However the disease is readily controlled by piped water supplies.

11.9 Protozoon and viral diseases

Malaria and yellow fever are sufficiently well documented elsewhere for them to receive but brief treatment here. Both the protozoon *Plasmodium* of malaria and the viral yellow and related fevers are carried by mosquitoes. In the former case by *Anopheles gambiae* and *A. funestus* in particular, and by *Aedes aegypti* in the latter. Malaria causes 2 million deaths per year from a case-load of 250 million, though yellow fever is a very serious disease only in white immigrants to the tropics. There are, however, many undescribed and uninvestigated viral 'fevers', some of them very serious, whose vectors are probably also mosquitoes.

11.10 Man-made tropical lakes, the balance of pros and cons

The larger man-made lake projects have attracted much criticism and scepticism that their overall costs would exceed the benefits they would confer. Much of this criticism has come from western observers, living in countries which have long enjoyed the benefits of abundant power and well-watered agriculture, and much of it, based on the early 'problem' period of formation of the new lake, has not been borne out once the lake has filled and reached some sort of equilibrium. Other criticism has been well founded, but it will be some time before a quantitative cost/benefit balance can be drawn up. Great endeavour will always attract carping as well as excessive optimism.

Some observations can be made, however. Without doubt the Volta scheme has provided, and will provide, abundant, cheap, electrical power for Ghana for many years. Its effects have yet been only felt in a minor way, for the loans of about £100 million (US $170 million) necessary to build the scheme are yet being paid off, and the non-Ghanaian Kaiser Aluminium, a company which financed the aluminium smelter which uses much of the generated power, will enjoy a pre-agreed low cost for its power for the rest of this century. Nonetheless, Tema, where the smelter is situated, has established much new industry to supplement the national income and to reduce dependence on imported goods. It was expected that 20 000 tons of fish would be harvested per year from the lake once the initial production surge had passed, but it seems that this was pessimistic for catches are currently running at 40 000 tons yr^{-1} after the peak of 60 000 yr^{-1} in the late 1960's. This has compensated for the reduced imports of beef protein from countries to the North affected by the Sahelian drought[180].

Looked at in the long term, the costs of the Volta scheme should decrease as the industrial benefits flourish. For the moment the huge lake has disrupted land communications, and water transport, still impeded by remaining submerged trees, has yet to replace them. There are signs that it will. A planned irrigation scheme has, for lack of finance, not yet been constructed, but cultivation of the wet mud flats left on drawdown of the water level allows cropping of maize, tomatoes, cow peas, sweet potatoes and others at times when lack of rain prevents growth elsewhere in the country. Schistosomiasis has certainly increased, but given political stability a progressively booming economy should eventually provide the public health measures necessary to break the parasite's cycle.

The Aswan High Dam in Egypt has provoked the greatest controversy and its advantages and disadvantages are on such a large scale that a balance cannot yet be drawn[442]. Again it has provided a constant supply of energy which will allow industrial development, but also it supplies an extra 19×10^9 m^3 of irrigation water for regular and numerous crops in a country almost exclusively dependent on agriculture but only 3% cultivable. The Aswan scheme has increased this by a factor of 1·6. Early fears that the lake would result in a net loss of water by seepage and evaporation have not been borne out.

The drawbacks of Aswan are more serious, however, than those of the Volta, which are largely those of delays in being able to take advantage of the opportunities offered. Egypt is arid and the evaporation of irrigation water is leading to increased soil salinity, which could threaten yields. The lack of fresh water flow to the Nile delta is also leading to encroachment of sea water further inland and soil salination. L. Nasser produces 10 000–13 000 tons of fish annually, but the detritus carried to the sea by the unimpeded Nile supported a sardine fishery of 15 000 tons yr^{-1} and probably other fisheries as well. The sardine fishing has declined completely, and the L. Nasser fish must be transported long distances to the centres of population near the coast. The fishery balance is certainly a negative one.

The uses of water and associated biological problems discussed in Chapters 10 and 11 have had strong utilitarian and economic undercurrents. The conservation of natural aquatic communities might, in contrast, seem something of a luxury that might be afforded by the richer societies, just as the arts might be supported only after food and shelter have been provided. To some extent this is true, but the costs of seriously disrupting natural fresh waters may outweigh the benefits of doing so. This thesis is discussed in the final chapter.

11.11 Further reading

McLachlan[339] is an excellent summary and key to much of the literature on tropical man-made lakes, and Petr[404] widens the essentially biological discussion of McLachlan to cover economic and social issues. Ackerman *et al.*[1] in particular, and also Lowe-McConnell[299] and Obeng[383] are symposium volumes packed with information on the problems of man-made lakes, and Rzóska[442, 443], Balon & Coche[19], Imevbore & Adegoke[248], and Davies *et al.*[89], Davies[88] and Bond *et al.*[39] respectively deal with Lakes Nasser-Nubia, Kariba, Kainji and Cabora Bassa. Sculthorpe[459], and a well-written article by Holm *et al.*[218] deal with the problems that growths of aquatic plants may cause. Muller[374a] is a fascinating and detailed account of tropical and temperate diseases caused by both flatworms and roundworms, and Peters & Gilles[401b] provide a profusely illustrated account of the whole range of tropical water-borne (and other) diseases.

CHAPTER 12
THE CONSERVATION OF
FRESHWATER ECOSYSTEMS

12.1 Introduction

Fresh waters are perhaps the most vulnerable of habitats and those most likely to be changed by the activities of man. This is for several reasons. Firstly, water is subconsciously disliked by many people, perhaps through fear of drowning, or past associations with once vast and inhospitable stretches of marsh. Secondly, lakes act as sinks for many of the products of human activity in their catchments, while the rivers are naturally-provided drains for the removal of waste to the sea. Thirdly, water is nevertheless a valuable and essential resource, often scarcest where it is required most and hence needing to be stored. Reservoirs have frequently replaced riverine habitats and their previously more diverse aquatic ecosystem and have caused problems with fish migration. Lastly, as with other ecosystems, man has had the notion that by some form of management he can 'improve' them.

Natural ecosystems represent the current equilibria in a continuous process of change and adjustment. All chemical systems (including living ones) tend always to adjust towards a state of maximum homeostasis when confronted with changing external conditions. This means that they develop mechanisms which tend to annul or minimize the effects of random external changes (e.g. short term weather), or to anticipate and perhaps make use of regularly predictable ones (e.g. tidal and photoperiodic changes). This is achieved in ecosystems through the continuous operation of natural selection on the organisms comprising them. It has resulted in production of arrays of species best fitted, in the prevailing environment, to maintain a system of maximum homeostasis while simultaneously retaining a capacity, through genetic mechanisms, to keep on adjusting to inevitable continuing changes in the natural physical and chemical environment. It is impossible to 'improve' an ecosystem formed and regulated in this way. Though we may not yet fully understand them there are always good reasons, in undisturbed ecosystems, why, for example, fish growth is moderate, or aquatic macrophyte beds are extensive. Attempts to increase the fish growth or clear the weed-beds will always result in a chain of repercussions which will inevitably have to be counteracted, perhaps at considerable expense. Geomorphological systems may also be viewed in the same terms. A river flood-plain may only be under water once every few years, but it is still part of the river bed, essential for the efficient and natural disposal of occasional high discharges. Attempts to change its use from this purpose to farming, for example, must mean

expensive engineering works to increase the height of the river banks to accommodate the flood water. Objective economic cost-benefit analyses have rarely
been produced for such schemes. In the long run one suspects that, so far as the
general public good is concerned, the balance might be very unfavourable.

Inevitably, however, aquatic ecosystems have been changed either from a
humanitarianly defensible need for water-borne disease control or for the
establishment of farmland, from simple ignorance of the consequences of pollution and eutrophication, or from the need to prevent flood damage to property
built on natural flood-plains through stupidity or greed. Conservation of natural
ecosystems can be defined formally in several ways. One is that it is a process of
maintaining, through whatever means necessary or available, the maximum
diversity of organisms and ecosystems; another is that it is 'the total management of the rural areas of this country for the fair and equal benefit of all groups
which have a direct interest in their use[543]. Both of these are compromise
definitions. What conservation of freshwater really amounts to is often making
the best of what is left after waterways and their catchments have been used for
the more obvious needs of society. It is true that most developed countries have
legislation permitting the reservation of particularly interesting areas, and that
attitudes towards the retention of as much natural habitat as possible are
changing favourably. It is also true that enlightened countries are developing
ways of restoring much changed lakes and swamps and of limiting water pollution, and that powerful allies may be found in the large numbers of people who
enjoy angling, wildfowling or birdwatching.

Nonetheless, on balance, more waterways are being degraded and lost than
created and restored. The agents of change are varied, but two, drainage and
river management[172] and eutrophication seem of widespread significance.
Drainage usually involves civil engineering works to move water to the sea as fast
as possible and manifests itself in river dredging and straightening, building of
flood banks protected with steel or concrete pilings, and the clearance of water
weeds. It may also include the filling in of small ponds and marshes (wetlands)
to create farm or building land. Eutrophication has already been mentioned
(see Chapter 10) in connection with water supply, but it also has many other
ramifications. Other threats are toxic pollution, overfishing, the introduction of
exotic species and recreation pressure. Most waterways are affected by a complex
of factors, often interconnected, which are best illustrated by actual cases. The
N. American Great Lakes, the Florida Everglades and the Norfolk Broads will
be discussed in detail.

12.2 The North American Great Lakes

The North American Laurentian Great Lakes stretch almost half way across
the continent, and are among the largest lakes in the world (see Fig. 2.2). They
drain a huge area, originally of conifer forest to the north and deciduous forest
to the south, and were once probably all well-oxygenated, clearwater lakes with

maximum total phosphorus levels probably less than $2\mu g$ P l^{-1}—that is towards the lower end of the fertility continuum (see Chapter 2). The waterway they provided from the Atlantic Coast to what is now the mid-west and plains region of the U.S.A. and Canada was a main route by which the continent was explored by Europeans in the seventeenth and eighteenth centuries. The discovery of minerals and cultivable land led to great increases in population so that, for example, the catchment of L. Erie, which in 1750 was a largely unexploited wilderness supporting perhaps 100 000 people, now contains some 12 millions with their associated industry and agriculture. L. Superior has changed the least in the last 200 years—its catchment still contains much intact forest—and changes have been greatest in the shallow L. Erie, and intermediate in the other three Great Lakes, Michigan, Huron and Ontario. The most apparent change has been eutrophication, though other changes have been equally significant.

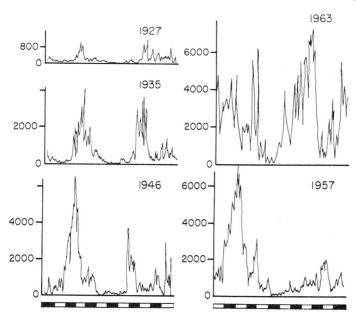

Fig. 12.1. Seasonal changes in total number per ml of phytoplankters in L. Erie during five representative years spanning the 1920's to 1960's. Counts were made daily on water pumped from the lake for use as a water supply. The steady increase consequent on the progressive eutrophication of the lake is shown. (Redrawn from Davis[91].)

In western L. Erie, for example, there has been a progressive change from late summer diatom populations to populations of blue-green algae since 1950, with an overall sevenfold increase in phytoplankton crop[91] (Fig. 12.1). The total solids concentration has increased by a half since 1920, and despite greater than 70% oxygen saturation at all depths in 1925–1930, levels of only 6·3% were recorded in deep water in 1969–1970. Populations in the sediments of the western basin of the oxygen-sensitive nymphs of mayflies *Hexagenia rigida* and

H. limbata occulta, which numbered around 400 individuals m⁻² until the early 1950's are now reduced to less than 1 m⁻². Similar increases in fertility, with consequent declines in diversity and in water quality, have been recorded in parts of Lakes Ontario, Huron and Michigan[250], particularly in partly enclosed bays and near to cities, though the offshore waters have changed to a smaller extent. Eutrophication has largely been a problem of these lakes in the twentieth century, yet changes in the lakes began in the eighteenth and nineteenth centuries. These changes are reflected in the commercial fish catches of the waterway[67, 465, 466].

Table 12.1 gives data on total commercial fish yields in the American Great Lakes over the past century. These have remained relatively constant in Lakes Superior, Michigan and Erie, but have declined in Lakes Huron and Ontario. The data mask the fact, however, that great declines in the fisheries for prized salmonid, coregonid and other fish have been compensated for, where the overall catch has not declined, by catches from a much less diverse fish community supported by increased fertility and production in the water. The reasons for the changes, in the order in which they became significant were: intensive selective fishing, modification of the tributary rivers, invasion or introduction of marine species, and lastly, eutrophication.

Table 12.1. Commercial fishery yields in the N. American Great Lakes in the nineteenth and twentieth centuries. Values are given in lb acre⁻¹. (From Smith[466].)

	Ontario	Erie	Huron	Michigan	Superior
Pre-1870	3·5–4·02				
1879–89	1·46	8·54	1·36	1·69	0·40
1890–99	1·08	9·42	1·62	2·62	0·44
1900–09	0·8	6·49	1·69	2·93	0·71
1910–19	0·94	10·04	1·39	1·89	0·77
1920–29	0·98	7·19	1·27	1·43	0·75
1930–39	0·79	6·86	1·45	1·66	0·94
1940–49	0·60	6·38	0·79	1·61	1·09
1950–59	0·51	8·37	0·65	1·97	0·83
1960–69	0·47	8·21	0·55	2·39	0·62

The changes began in L. Ontario, the lowest basin on the waterway and the earliest of the lakes settled by Europeans. Progressively the changes have spread upstream so that the state of the upper Great Lakes fisheries in 1970 was approximately that of L. Ontario in 1900. Atlantic salmon, *Salmo salar*, were only ever present in L. Ontario in the Great Lakes system, for their upstream spawning migration from the sea was blocked by the Niagara Falls. Fishing for salmon began in the 1700's, was reduced by 1880, and had ceased altogether by 1900. The fishing was intensive, but the species survived it for many decades, and seems to have disappeared because of changes in the streams flowing into the lake where it moved to spawn. In the nineteenth century, forests were cleared over much of the catchment, dams were built on the streams so as to operate water-power for the saw-mills, and much waterlogged sawdust floored the streams themselves.

Clear felling of trees has two effects on drainage streams, other than chemical ones (see Chapters 1 & 2). It reduces their flow in summer because more water evaporates than previously, and it increases their temperature, because the streams are no longer shaded. Repeated attempts to re-establish a salmon fishery in L. Ontario have failed because the spawning stream waters are now too warm and the flows insufficient to maintain the cool, oxygenated water and gravel bottom which the salmon require.

In the upper lakes and in L. Ontario, after the salmon declined, whitefish (*Coregonus clupeiformis* DeKay) and lake trout (*Salvelinus namaycush* (Walbaum) were both heavily overfished, and fisheries for them had declined in L. Erie by 1940 and L. Huron by the 1950's. A trout fishery is now maintained by annual stocking. Gradually, however, as the trout and coregonine fisheries became less profitable, lake herring, or cisco (*Leucichthys artedi* Le Sueur) and deep-water ciscoes (*Leucichthys* spp.) were progressively fished. Currently, the lake fisheries depend on percids and other fish which have been favoured by different changes taking place in the lakes.

Overfishing for these salmonid and coregonid species was probably not the only cause of decline; the sturgeon (*Acipenser fulvescens* Rafinesque), although a valuable commercial fish, was deliberately removed because of the damage it did to nets. By 1890–1910 it had almost disappeared, partly from overfishing, as its valuable by-products (gelatin, isinglass, a bladder extract used in clarifying beverages and sizing textiles, and, of course, caviar) were prized, and partly from ruination of its spawning habitat (see below). It is a particularly vulnerable fish, with a low growth rate and late sexual maturity. Even following a ban on commercial exploitation, it is now common only in parts of L. Huron.

The changes in tributary streams which so affected salmon reproduction in L. Ontario became widespread elsewhere in the late nineteenth and early twentieth centuries. Most of the commercially exploited fish were those of shallow water, which entered streams to spawn. Although the physical environment for spawning may not have changed too seriously for many of the species, congregation of sturgeon, coregonids, and percids in the water below mill dams provided easy fishing with seines, dipnets, and even spears. In this way two agents of change may have combined to produce declines. In the case of the sturgeon, drainage of swamps and marshes associated with the headwaters removed a favoured breeding habitat.

Between 1860 and 1880, two marine species, the sea lamprey (*Petromyzon marinus* L.) and the alewife (*Alosa pseudoharengus* (Wilson)) entered L. Ontario. They may have come up the St. Laurence River, which opens out at the northern edge of their ranges, or via the canal built in the early 1800's between the Hudson River and L. Ontario. The Hudson River enters the sea at New York. The lamprey is not a fish, but a cyclostome, which is parasitic on fish. It feeds by rasping fish flesh with a tongue after it has attached by a sucker which forms its jawless mouth. Both species could have entered L. Ontario at any time in the previous centuries, but if they did, they were unable to establish significant populations. Possibly the community changes caused by fishing and stream modification provided suitable niches for them. The Erie and Welland canals, both of which, for navigation purposes, by-pass the Niagara Falls, removed an otherwise impassable barrier to migration of these species upstream to the upper lakes. The lamprey reached L. Erie by 1921, L. Huron and L. Michigan in the early 1930's and L. Superior in 1946. The alewife generally lagged behind, reaching Erie and Huron in 1931–1933, Michigan in 1949, and Superior in 1953.

Both immigrants like deep water for parts of their life cycles, and are neither abundant nor problematic in the relatively shallow L. Erie. In the other lakes they have caused major changes. The pattern appears to have been one of parasitism by the lamprey, firstly of the larger deep water carnivores, such as lake trout, burbot (*Lota lota* L.), and deep water ciscoes, then of smaller species, until the lamprey population itself declines. Reduction of the large piscivores then allows increase of the alewife, predation on which is reduced, while the alewife feeds aggressively on large zooplankters and benthic Crustacea, such as *Pontoporeia*. These are also the main food sources of the young of the piscivores, whose populations cannot then recover from the lamprey depredations. Alewives may even increase their competitive advantage by also feeding on the large piscivore young, and are now among the commonest Great Lakes fish.

With all of these influences it is difficult to separate out the effects of progressive eutrophication this century. Increasing nutrient enrichment initially leads to increases in littoral aquatic macrophyte growth as well as to increased phytoplankton and zooplankton production. The latter may increase the growth rates of indigenous fish, while the former, by providing extra habitat for snails, may increase the incidence of parasitism by trematodes using snails and fish as successive hosts. As lakes become more fertile there is a decline in the diversity of the fish community, while some tolerant species may become extremely productive.

The mechanisms by which certain species disappear on eutrophication are not fully known. They may involve loss of gravelly spawning habitat as increased sedimentation covers it with organic deposits which easily become too deoxygenated for salmonid eggs to survive; the burbot lives and spawns in the deepest parts of well-oxygenated lakes, from where hypolimnial deoxygenation may force it at an early stage. Extensions of marginal weed-beds, in providing cover, spawning habitat and abundant invertebrate food may favour some species unable previously to compete successfully. Because of the complexity of fish biology, however, it is impossible yet to state quantitatively at what stage of enrichment major changes in fish community will occur. Fig. 12.2 gives a qualitative schema which may have general relevance in temperate regions[70]. What can be said is that conditions leading to summer total phosphorus or chlorophyll *a* levels in the epilimnion of the order of 20–30 μg l^{-1} have led, in L. Erie and in similar large lakes such as the European Bodensee, to predominance of percid and cyprinid fish. Extreme eutrophication, which may result in total phosphorus levels ten times as high, leads to little further change in the fish communities,

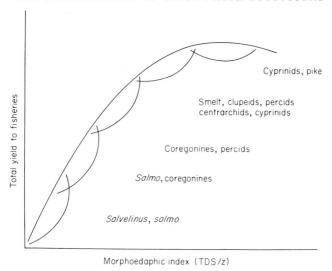

Fig. 12.2. A suggested relationship between fisheries yield and nature of the catch in progressively more fertile lakes. The use of the morphoedaphic index is discussed in Chapter 2. (Redrawn from Colby *et al.*[70].)

compared with that which increases them from less than 10 μg l^{-1} to more than about 20 μg l^{-1}.

This account of the N. American Great Lakes has been much simplified and more details will be found in Scott & Crossman[457] as well as the articles referred to. Such large lakes are not uniform and the diversity of their biota and the complexity of its interactions is still very great. Attempts are being made to reverse the changes that have occurred by use of phosphate-stripping (see Chapter 10) to limit eutrophication, and restriction (ultimately it is hoped a complete ban) of detergent phosphate use, by restocking of some fish and the use of lampricides. But the very long turnover times (flushing times) of the water masses of the larger lakes delay the effects of remedial action. Long-term planning and foresight are thus essential. The fact that four of the lakes straddle two countries, the U.S.A. and Canada, adds administrative complications, which are, apparently, being overcome with alacrity. The sheer volumes of the water masses concerned have fortunately buffered the changes in this case, but not in the following two that will be considered.

12.3 The Florida Everglades

The very name of the Everglades conjures an impression of perpetual greenness, and many people imagine the area as a tangle of dark and dripping woodland. In fact the Everglades is a complex of ecosystems based on a shallow river almost as wide, up to 100 km, as it is long, 180 km, slowly flowing from the lowlands around L. Okeechobee to the sea at the south western tip of Florida, U.S.A.

(Fig. 12.3). The flow has never been rapid—only a few hundred metres per day —and the river is at most 30 cm deep over much of the area. The main fresh-water community is of emergent aquatic macrophytes, dominated by the saw-sedge, *Cladium jamaicensis* Crantz., which grows several metres tall, so that the description 'River of Grass' is embodied in the title of a well-known book on the area by Marjory Stoneman Douglas[109].

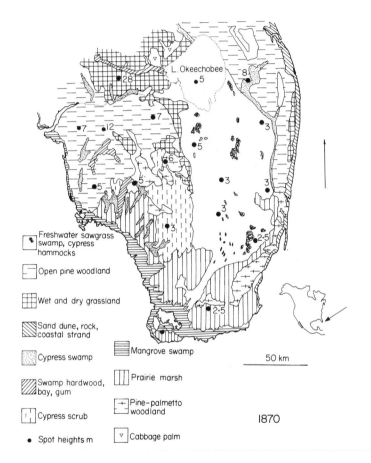

Fig. 12.3. Natural vegetation of the Florida Everglades area as it was in 1870. (Modified from Caulfield[66].)

The southern tip of Florida is floored by a porous limestone (oolite) and is very flat. Running along the eastern coast is a ridge·of rock some 6 m above mean sea level, and the west coast also bears a slightly wider but still narrow and subdued 'upland'. In the basin between the two lie the Everglades, with the land dipping only 7 m from L. Okeechobee towards the sea. Rainfall is very seasonal in the area; most of the 200 cm or so falls in June and July, sometimes in heavy showers associated with hurricanes which also occur in September. The winter months are dry and the natural water supply to the Everglades is then least.

To the south of L. Okeechobee was an area of several thousand hectares of peat, up to 4 m deep, laid down over several thousand years in swamps associated with the lake. This acted as a sponge, taking up much water in wet periods and releasing it steadily all the year around. The flow of water penetrated the Everglades along three main water-courses or sloughs, running south or south-westwards—the Shark River slough, the Lostman's River slough, and the Taylor slough. As the water reached the sea the saw-grass community merged with transitional vegetation and was then replaced by a dense, tangled forest of mangroves up to 30 m high. Mangrove forest is the sub-tropical and tropical equivalent of salt marsh and is subjected to much the same intertidal regime. The steady flow of fresh water confined the mangrove to the coast by stopping penetration of salt water inland and mangrove trees were not found beyond a few km from the river mouths.

Dotted among the sawgrass vegetation on islands of oolite penetrating a few centimetres above the general basin level are woods of slash pine (*Pinus caribaea*) and palmetto (*Serenoa serrulata*) and sometimes clumps of hardwood trees. Small groves of other trees, swamp and pond cypresses (*Taxodium* spp.), occupy depressions in the oolite where peat has accumulated. The saw grass does not undergo succession to drier forest (see Chapter 8) because light fires, begun by lightning at the start of the rainy season, have occasionally burnt the surface vegetation and litter, though left undamaged the deeper peat and rhizomes of the saw-grass[434]. The pine woodland is also fire-resistant and is replaced by hard-wood only after a long period when fires have not intervened.

This complex of plant communities in turn supports a rich fauna, of which the birds, some 300 species of them, are best known, for the area is at the junction of several migration flyways[230]. Huge flocks were described by the naturalist John James Audubon in 1832 and it is still possible to see flocks of 100 000 ducks. Ibis, spoonbills, herons, pelicans, coots, plovers, gulls, terns, storks, and cranes are also to be seen in abundance. Other vertebrates form a no less exciting collection. The alligator is perhaps best known, but the American crocodile lives in the mangrove swamps, and snakes and turtles are common. Some 25 mammal species—opossums, raccoons, wildcat, otter, white-tailed deer, mountain lion, black bear, and, offshore, the manatee and dolphin—have been recorded, and half of these depend significantly on the freshwater communities. The fish and invertebrate faunas are even richer.

Not surprisingly, the first threat, now much reduced, to the Everglades came from poaching of the rich fauna. In the 1890's millions of feathers of egrets and other Florida birds formed the raw materials of a thriving millinery trade, and by 1930 100 000 alligator hides were being processed into leather each year in Florida tanneries. It has been estimated that the alligator population was reduced to only 1 % of its original level by the 1960's, but it is now recovering owing to stringent protection laws[157]. The real danger for the Everglades' ecosystem, and also for the whole of southern Florida, comes from interference with the natural drainage patterns.

As early as 1882 it was realized that the peatlands around L. Okeechobee were extremely fertile, could they be drained of the standing water that covered them for eight months of the year. A canal was dredged between the Caloosahatchee River and L. Okeechobee (Fig. 12.4). Sugar cane and winter vegetables thrived on the areas drained.

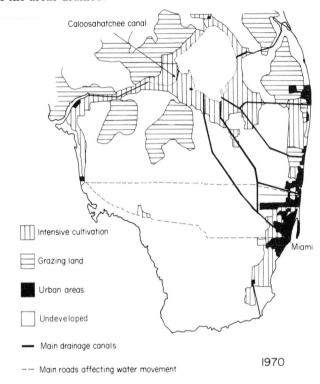

Fig. 12.4. The Florida Everglades area in 1970. (Modified from Caulfield[66].)

In 1925, and particularly in 1928, hurricanes were severe enough to cause flooding of water from L. Okeechobee into adjacent drained areas which had then been settled as fertile farmland. These areas had formed the natural wetland of the lake, normally accommodating flood levels. In 1928, between 1500 and 2500 people were drowned in the floods. An embankment was constructed around the lake and a period of digging of further drainage canals began so that future potential flood water could be drained rapidly to the sea. The canals (Fig. 12.4) emerge at the coast among the built-up areas of the eastern oolite ridge and effectively divert the waterflow to the east from its original southward progress through the Everglades. Pressure for drainage has been stimulated not only by farming interests but also by the warm Florida climate. For several decades the area has been promoted as a retirement and holiday haven and the demand for building land, otherwise confined to the coastal ridges, has been aggressive.

The canal system incorporates large, shallow reservoirs called water conservation areas. These can be used for temporary storage of water released from L. Okeechobee. They are also necessary to delay the flow to the sea sufficiently for the groundwater aquifer from which the coastal cities derive their fresh water supply to be recharged. They have had to be built because the natural peat regulator south of L. Okeechobee, which bore exactly these functions has been much reduced. Drainage results in rapid oxidation of the peat, which, once 4 m deep, now disappears at the rate of 2 cm per year.

A main road, the Tamiami Trail, skirts the southernmost of the reservoir areas, and water may be released through culverts under it to the remaining southern portion of the Everglades. The problem is that the water supply is now insufficient—a high proportion is diverted to the east coast—and the natural seasonal rhythm of supply is not necessarily maintained. Water is released to the Everglades largely when it is convenient for the drainage system.

In the pristine Everglades the seasonality of the water supply was something the ecosystem had adapted to. In the dry season animals congregated in the deeper parts of the sloughs and breeding cycles were related to this. The concentration of invertebrates at the edges of the receding water, for example, provided the wood ibis, which is a wading bird seizing its prey after contact as it sweeps its bill through the water, with a rich food supply during the fledgling season. Nests of birds built on the swamp floor in the dry season may now be destroyed by unseasonal inputs of water.

The overall lack of water is probably most crucial, however. The Everglades have always been subject to light fires, indeed the diversity of their vegetation depends on fire to prevent succession in the *Cladium* swamps. But the extreme drought now caused by diversion of the water supply, particularly in years of low rainfall such as October–April 1970–71 when only a third of the expected 35 cm fell, has led to especially destructive fires. These have bitten deep into the sub-surface peat, setting hundreds of hectares on fire at times, and damaging the vegetation perhaps irreversibly. Uncontrollable fires such as these inevitably also threaten adjacent urban areas, as well as causing smoke pollution for long periods. The lack of water in the Everglades is also reducing the areas of water which persist through the dry season. Since fish and reptiles congregate in these pools this has led to heavier than normal predation and mass fish deaths due to deoxygenation. The lack of water is affecting the cities also. The groundwater aquifer is not being recharged rapidly enough to prevent sea water moving into the oolite and contaminating the drinking-water wells. In recent years, new wells have had to be drilled further inland for East coast cities.

When the costs of the drainage and civil engineering works are added up, including the inevitable future costs of protecting water supplies, repairing and preventing fire damage, it seems unlikely that they will be balanced by benefits which can be construed as real. The drained peat, as it oxidizes away, will not indefinitely form good farmland, and the retirement colonies of southern Florida are not without sociological problems. The cost of the damage to the Everglades is impossible to estimate. Suffice it to say that even without taking into account the conservation value, the costs of a proposed barge canal across a swampy area of central Florida were found to exceed the benefits and the project has been stopped by Government intervention.

12.4 The Norfolk Broads

In the local dialect of eastern England, 'broad' means a broadening or widening of a river. The Norfolk Broads are a collection of small lakes usually to one side

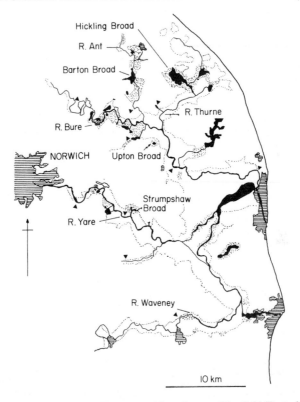

Fig. 12.5. The Norfolk Broadland. (For general location see Fig. 8.6.) Dotted outline shows the original extent of wetland, and stipple the remaining areas which have not been converted by drainage to agricultural land. Horizontal shading shows the main towns, and triangles the main sewage effluent outfalls.

of the lowland rivers which drain eastern Norfolk, but sometimes astride the river or even isolated in the valley a few km from the river (Fig. 12.5). The lakes all lie in peat deposits and their basins were excavated by man in a few hundred years prior to the fourteenth and fifteenth centuries[278]. The priory records of Norwich Cathedral and other documents reveal that large quantities of peat were burned for fuel in the twelfth century and peat borings show clear evidence that the basins were dug out rather than naturally formed (Fig. 12.6). The sides are vertical and long islands left in the middle, originally separating the workings, show similar peat stratigraphy to that of the 'mainland'.

The workings were abandoned in the fourteenth and fifteenth centuries when changes in the relative levels of land and sea, and probably a period of relatively wetter climate, led to flooding of them. The surrounding marshes and fens continued to be used for the hunting of wildfowl, for fishing and for the cropping of reed (*Phragmites australis*) and sedge (*Cladium mariscus*) as roofing materials. This led to progressive digging out of channels or dykes connecting the lake basins with the rivers. In this way the marsh products could easily be transported by boat to the villages and towns.

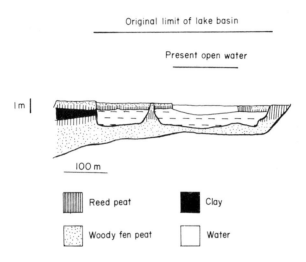

Fig. 12.6. Cross-section, reconstructed from peat borings, through a typical Norfolk Broad (Hoveton Great Broad). Interrupted lines show lake sediment laid down after excavation of the broad in the original fen peats of the valley. (Redrawn from Ellis[127].)

Undoubtedly, the Broadland was, for several centuries, an area of immense ecological interest for the waters are naturally fertile and a number of water plants, aquatic and fen invertebrates have been recorded in the British Isles only from this area. Cutting of the fens for reed prevented their succession to alderwood (*Alnus glutinosa* woodland), and helped maintain a diversity of communities.

The higher land of the valley slopes was agriculturally fertile and, with the expansion in farming brought about by the earliest agriculturalists and by the Enclosure Acts of the late eighteenth and early nineteenth centuries, attention was directed also to the fertile alkaline peats of the valley bottoms. Drainage was first carried out by wind pumps, which, though mostly derelict, still grace the Broadland scene, and more dykes were cut to carry the drainage water. Drainage by this means was not so efficient that it severely changed the Broadland eco-system however, and the area was still noted as rich and varied in the early part of this century when the conversion of the wind pumps to more powerful steam pumps began.

During the late nineteenth century, Broadland was 'discovered' as a holiday area and a thriving tourist industry began on the waterways. At the start all boats hired out were sail powered, but motor boats gradually began to dominate the hire-fleets in the twentieth century. During the same period the resident population also began to increase greatly, particularly in the towns, and in the early decades of the twentieth century the first sewage treatment works began to discharge effluent to the rivers. Palaeolimnological studies have shown that these changes, started early but then intensified, have culminated in the present greatly-altered and much impoverished wetland. The most serious changes have occurred variously in the past 30 years and have been associated with steadily increasing nutrient loading which has caused loss of the original clear and sparkling waters with phytoplankton crops of perhaps 10–20 μg chlorophyll a l^{-1} and with rich aquatic macrophyte beds of *Chara*, *Najas* and other genera. These have progressively been replaced by dense tangles of ranker, taller, water weeds like *Potamogeton pectinatus*, *Myriophyllum spicatum* and *Ceratophyllum demersum*, and then by dense phytoplankton populations with crops of 300 μg chlorophyll a l^{-1} or more. Even in water only 1–2 m deep the bottom can no longer be seen in summer and rich stands of even the more tolerant macrophytes are found only in four out of more than forty broads, mostly isolated from the main waterway, and in the drainage dykes on the farmland. A whole syndrome of effects of eutrophication, some of them of considerable economic importance, has arisen.

The course of the eutrophication has been traced through sediment core analyses. Table 12.2 gives some details of the course of events in Strumpshaw Broad on the R. Yare, downstream from Norwich, the only substantial city in the area. Increased fertility of the broads in general has its origins in the agri-cultural expansion of the nineteenth century. Cultivation and fertilization of the land at first led to increased growth of aquatic macrophytes and the epiphytic algae attached to their surfaces. At Strumpshaw Broad, however, agricultural influence was dominated by the greater source of nutrients from the city of Norwich sewage, even in the earliest stages. Table 12.2 shows how a major change occurred around 1912 as the city sewerage system was replanned and septic tanks in many areas were replaced by mains drainage with an effluent discharging to the river. This led to a forty year phase of rank macrophyte growth, though from the 1950's even this was replaced by phytoplankton as the sewage works was greatly expanded.

The broads which still have abundant macrophytes receive only water running off the land and do not receive sewage effluent. Alone, the land drainage water is insufficiently fertile to support large enough plankton and epiphyte crops to deprive the macrophytes of sufficient light for their growth. Very large quantities of nitrate ions are washed into the waterway from land fertilization,

Table 12.2. Summary of events in Strumpshaw broad deduced from analysis of a sediment core. (From Moss[369].)

Date and Zone	Nutrient input	Events
A pre-1800	Low input though river relatively fertile	Moderate crops of bottom-growing macrophytes. Clear water, marl deposition. Plankton virtually absent
B 1800–1900	Increased input with growth of City of Norwich	Blanketing of low-growing macrophytes (*Chara*) with filamentous algae. Increases in epiphytes. Plankton scarce. Snail numbers and rate of sedimentation increasing
C 1900–1912	Increased input with growth of city	Short transition zone as *Chara* flora replaced by tall-growing macrophytes. Water more turbid as plankton starts to increase
D 1912–1950	Increasing more rapidly with population and conversion of septic tanks to main sewerage. River organically very polluted	Plankton increase to maximum. Aquatic plants and epiphytes in steady decline. Initially abundant tall-growing macrophytes; markedly increased organic sedimentation rates but then declining. Macrophytes, when present, act as filter for river sediment and minimize effects of tidal flushing. Initial epiphyte maxima, snails plentiful. Marl deposition reduced
E 1950–1974	Post-war agricultural fertilization coupled with major increases in effluent loading. Use of phosphate detergents. Expansion of sewage works and extent of area served by it	Decreasing volume of broad and increased flushing rate leads to little deposition of plankton diatoms. Loss of macrophytes complete. Inorganic sedimentation rate declines as marl source is lost but more material is washed in from the river. Organic sedimentation variable and mainly of reed and river material. Sediments watery, not stabilized by macrophytes
F post-1974	Continued very high	Much of broad a bare mud flat at low tide. Net deposition of sediment probably has ceased

but the phosphorus input is only moderate from this source and can support maximum summer total phosphorus levels of only about 70 μg P l^{-1}, of which perhaps only half is available to phytoplankton, the remainder being adsorbed to clay colloids.

Where enrichment has been sufficient to prevent macrophyte growth, the summer total phosphorus levels of around 200 μg P l^{-1} have come from faecal and excretory sources. Mostly they have come from human sewage effluent, and phosphorus budgets for Barton Broad (Table 12.3) show the present situation compared with that at previous times. Major changes in the broad were evident in almost complete loss of submerged macrophytes in the 1960's, but had their origins in the opening of the first of two sewage works on the R. Ant in 1924.

Table 12.3. Past phosphorus budgets for Barton broad. (From Moss[370].) Past budgets have been reconstructed from analyses of sediment and application of the current % retention of phosphorus in the sediment to estimate past loading. The components of the load have been calculated from historical information on sewage disposal.

	Retention of phosphorus in sediment gPm⁻² yr⁻¹	Calculated total P loading gPm⁻² yr⁻¹	Calculated [$^m{}_w$] μgPl⁻¹	Loading from sewage effluent gPm⁻² yr⁻¹	Loading from land drainage gPm⁻² yr⁻¹	Partial [$^m{}_w$] sewage μgPl⁻¹	Partial [$^m{}_w$] land drainage μgPl⁻¹
1800	0·08	0·4	13·3	0·0	0·4	0·0	13·3
1900	0·31	1·55	52·0				
1920	0·43	2·15	72·0				
1940	0·71	3·55	119·0	2·15	1·40	72·0	47·0
1974		10·83*	329·0*	8·6*	2·23*	72·0	47·0
/76 (mean)			361·0			287·0	74·0

* Measured (Osborne[390].)

[$^m{}_w$] is the concentration of total phosphorus.

The upper part of the R. Thurne, with Hickling Broad (Fig. 12.5), a National Nature Reserve and the largest of the lakes, together with Horsey Mere and Martham Broad, is of particular interest. There are no effluent discharges into it and although Martham Broad retains its macrophytes and reasonably clear water, Hickling Broad and Horsey Mere do not. Both have dense phytoplankton populations and the swards of *Chara*, *Najas marina* and other aquatic macrophytes for which Hickling Broad in particular was famous until the early 1970's are now replaced by isolated beds or clumps of *Potamogeton pectinatus*, *Hippuris vulgaris* and *Myriophyllum spicatum*.

The eutrophication of Hickling Broad seems to have come from a major expansion in the size of a roost of black-headed gulls (*Larus ridibundus*). A maximum of only 25 000 birds was present in the 1950's, and something less than this in 1921. In 1976 the maximum was estimated as 250 000 birds and in 1977 as 100 000. They fly in from their breeding grounds, mostly around the Baltic Sea, in autumn, and feed on the rich invertebrate populations of newly-ploughed fields and on edible waste at town rubbish tips during the winter, before migrating back to mainland Europe in January and February. Their excretion of some 40 mg total phosphorus per bird per day amply provides sufficient phosphorus to support the algal growth of Hickling Broad and Horsey Mere, even if only half of it is deposited in roosting hours. Most of the gull excreta falls to the surface sediment of Hickling Broad, and phosphate is released from it as it decomposes with rising temperatures in spring and early summer. The slight ebb and flow of tide in the R. Thurne causes a mixing of water between Hickling Broad and Horsey Mere, so the latter is heavily fertilized though many gulls do not roost on it. Martham Broad receives little water from Hickling Broad, remains only moderately enriched, from farming activities, and retains a rich macrophyte flora. The gull problem is probably not a natural one, but ultimately one induced by man. All common species of gulls have been increasing in numbers in Europe during this century for reasons still obscure, but thought to be related to increasing availability of edible waste on rubbish tips, greater conversion of land to arable, and changes in gull habits necessitated by the decline in numbers of inshore fishing boats and the waste offal thrown overboard from them. Legislation protecting the breeding birds in mainland Europe has probably also been important.

The mechanism by which increased enrichment causes the decline of submerged macrophytes is not just a simple one of shading by increased phytoplankton crops (Fig. 12.7). There are palaeolimnological evidence and experimental studies[406] which show that increases in epiphytic algae and blanketing filamentous algae first occur. Indeed, where the problem of moderately fertilized rivers and lakes is not one of too much choking macrophyte growth, it tends to be one of blanket weed, caused for example by the filamentous alga, *Cladophora*. In the Broads as they were prior to the recent problem period, incoming nutrients were largely taken up by the macrophytes and their epiphytes. The phytoplankton was restricted in its development through competition for nutrients and possibly also because the macrophytes secreted organic inhibitors. As nutrient

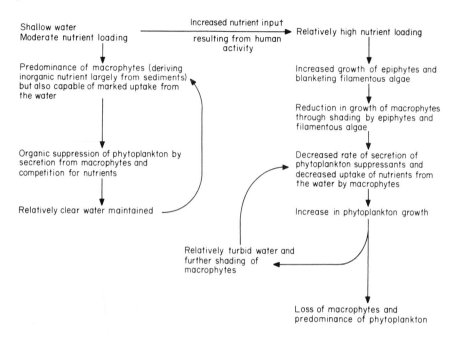

Fig. 12.7. A hypothesis to account for the decline of aquatic macrophyte populations in progressively eutrophicated fresh waters. (Redrawn from Phillips *et al*.[406].)

loading increased, the epiphytic and littoral filamentous algae increased their growth first, being, through selection for their particular microhabitat close to the macrophytes, more tolerant of any inhibitors secreted than the phytoplankters. Progressively they restricted light to the macrophytes—even a moderate epiphytic community absorbs as much light as several metres of water—and reduced macrophyte growth. At this stage less energy was available to the macrophytes for production of plankton inhibitors, and less biomass was present to provide substantial competition for nutrients and the plankton began more fully to exploit the increased nutrient load.

The secondary consequences of the extreme eutrophication that has occurred in Broadland have been varied. Increased macrophyte growth itself was a problem, particularly from the 1940's to the 1960's when enrichment from agricultural changes following World War II started to affect broads not already moderately fertilized by sewage effluent. Navigation channels had to be cut once or twice a year in some of the rivers, but this was a relatively inexpensive problem. Now, loss of the macrophytes has meant greater access for boats to sheltered bays and the edges of the broads and rivers. Previously weed tangles kept the boats, which have increased greatly in number in the post-war years, to the main channels. At the edges of the broads and along the river banks, floating leaved macrophytes, like the water lilies, *Nuphar lutea* and *Nymphaea alba* have been so physically damaged by boats that they too have disappeared except in sheltered areas, and the reed fringes behind the water lilies have also suffered. The banks are now less buffered from the erosion caused by boat wash and waves and have been receding to an extent which has threatened the raised floodbanks which protect the drained farmland in the valleys. The banks in many areas have had to be protected with expensive and unaesthetic steel, concrete and wood piling. The total sums involved now run into millions of pounds.

Sedimentation rates have also increased wherever there has been an increase in nutrient loading. Partly the increase has been through production of more autochthonous matter, partly through movement of sediment out of the marshes and down the river, after macrophytes, which are known to stabilize sediments, have disappeared. The use of more powerful land drainage pumps, now operated by electricity, also speeds the passage of water from land to the waterways giving less time for particles to settle out in the dykes and therefore increasing the particle load on the main waterway. This has meant that some broads, situated alongside channels, now have either a reduced water depth (< 25 cm), which itself seems to inhibit recolonization of macrophytes, or bare mud flats (e.g. Strumpshaw Broad, see Table 12.2) at low tide. It also means that the costs of dredging the navigation channels have increased and must continue to do so.

In the Thurne broads (Fig. 12.5), which, because of their nearness to the sea in an area of permeable sandy soils, are brackish (up to 10% of sea water salinity), there has been an increase in the population sizes of a brackish phytoplankter, *Prymnesium parvum* Carter, which kills fish. *Prymnesium* is a relatively slow-growing species, requiring quite complex media to grow in the laboratory. Probably it does not compete well with other species, and until the period of extreme enrichment by seagulls in Hickling Broad and Horsey Mere it rarely formed populations large enough (more than about 10^4 cells ml^{-1}) to kill fish. Since 1969, however, such lethal populations have been an almost annual feature and in 1969 some 250 000 fish were killed. It was not financially possible (nor indeed sensible) to restock these fish and the once famous coarse fishery of Hickling Broad has not recovered. *Prymnesium* secretes several toxins, some of which are phospholipids. Ichthyotoxins change the permeability of the gill surfaces of fish and the respiratory surfaces of invertebrates such that haemolysins and cytotoxins can enter. In commercial, fertilized, saline fish ponds in Israel, *Prymnesium* has been a problem, now controlled by the use of ammonia as an algicide. To some extent ammonia levels in the Thurne Broads determine when *Prymnesium* populations may develop, but addition of even more combined nitrogen to the water to control *Prymnesium* would exacerbate the eutrophication problem, for Hickling Broad water is now so enriched with phosphorus that there is a shortage of nitrogen in the summer.

Birds, as well as fish have been killed; in this case an anaerobic bacterium, *Clostridium botulinum* type C, has been responsible. Less is known of this problem than of others but it seems that *Clostridium*, present ubiquitously but inertly as spores in aerated habitats, now finds conditions in the deoxygenated surface sediments more frequently suitable for its germination as an active bacterial cell. The cells secrete a powerful botulism toxin which water birds absorb from their drinking-water, or more likely food if this includes mud particles or decaying vegetation as they forage for bottom invertebrates. The toxin is usually fatal, and several hundreds or thousands of birds are now annually killed, particularly in warm summers when bacterial activity in the sediment is greatest.

All of the eutrophication problems of Broadland have been recorded, individually or collectively, elsewhere and where they have been tackled they have been found to be reversible. As discussed in Chapter 10, it is always cheaper, in the long term, to tackle a problem at its source than to attempt continually to treat its symptoms. In much of Broadland it will be necessary to reduce the nutrient loading from sewage effluent by diversion of it to the sea or by phosphate stripping. Isolation of small areas of water in dykes to the sides of rivers so that they receive only land drainage water has been shown to result in clarification of the water and abundant recolonization of macrophytes. Nutrient budgeting has shown that phosphate stripping, if done efficiently, would reduce the total phosphorus level of the main broads sufficiently to allow macrophyte regrowth. Little can be done to mitigate agricultural fertilization, which will lead, in the absence of effluent, to perhaps greater water weed biomass than thought desirable by those responsible for maintaining navigation channels, but on balance this problem will be less expensive than the present ones.

Where phosphate loading has been reduced in deep lakes, such as L. Washington, the results have been rapidly seen[118]. In shallow lakes, release of phosphate from the sediment may still occur as labile organic matter now being laid down still renders the sediment surface anaerobic in summer (see Chapter 5). This too should decline as the external loading of phosphorus is reduced, for the released phosphorus is removed down river by the general fast flushing of the broads. In those broads where sediment has accumulated to within centimetres of the water surface expensive rehabilitation by pumping out of the mud will be necessary, and since these broads are mostly not on main navigation routes the costs can only be balanced by the less quantifiable nature conservation advantages. Funds for this may not be so readily forthcoming.

The seagull problem at Hickling Broad is also a thorny one. No one knows just why seagull numbers have increased overall and killing of seagulls on the scale that would be necessary is probably unacceptable on aesthetic grounds, and potentially useless on anything but a continuing basis. The philosophy of tackling a problem at source must again be followed and the reasons why seagull numbers have increased overall in Europe must be investigated properly. For the moment such a large roost is itself not uninteresting and for a limnologist, at least, guanotrophic lakes are as interesting as clear mountain tarns!

12.5 Postscript

The three examples of conservation problems outlined above are less than half the whole story in each case. It is now relatively easy for a limnologist to diagnose problems and suggest solutions. It is relatively difficult to have these solutions implemented, for implementation is usually a political and social matter. Lobbies and factions will always be involved and decisions may rest on personalities or be made by lay bodies unable or unwilling to grasp the technical issues and the long-term economic implications. The accounts of the Great Lakes, Everglades and Broads are incomplete without a study of the politics and sociology which permeate them. This is outside the scope of this book, but it should not be outside the scope of a scientist concerned with the freshwater, or any other, environment. The end of this book is really a starting point.

12.6 Further reading

There is no shortage of conservation problems involving fresh water, and the examples given here are merely representative. The reader will readily find, if he or she wishes, some issue local to him or her, with its own literature, though often in mimeographed or other ephemeral form and not readily available outside the area of immediate interest.

General works including information on aquatic conservation include Duffey & Watt[111], Morgan[352] and many semi-popular works. Aron & Smith[15] deals with the problems caused by ship canals, and Bellamy & Pritchard[29] and Moore & Bellamy[350] with peatlands. An issue of the *Journal of the Fisheries Research Board of Canada* (1975, **32** (1)) deals with the problems of hydroelectric projects, and Darnell[87], Goldstein[175], Good *et al.*[179], and IUCN[250] are all concerned with the conservation of wetlands. Armillas[13] tells of early uses of swamps by the Aztecs, who seem to have been considerably less short-sighted than we. *Biological Conservation* and *Environmental Conservation* publish relevant articles.

For work on the American Great Lakes, the *Journal of the Fisheries Research Board of Canada* in the last ten years has been a veritable gold mine. For example, major symposia were produced in 1972 (**29**(6)) on fish communities and in 1976 (**33**(3)) on L. Erie. The Proceedings of the Conference on Great Lakes Research (University of Michigan, Ann Arbor) and Marshall[331] are also useful starting points for the very extensive literature which has resulted from the great concern shown by Canada and the U.S.A. for these very valuable resources.

Caulfield[66] and Douglas[109] are good general works on the Everglades, the former having magnificent photographs and the latter a very full account of the political background. A basic work on the natural history is Robertson[435] and more technical accounts, largely of the vegetation are Davis[92, 93], Egler[124], Loveless[298] and Shelford[463].

The Broads are now well served with technical literature which is widely available—Ellis[127], George[163], Holdway *et al.*[217], Leah *et al.*[281], Mason & Bryant[333], Moss[366, 367], Nature Conservancy[380], Osborne & Moss [391], Osborne & Phillips[392], Phillips[405]—but as always, the equally important non-technical side to the problems is covered in ephemera only locally available.

REFERENCES

1 ACKERMANN W.C., WHITE G.F. & WORTHINGTON E.B. (1973) *Man-made lakes, their problems and environmental effects*. Geophys. Un. Monogr., **17**.
2 ALEXANDER R.M. (1967) *Functional design in fishes*. Hutchinson, London.
3 ALGEUS S. (1950) The utilization of aspartic acid, succinamide and asparagine by *Scenedesmus obliquous*. *Physiol. Plant.*, **3**, 225–35.
4 ALLAN J.D. (1976) Life history patterns in zooplankton. *Am. Nat.*, **110**, 165–80.
5 ALLANSON B.R. (1973) The fine structure of the periphyton of *Chara* sp. and *Potamogeton natans* from Wytham Pond, Oxford and its significance to the macrophyte—periphyton metabolic model of R. G. Wetzel and H. L. Allen. *Freshwat. Biol.*, **3**, 535–42.
6 ALLEN H.L. (1967) Acetate utilization by heterotrophic bacteria in a pond. *Hidrologiai Kozlony*, **1967**, 295–7.
7 ALLEN H.L. (1971) Primary productivity, chemo-organotrophy, and nutritional interactions of epiphytic algae and bacteria on macrophytes in the littoral of a lake *Ecol. Monogr.*, **41**, 97–127.
8 ALLEN K.R. (1951) The Horokiwi stream. A study of a trout population. *Fish. Bull. N.Z. Marine Dept.*, **10**, 1–231.
9 AMERICAN PUBLIC HEALTH ASSOCIATION (1971) *Standard methods for the examination of water and wastewater*. 13th edn. 874 pp. Washington.
10 AMERICAN PUBLIC HEALTH ASSOCIATION (1976) *Standard methods for the examination of water and wastewater*. 14th edn. 1193 pp. A.P.H.A., Washington.
11 ANTIA N.J. (1977) A critical appraisal of Lewin's Prochlorophyta. *Br. phyc. J.*, **12**, 271–6.
12 APPLEBY P.G. & OLDFIELD F. (1978) The calculation of lead—210 dates assuming a constant rate of supply of unsupported ^{210}Pb to the sediment. *Catena*, **5**, 1–8.
13 ARMILLAS P. (1971) Gardens on swamps. *Science*, **174**, 653–61.

14 ARNOLD D.E. (1971) Ingestion, assimilation, survival, and reproduction by *Daphnia pulex* fed seven species of blue-green algae. Limnol. Oceanogr., **16**, 906–20.
15 ARON W.I. & SMITH S.H. (1971) Ship canals and aquatic ecosystems. *Science*, **174**, 13–20.
16 BAGENAL T.B. (1978) *Methods for assessment of fish production in fresh waters*. IBP Handbook No. 3. Third Edition. 384 pp. Blackwell Scientific Publications, Oxford.
17 BAILEY-WATTS A.E. (1976) Planktonic diatoms and some diatom-silica relations in a shallow eutrophic Scottish loch. *Freshwat. Biol.*, **6**, 69–80.
18 BALINSKY B.I. & JAMES G.V. (1960) Explosive reproduction of organisms in Kariba Lake. *S. Afr. J. Sci.*, **56**, 101–4.
19 BALON E.K. & COCHE A.G. (eds) (1974) *Lake Kariba, a Man-made tropical ecosystem in Central Africa*. 767 pp. Dr. W. Junk, The Hague.
20 BARKO J.H., MURPHY P.G. & WETZEL R.G. (1977) An investigation of primary production and ecosystem metabolism in a Lake Michigan dune pond. *Arch. Hydrobiol.*, **81**, 155–87.
21 BÄRLOCHER F. & KENDRICK B. (1973) Fungi and food preferences of *Gammarus pseudolimnaeus*. *Arch. Hydrobiol.*, **72**, 501–16.
22 BÄRLOCHER F. & KENDRICK B. (1973) Fungi in the diet of *Gammarus pseudolimnaeus* (Amphipoda). *Oikos*, **24**, 295–300.
23 BAYLY I.A.E. & WILLIAMS W.D. (1973) *Inland waters and their ecology*. Longman, London.
24 BEADLE L.C. (1974) *The inland waters of tropical Africa*. 265 pp. Longman, London.
25 BEAUCHAMP R.S.A. (1964) The rift valley lakes of Africa. *Verh. int. Verein theor. angew. Limnol.*, **15**, 91–9.
25a BEETON A.M. (1969) Changes in environment and biota of the Great Lakes.

In *Eutrophication: Causes, Consequences, Correctives*, pp. 150–87. National Academy of Sciences, Washington, D.C.

26 BEHRENDT A. (1977) *The management of angling waters*. 205 pp. André Deutsch, London.

27 BELCHER J.H., SWALE E.M.F. & HERON J. (1966) Ecological and morphological observations on a population of *Cyclotella pseudostelligera* Hustedt. *J. Ecol.*, 54, 335–40.

28 BELLAIRS A. d'A. (1969) *The life of reptiles*, (2 vols.), Weidenfeld, London.

29 BELLAMY D.J. & PRITCHARD T. (1973) Project 'Telma': A scientific framework for conserving the world's peatlands. *Biol. Conserv.*, 5, 33–40.

30 BELLAMY L.S. & REYNOLDSON T.B. (1974) Behaviour in competition for food amongst lake-dwelling triclads. *Oikos*, 25, 356–64.

31 BERG K. (1938) Studies on the bottom animals of Esrom Lake. *K. danske vidensk. Selsk. skr.*, 7, 1–255.

32 BERG K. & JONASSON P.M. (1965) Oxygen consumption of profundal lake animals at low oxygen content of the water. *Hydrobiologia*, 26, 131–43.

33 BERMAN T. (1970) Alkaline phosphatases and phosphorus availability in Lake Kinneret. *Limnol. Oceanogr.*, 15, 663–74

34 BILLAUD V.A. (1968) Nitrogen fixation and the utilization of other inorganic nitrogen sources in a subarctic lake. *J. Fish. Res. Bd. Can.*, 25, 2101–11.

35 BIRDSEY E.C. & LYNCH V.H. (1962) Utilization of nitrogen compounds by unicellular algae. *Science, N.Y.*, 137, 763–4.

36 BIRKS H.J.B. & WEST R.G. (1973) *Quaternary Plant Ecology*. 321 pp. Blackwell Scientific Publications, Oxford.

37 BISHOP J.E. (1973) *Limnology of a small Malayan River, Sungai Gombak*. Dr. W. Junk, The Hague.

38 BOLD H.C. & WYNNE M.J. (1978) *Introduction to the Algae*. 706 pp. Prentice-Hall Inc., Englewood Cliffs.

39 BOND W.J., COE N., JACKSON P.B.N. & ROGERS K.H. (1978) The limnology of Cabora Bassa, Moçambique, during its first year. *Freshwat. Biol.*, 8, 433–48.

40 BOYCOTT A.E. (1936) The habitats of freshwater Mollusca in Britain. *J. Anim. Ecol.*, 5, 116–86.

41 BRINKHURST R.O. (1974) *The benthos of lakes*. 190 pp. Macmillan, London.

42 BRINKHURST R.O. & JAMIESON B.G.M. (1971) *The aquatic oligochaeta of the world*. 860 pp. Oliver & Boyd, Edinburgh.

43 BRINSON M.M. (1977) Decomposition and nutrient exchange of litter in an alluvial swamp forest. *Ecology*, 58, 601–9.

44 BRISTOW J.M. & WHITCOMBE M. (1971) The role of roots in the nutrition of aquatic vascular plants. *Am. J. Bot.*, 58, 8–13.

45 BROCK T.D. (1967) Life at high temperatures. *Science*, 158, 1012–19.

46 BROCK T.D. (1970) *Biology of microorganisms*. 738 pp. Prentice-Hall, Englewood Cliffs.

47 BROCK T.D. & DARLAND G.K. (1970) Limits of microbial existence. *Science*, 169, 1316–18.

48 BROOKS J.L. (1965) Predation and relative helmet size in cyclomorphic *Daphnia*. *Proc. nat. Acad. Sci. U.S.*, 53, 119–26.

49 BROOKS J.L. & DODSON S.I. (1965) Predation, body size, and composition of plankton. *Science, N.Y.*, 150, 28–35.

50 BROWN M.E. (ed.) (1957) *The physiology of fishes*. Vol. 1 *Metabolism*, Vol. 2 *Behaviour*. Academic Press, London.

51 BROWN S. & COLMAN B. (1963) Oscillaxanthin in lake sediments. *Limnol. Oceanogr.*, 8, 352–3.

52 BRYLINSKI M. & MANN K.H. (1973) An analysis of factors governing productivity in lakes and reservoirs. *Limnol. Oceanogr.*, 18, 1–14.

53 BUNT J.S. (1968) Some characteristics of micro algae isolated from Antarctic sea ice. *Ant. Res. Ser.*, 11, 1–14.

54 BURDEN E.H.W.J. (1961) The toxicity of nitrates and nitrites with particular reference to the potability of water supplies. *Analyst.* 86, 429–33.

55 BURNS C.W. (1968) The relationship between body-size of filter-feeding Cladocera and the maximum size of particle ingested. *Limnol. Oceanogr.*, 13, 675–8.

56 BURNS C.W. & RIGLER F.H. (1967) Comparison of filtering rates of *Daphnia rosea* in lake water and in suspensions of yeast. *Limnol. Oceanogr.*, 12, 492–502.

57 BURRIS R.H., EPPLING F.J., WAHLIN H.B. & WILSON P.W. (1943) Detection of nitrogen fixation with isotopic nitrogen. *J. Biol. Chem.*, 148, 349–57.

58 CALOW P. (1973) The food of *Ancylus fluviatilis* Mull; a littoral, stone-dwelling herbivore. *Oecologia*, 13, 113–33.

59 CALOW P. & FLETCHER C.R. (1972) A new radiotracer technique involving ^{14}C and ^{51}Cr for estimating the assimila-

tion efficiencies of aquatic primary consumers. *Oecologia (Berl.)*, **9**, 155–70.

60 CAMPBELL J.I. & MEADOWS P.S. (1972) An analysis of aggregation formed by the Caddis fly larva *Potamophylax latipennis* in its natural habitat. *J. Zool Lond.*, **167**, 133–47.

61 CANTER H.M. & LUND J.W.G. (1968) The importance of protozoa in controlling the abundance of planktonic algae in lakes. *Proc. Linn. Soc. Lond.*, **179**, 203–19.

62 CARLANDER K.D. (1969) *Handbook of freshwater fishery biology*. 720 pp. 3rd edn. Iowa State U.P. Ames.

63 CARTER G.S. (1955) *The papyrus swamps of Uganda*. 15 pp. Heffer, Cambridge.

64 CASTENHOLZ R.W. (1969) Thermophilic blue-green algae and the thermal environment. *Bact. Rev.*, **33**, 476–504.

65 CASTENHOLZ R.W. & WICKSTROM C.E. (1975) Thermal Streams. In *River Ecology*, pp. 264–85, in (ed.) Whitton B.A. Blackwell Scientific Publications, Oxford.

66 CAULFIELD P. (1971) *Everglades*. 143 pp. Sierra Club, Ballantyne.

66a CHAMBERLAIN W. (1968) *A preliminary investigation of the nature and importance of soluble organic phosphorus in the phosphorus cycle of lakes*. 232 pp. Ph.D. thesis Univ. of Toronto, Ontario.

67 CHRISTIE W.J. (1974) Changes in the fish species composition of the Great Lakes. *J. Fish. Res. Bd. Can.*, **31**, 827–54.

68 CLEGG J. (1959) *The freshwater life of the British Isles*. 352 pp. Frederick Warne, London.

69 CLEGG J. (1965) *The freshwater life of the British Isles*. (2nd edn.) 352 pp. Warne, London.

70 COLBY P.J., SPANGLER G.R., HURLEY D.A. & McCOMBIE A.M. (1972) Effects of eutrophication on salmonid communities in oligotrophic lakes. *J. Fish. Res. Bd. Can.*, **29**, 975–83.

71 COLEMAN M.J. & HYNES H.B.N. (1970) The vertical distribution of the invertebrate fauna in the bed of a stream. *Limnol. Oceanogr.*, **15**, 31–40.

72 COLLINGWOOD R.W. (1977) A survey of eutrophication in Britain and its effects on water supplies. *Wat. Resour. Cent. Tech. Rept.*, **40**, 41 pp.

73 COOK C.D.K., GUT B.J., RIX E.M., SCHNELLER J. & SEITZ M. (1974) *Water plants of the world*. 561 pp. Dr. W. Junk, The Hague.

74 COTT H.B. (1961) Scientific results of an enquiry into the ecology and economic status of the Nile crocodile (*Crocodilus niloticus*) in Uganda & Northern Rhodesia. *Trans. zool. Soc. Lond.*, **29**, 211–392.

75 CRANWELL P.A. (1973) Branched chain and cyclopropanoid acids in a recent sediment. *Chem. Geol.*, **11**, 307–13.

76 CRANWELL P.A. (1974) Monocarboxylic acids in lake sediments: indicators derived from terrestrial and aquatic biota of palaeoenvironmental trophic levels. *Chem. Geol.*, **14**, 1–14.

77 CRANWELL P.A. (1976) Organic geochemistry of lake sediments. In *Environmental Biogeochemistry Vol. 1 Carbon, nitrogen, phosphorus, sulphur and selenium cycles*, pp. 75–88 (ed.) Nriagu, J.O. Ann Arbor Science, Ann Arbor, Michigan.

78 CRAWFORD R.M. (1966) The control of anaerobic respiration as a determining factor in the distribution of the genus *Senecio*. *J. Ecol.*, **54**, 403–13.

79 CREER K.M., THOMPSON R., MOLYNEUX L. & MACKERETH F.J.H. (1972) Geomagnetic secular variation recorded in the stable magnetic remanence of recent sediments. *Earth. Planet. Sci. Letters*, **14**, 115–27.

80 CRISP D.T. (1970) Input and output of minerals for a small watercress bed fed by chalk water. *J. appl. Ecol.*, **1**, 117–40.

81 CUMMINS K.W. (1974) Structure and function of stream ecosystems. *BioScience*, **24**, 631–41.

82 CUMMINS K.W., PETERSEN R.C., HOWARD F.O., WUYCHECK J.C. & HOLT V.I. (1973) The utilization of leaf litter by stream detritivores. *Ecology*, **54**, 336–45.

83 CUSHING D.H. (1968) *Fisheries Biology*. 200 pp. University of Wisconsin Press.

84 CUSHING D.H. (1975) *Marine ecology and fisheries*. Cambridge University Press, Cambridge & London.

85 CUSHING D.H. (1977) *Science and the fisheries*. 60 pp. Edward Arnold, London.

86 CUSHING E.J. & WRIGHT H.E. (1967) *Quaternary Palaeoecology*. 433 pp. Yale University Press, London & New Haven.

87 DARNELL R.M. (1976) *Impacts of construction activities in wetlands of the United States*. 392 pp. Report of Contract 68-01-2452. U.S. Environmental Protection Agency. Corvallis, Oregon.

88 DAVIES B.R. (1975) Cabora Bassa hazards. *Nature*, Lond., **254**, 477–8.

89 DAVIES B.R., HALL A. & JACKSON P.B.N. (1975) Some ecological aspects of the Cabora Bassa dam. *Biol. Cons* **8**, 189–201.

90 DAVIES R.W. & REYNOLDSON T.B. (1971)

The incidence and intensity of predation on lake-dwelling triclads in the field. *J. anim. Ecol.*, **40**, 191–214.

91 DAVIS C.C. (1964) Evidence for the eutrophication of Lake Erie from phytoplankton records. *Limnol. Oceanogr.*, **9**, 275–83.

92 DAVIS J.H. (1943) The natural features of southern Florida. *Fla. Geol. Surv. Biol. Bull.*, **25**, 311 pp.

93 DAVIS J.H. (1946) The peat deposits of Florida, their occurrence, development and uses. *Fla. Geol. Surv. Geol. Bull.*, **30**, 247 pp.

94 DAVIS M.B. (1968) Pollen grains in lake sediments: redeposition caused by seasonal water circulation. *Science, N.Y.*, **162**, 796–9.

95 DAVIS R.B. (1974) Stratigraphic effects of tubificids in profundal lake sediments. *Limnol. Oceanogr.*, **19**, 466–88.

96 DEEVEY E.S. (1941) Limnological studies in Connecticut VI. The quantity and composition of the bottom fauna of thirty six Connecticut and New York lakes. *Ecol. Monogr.*, **11**, 413–55.

97 DEEVEY E.S. (1942) The biostratonomy of Linsley Pond. *Am. J. Sci.*, **240**, 233–64, 313–24.

98 DEEVEY E.S. (1969) Cladoceran populations of Rogers Lake, Connecticut during late- and post-glacial time. *Mitt. int. Ver. theor. angew. Limnol.*, **17**, 56–63.

99 DEMARTE J.A. & HARTMAN R.T. (1974) Studies on absorption of ^{32}P ^{59}Fe and ^{45}Ca by water-milfoil (*Myriophyllum exalbescens* Fernald) *Ecology*, **55**,188–94.

100 DENNY P. (1972) Sites of nutrient absorption in aquatic macrophytes. *J. Ecol.*, **60**, 819–29.

101 DILLON P.J. & RIGLER F.H. (1974) A test of a simple nutrient budget model predicting the phosphorus concentration in lake water. *J. Fish. Res. Bd. Can.*, **31**, 1771–8.

102 DILLON P.J. & RIGLER F.H. (1975) A simple method for predicting the capacity of a lake for development based on lake trophic status. *J. Fish. Res. Bd. Can.*, **32**, 1519–31.

103 DILWORTH M.J. (1966) Acetylene reduction by nitrogen fixing preparations from *Clostridium pasteurianum*. *Biochim. Biophys. Acta.*, **127**, 285–94.

104 DODGE J.D. (1973) *The fine structure of algal cells*. 261 pp. Academic Press, London.

105 DODSON S.I. (1974) Zooplankton competition and predation: an experimental test of the size–efficiency hypothesis. *Ecology*, **55**, 605–13.

106 DOEMEL W.N. & BROCK T.D. (1970) The upper temperature limit of *Cyanidium caldarium*. *Arch. Mikrobiol.*, **72**, 326–32.

107 DONNER J. (1966) *Rotifers*. 80 pp. Warne, London.

108 DOUGLAS I. (1968) The effects of precipitation chemistry and catchment area lithology on the quality of river water in selected catchments in Eastern Australia. *Earth Sci. J.* **2**, 126–44.

109 DOUGLAS M.S. (1947) *The Everglades: River of Grass.* 308 pp. Ballantine Books, New York.

110 DOWNING K.M. & MERKENS J.C. (1957) The influence of temperature on the survival of several species of fish in low tensions of dissolved oxygen. *Ann. Appl. Biol.*, **45**, 261–7.

111 DUFFEY E. & WATT A.S. (eds.) (1971) *The scientific management of animal and plant communities for conservation.* 625 pp. Blackwell Scientific Publications, Oxford.

112 DUNN I.G. (1972) The commercial fishery of L. George, Uganda (E. Africa). *Afr. J. Trop. Hydrobiol. Fish.*, **2**, 109–20.

113 DUTHIE J.R. (1972) Detergents: nutrient considerations and total assessment. In *Nutrients & Eutrophication*, pp. 205–16. Spec. Symp. 1. American Society of Limnology & Oceanography. Lawrence, Kansas.

114 DVORAK J. (1970) Horizontal zonation of macrovegetation, water properties and macrofauna in a littoral stand of *Glyceria aquatica* (L.) Wahlb. in a pond in South Bohemia. *Hydrobiologia*, **35**, 17–30.

115 EDMONDSON W.T. (1957) Trophic relations of the zooplankton. *Trans. Am. Micro. Soc.*, **76**, 225–46.

116 EDMONDSON W.T. (ed.) (1959) H.B. Ward & G.C. Whipple. *Fresh-water Biology*. 2nd edn. 1248 pp. Wiley, New York.

117 EDMONDSON W.T. (1965) Reproductive rate of planktonic rotifers as related to food and temperature in nature. *Ecol. Monogr.*, **35**, 61–111.

118 EDMONDSON W.T. (1970) Phosphorus, nitrogen, and algae in Lake Washington after diversion of sewage. *Science, N.Y.*, **169**, 690–1.

119 EDMONDSON W.T. (1974) Secondary production. *Mitt. int. verein. theor. angew. Limnol.*, **20**, 229–72.

120 EDMONDSON W.T. & ALLISON D.E.

(1970) Recording densitometry of X-radiographs for the study of cryptic laminations in the sediment of Lake Washington. *Limnol. Oceanogr.*, **15**, 138–44.

121 EDMONDSON W.T. & WINBERG G.G. (eds.) (1971) *A manual on methods for the assessment of secondary productivity in freshwaters.* 368 pp. Blackwell Scientific Publications, Oxford.

122 EDWARDS A.M.C. (1971) *Aspects of the chemistry of four East Anglian rivers.* Ph.D. Thesis. University of East Anglia, Norwich.

123 EGGLISHAW H.J. (1970) Production of salmon and trout in a stream in Scotland. *J. Fish. Biol.*, **2**, 117–36.

124 EGLER F.E. (1952) Southeast saline everglades vegetation, Florida, and its management. *Vegetatio*, **3**, 213–65.

125 ELLIOTT J.M. (1971) The distances travelled by drifting invertebrates in a Lake District stream. *Oecologia (Berl.)*, **6**, 350–79.

126 ELLIOTT J.M. (1977) *Some methods for the statistical analysis of samples of benthic invertebrates.* 2nd edn. *Sci. Pub. Freshwat. Bio.. Ass.*, **25**, 160 pp.

127 ELLIS E.A. (1965) *The Broads.* Collins, London.

128 EMINSON D.F. (1978) *The ecology of epiphytic diatoms in Broadland.* Ph.D. thesis. Univ. of East Anglia, Norwich.

129 ESCH G.W. & MCFARLANE R.W. (1976) *Thermal ecology* II. ERDA Technical Information Center, CONF—750425/RAS 404 pp.

130 FASSETT N.C. (1957) *A manual of aquatic plants.* 2nd edn. 405 pp. University of Wisconsin Press, Madison.

131 FAY P., STEWART W.D.P., WALSBY A.E. & FOGG G.E. (1968) Is the heterocyst the site of nitrogen fixation in blue-green algae? *Nature, Lond.*, **220**, 810–12.

132 FISH G.R. (1956) Some aspects of the respiration of six species of fish from Uganda. *J. exp. Biol.*, **33**, 186–95.

133 FISHER S.G. & LIKENS G.E. (1973) Energy flow in Bear Brook, New Hampshire: an integrative approach to stream ecosystem metabolism. *Ecol. Monogr.*, **43**, 421–39.

134 FITTKAU E.J. (1970) Role of caimans in the nutrient regime of Amazon affluents (an hypothesis). *Biotropica*, **2**, 138–42.

135 FITZGERALD G.P. (1969) Some factors in the competition or antagonism among bacteria, algae & aquatic weeds. *J. Phycol.*, **5**, 351–9.

136 FOGG G.E. (1971) Extracellular products of algae in freshwater. *Arch. Hydrobiol. Beih. Ergebn. Limnol.*, **5**, 1–25.

137 FOGG G.E. (1975) *Algal cultures and phytoplankton ecology.* 2nd edn. 175 pp. University of Wisconsin Press, Madison.

138 FOGG G.E., STEWART W.D.P., FAY P. & WALSBY A.E. (1973) *The Blue green algae.* 459 pp. Academic Press, London.

139 FOGG G.E. & WESTLAKE D.F. (1955) The importance of extracellular products of algae in freshwater. *Verh. int Verein theor. angew. Limnol.*, **12**, 219–32.

140 FORBES S.A. (1887) The lake as a microcosm. *Bull. Peoria (Ill) Sci. Ass.* (Reprinted in *Bull. Ill. nat. Hist. Surv.*, **15**, 537–50 (1925)).

141 FRANTZ T.C. & CORDONE A.J. (1967) Observations on deepwater plants in Lake Tahoe, California & Nevada. *Ecology*, **48**, 709–14.

142 FREY D.G. (1969) (ed.) *Symposium on palaeolimnology. Mitt. int. Ver. Limnol.*, **17**, 448 pp.

143 FRITSCH F.E. (1948) *The structure and reproduction of the algae.* Vols. 1 & 2. Cambridge University Press, Cambridge and London.

144 FROST W.E. & BROWN M.E. (1967) *The trout.* Collins, London.

145 FRYER G. (1973) The Lake Victoria fisheries: some facts and fallacies. *Biol Conserv.*, **5**, 304–8.

146 FRYER G. (1977) The atyid prawns of Dominica. *A. Rept. Freshwat. Biol. Ass.*, **45**, 48–54.

147 FRYER G. (1977) Studies on the functional morphology and ecology of atyid prawns of Dominica. *Phil. Trans. R. Soc. (B.)*, **277**, 57–129.

148 FRYER G. & ILES T.D. (1969) Alternative routes to evolutionary success as exhibited by African cichlid fishes of the genus *Tilapia* and the species flocks of the Great Lakes. *Evolution*, **23**, 359–69.

149 FRYER G. & ILES T.D. (1972) *The cichlid fishes of the Great Lakes of Africa.* 641 pp. Oliver and Boyd, Edinburgh.

150 FURSE M., KIRK R.C., MORGAN P.R. & TWEDDLE D. (1979) Fishes: Diversity in the swamps and abundance in the lake, pp. 175–208 in Kalk M. (ed.) L. Chilwa. Dr. W. Junk. b.v, The Hague.

151 GAARDER T. & GRAN H.H. (1927) Investigations of the production of plankton in the Oslo Fjord. *J. Cons. perm. int. Explor. Mer*, **42**, 1–48.

152 GAJEVSKAYA N.S. (1969) *The role of higher aquatic plants in the nutrition of animals of freshwater basins.* Translated

by D. G. Maitland Muller, (ed.) Mann K.H. National Lending Library for Science & Technology, U.K.

153 GANF G.G. (1974) Phytoplankton biomass & distribution in a shallow eutrophic lake (Lake George, Uganda). *Oecologia*, **16**, 9–29.

154 GANF G.G. & BLAZKA P. (1974) Oxygen uptake, ammonia and phosphate excretion by zooplankton of a shallow equatorial lake (Lake George, Uganda). *Limnol. Oceanogr.*, **19**, 313–25.

155 GANF G.G. & VINER A.B. (1973) Ecological stability in a shallow equatorial lake (Lake George, Uganda). *Proc. roy. Soc. (B)*, **184**, 321–46.

156 GARDNER W.S. & LEE G.F. (1975) The role of amino acids in the nitrogen cycle of lake Mendota. *Limnol. Oceanogr.*, **20**, 379–88.

157 GARRICK L.D. & LANG J.W. (1977) The alligator revealed. *Natural History*, **86**, 54–61.

158 GARROD D.J. (1961) The rational exploitation of the *Tilapia esculenta* stock of the North Buvuma island area, Lake Victoria. *East Afr. Agric. Forestry J.*, **27**, 69–76.

159 GARROD D.J. (1961) The history of the fishing industry of Lake Victoria, East Africa, in relation to the expansion of marketing facilities. *East Afr. Agr. Forest. J.*, **27**, 95–9.

160 GAUDET J.J. (1977) Uptake, accumulation, and loss of nutrients by papyrus in tropical swamps. *Ecology*, **58**, 415–22.

161 GENTNER S.R. (1977) Uptake and transport of iron and phosphate by *Vallisneria spiralis* L. *Aq. Bot.*, **3**, 267–72.

162 GEORGE E.A. (1976) A guide to algal keys (excluding seaweeds). *Br. phycol. J.*, **11**, 49–55.

163 GEORGE M. (1976) Land use and nature conservation in Broadland. *Geography*, **61**, 137–42.

164 GERHART D.Z. & LIKENS G.E. (1975) Enrichment experiments for determining nutrient limitation: four methods compared. *Limnol. Oceanogr.*, **20**, 649–53.

165 GERKING S.D. (1957) A method of sampling the littoral macrofauna and its application. *Ecology*, **38**, 219–26.

166 GERKING S.D. (1978) *Ecology of freshwater fish production.* 504 pp. Blackwell Scientific Publications, Oxford.

167 GERLOFF G.C. & KROMBHOLZ P.H. (1966) Tissue analysis as a measure of nutrient availability for the growth of aquatic plants. *Limnol. Oceanogr.*, **11**, 529–37.

168 GESSNER F. (1952) Der Druck in seiner Bedeutung für das Wachstum submerser Wasserpflanzen. *Planta*, **40**, 391–7.

169 GESSNER F. (1955) *Hydrobotanik I.* Energiehaushalt (Hochschulbücher für Biologie Band 3). 517 pp. VEB Deutscher Verlag der Wissenschaften, Berlin.

170 GESSNER F. (1959) *Hydrobotanik II* Stoffhaushalt (Hochschulbücher für Biologie Band 3). 701 pp. VEB Deutscher Verlag der Wissenschaften, Berlin.

171 GIBBONS J.W. & SHARITZ R.R. (1974) *Thermal ecology.* 670 pp. U.S. Atomic Energy Commission, Augusta, Geo.

172 GILETTE R. (1972) Stream channelization: conflict between ditchers, conservationists. *Science*, **176**, 890–3.

173 GODWIN H. (1975) *The history of the British flora.* 2nd edn. 541 pp. Cambridge University Press, Cambridge & London.

174 GODWIN H. (1978) *Fenland: its ancient past and uncertain future.* 196 pp. Cambridge University Press, Cambridge & London.

175 GOLDSTEIN J.H. (1971) *Competition for wetlands in the mid-west, an economic analysis.* 105 pp. Resources for the Future Inc. Baltimore, The John Hopkins Press.

176 GOLTERMANN H.L. (1975) *Physiological limnology.* 489 pp. Elsevier, Amsterdam.

177 GOLTERMAN H.L. (ed.) (1977) *Interactions between sediments and freshwater.* 473 pp. Dr. W. Junk, The Hague.

178 GOLTERMAN H.L., CLYMO R.S. & OHNSTAD M.A.M. (1978) *Methods for physical and chemical analysis of fresh waters.* 2nd Edition. 213 pp. Blackwell Scientific Publications, Oxford.

179 GOOD R.E., WHIGHAM D.F. & SIMPSON R.L. (1978) *Freshwater wetlands.* 378 pp. Academic Press, London.

180 GOODWIN P. (1976) Volta ten years on. *New Scientist*, Sept. 16 1976, 596–7.

181 GOULDEN C.E. (1964) The history of the Cladoceran fauna of Esthwaite Water (England) and its limnological significance. *Arch. Hydrobiol.*, **60**, 1–52.

182 GRIFFITHS M. & EDMONDSON W.T. (1975) Burial of oscillaxanthin in the sediment of Lake Washington. *Limnol. Oceanogr.*, **20**, 945–52.

183 GRIFFITHS M., PERROTT P.S. & EDMONDSON W.T. (1969) Oscillaxanthin in the sediment of Lake Washington. *Limnol. Oceanogr.*, **14**, 317–26.

184 GUGGISBERG C.A.W. (1972) *Crocodiles their natural history, folklore & conservation.* David & Charles, Newton Abbot.

185 GULLAND J.A. (1969) *Manual of methods for fish stock assessment. Part* I. *Fish population analysis.* 153 pp. FAO, Rome.

186 GUNNISON D. & ALEXANDER M. (1975) Resistance and susceptibility of algae to decomposition by natural microbial communities. *Limnol. Oceanogr.,* **20,** 64–70.

187 HALL C.A.S. & MOLL R. (1975) Methods of assessing aquatic primary productivity. In *Primary productivity of the biosphere,* (eds.) Lieth H. & Whittaker R.H. pp. 19–54. 339 pp. Springer-Verlag, New York, Berlin.

188 HALL D.J. (1964) An experimental approach to the dynamics of a natural population of *Daphnia galeata mendotae. Ecology,* **45,** 94–111.

189 HALL D.J., COOPER W.E. & WERNER E.E. (1970) An experimental approach to the production dynamics and structure of freshwater animal communities. *Limnol. Oceanogr.,* **15,** 829–928.

190 HALL D.J., THRELKELD S.T., BURNS C.W. & CRAWLEY P.H. (1976) The size & efficiency hypothesis and the size structure of zooplankton communities. *A. Rev. Ecol. Syst.,* **7,** 177–208.

191 HALL K.J. & HYATT K.D. (1974) Marion Lake (IBP)—from bacteria to fish. *J. Fish. Res. Bd. Can.,* **31,** 893–911.

192 HANEY J.F. An *in situ* method for the measurement of zooplankton grazing rates. *Limnol. Oceanogr.,* **16,** 970–6.

192a HARDING D. (1966) Lake Kariba, the hydrology and development of fisheries. In *Man-Made Lakes,* pp. 7-20 (ed.) Lowe-McConnell R. Institute of Biology, London.

193 HARDING J.P. & SMITH W.A. (1974) A key to the British freshwater Cyclopoid and Calanoid Copepods. *Sci. Pub. Freshwat. Biol. Ass.,* **18,** 1–55.

194 HARLIN M.M. (1975) Epiphyte-host relations in seagrass communities. *Aquat. Botany,* **1,** 125–31.

195 HARMSWORTH R.W. (1968) The developmental history of Blelham Tarn (England) as shown by animal microfossils, with special reference to the Cladocera. *Ecol. Monogr.,* **38,** 223–41.

196 HART C.W. & SAMUEL L.H. (eds.) (1974) *Pollution ecology of freshwater invertebrates.* 389 pp. Academic Press, New York.

197 HARTLEY P.H.T. (1948) Food and feeding relationships in a community of freshwater fishes. *J. anim. Ecol.,* **17,** 1–14.

198 HARTMAN R.T. & BROWN D.L. (1967) Changes in internal atmosphere of sub-mersed vascular hydrophytes in relation to photosynthesis. *Ecology,* **48,** 252–8.

199 HASLAM S.M. (1978) *River plants.* Cambridge University Press, Cambridge and London.

200 HASLAM S.M., SINKER C.A. & WOLSELEY P.A. (1975) British water plants. *Field Studies.* **4,** 243–351.

201 HASLER A.D. (1966) *Underwater guideposts: homing of salmon.* University Press, Wisconsin, Madison.

202 HASLER A.D. (1975) *Coupling of land and water systems.* 309 pp. Springer-Verlag, Berlin.

203 HASLER A.D. & JONES E. (1949) Demonstration of the antagonistic action of large aquatic plants on algae and rotifers. *Ecology,* **30,** 359–64.

204 HAWORTH E.Y. (1969) The diatoms of a sediment core from Blea Tarn, Langdale. *J. Ecol.,* **57,** 429–39.

205 HAWORTH E.Y. (1972) Diatom succession in a core from Pickerel Lake, North eastern South Dakota. *Geol. Soc. Am. Bull.,* **83,** 157–72.

206 HEBERT P.D.N. (1978) The adaptive significance of cyclomorphosis in *Daphnia*: more possibilities. *Freshwat. Biol.,* **8,** 313–20.

207 HECKY R.E. & KILHAM P. (1973) Diatoms in alkaline, saline lakes: ecology and geochemical implications. *Limnol. Oceanogr.,* **18,** 53–71.

208 HENDERSON H.F., RYDER R.A. & KUDHONGANIA A.W. (1973) Assessing fishery potentials of lakes and reservoirs. *J. Fish Res. Bd. Can.,* **30,** 2000–9.

209 HENSHAW G.G., COULT D.A. & BOULTER D. (1961) Cytochrome-c Oxidase, the terminal oxidase of *Iris pseudacorus* L. *Nature, Lond.,* **197,** 579.

210 HER MAJESTY'S STATIONERY OFFICE (H.M.S.O.) (1976) *Agriculture and Water Quality.* 469 pp.

211 HICKLING C.F. (1961) *Tropical inland fisheries.* 287 pp. Longman, London.

212 HICKLING C.F. (1966) On the feeding processes of the white amur *Ctenopharyngodon idella* Val. *J. Zool.,* **148,** 408–19.

213 HILDEBRAND S.G. (1974) The relation of drift to benthos density and food level in an artificial stream. *Limnol. Oceanogr.,* **19,** 951–7.

214 HILDREW A.G. & TOWNSEND C.R. (1976) The distribution of two predators and their prey in an iron-rich stream. *J. anim. Ecol.,* **45,** 41–57.

215 HOBBIE J.E., CRAWFORD C.C. & WEBB K.L. (1968) Amino acid flux in an estuary. *Science,* **159,** 1463–4.

216 HOLČIK J. (1970) Standing crop, abundance, production and some ecological aspects of fish populations in some inland waters of Cuba. *Vestnik. Cs. spol. zool.* (*Acta. soc. zool. Bohemoslov.*), **34**, 184–201.

217 HOLDWAY P.A., WATSON R.A. & MOSS B. (1978) Aspects of the ecology of *Prymnesium parvum*. (Haptophyta) and water chemistry in the Norfolk Broads, England. *Freshwat. Biol.*, **8**, 295–311.

218 HOLM L.G., WELDON L.W. & BLACKBURN R.D. (1969) Aquatic weeds. *Science, N.Y.*, **166**, 699–708.

219 HOLME N.A. & MCINTYRE A.D. (1971) (eds.) *Methods for the study of the marine benthos.* IBP Handbook No. 17. Blackwell Scientific Publications, Oxford.

220 HOLT S.J. & TALBOT L.M. 1978. New principles for the conservation of wild living resources. *J. Wildl. Man.* **43** (Suppl.) Wildlife Monographs **59**, 33 pp.

221 HOPSON A.J. (1969) A description of the pelagic embryos and larval stages of *Lates niloticus* (L.) (Pisces: centropomidae)) from Lake Chad, with a review of early development in lower percoid fishes. Zool. J. Linn. Soc. **48**, 117–34.

222 HOPSON A.J. (1972) A study of the Nile perch in L. Chad. Overseas Res. Publ. **19**, H.M.S.O., London.

223 HORIE S. (1977) *Paleolimnology of Lake Biwa and the Japanese Pleistocene.* Contributions of the L. Biwa Research Project Otsu, Japan.

224 HORNE A.J. & FOGG G.E. (1970) Nitrogen fixation in some English lakes. *Proc. roy. Soc. B.*, **175**, 351–66.

225 HORNE A.J. & GOLDMAN C.R. (1972) Nitrogen fixation in Clear Lake, California. I. Seasonal variation and the role of heterocysts. *Limnol. Oceanogr.*, **17**, 678–92.

226 HOWARD-WILLIAMS C. (1972) Limnological studies in an African swamp: seasonal and spatial changes in the swamps of Lake Chilwa, Malawi. *Arch. Hydrobiol.*, **70**, 379–91.

227 HOWARD-WILLIAMS C. (1975) Vegetation changes in a shallow African lake: response of the vegetation to a recent dry period. *Hydrobiologia*, **47**, 381–98.

228 HOWARD-WILLIAMS C. (1977) Swamp ecosystems. *Malay. Nat. J.* **31**, 113–25.

229 HOWARD-WILLIAMS C. & LENTON G.M. (1975) The role of the littoral zone in the functioning of a shallow tropical lake ecosystem. *Freshwat. Biol.*, **5**, 445–59.

230 HOWELL A.D. (1932) *Florida Bird Life.* 577 pp. Florida Dept. of Game and Fresh Water Fish, Talahassee.

231 HUBER-PESTALOZZI G. (1938—), *Das Phytoplanktons der Susswassers. Systematik und Biologie,* (a continuing series). E. Schweitzer' bartische Verlagsbuchlandlung, Stuttgart.

232 HUGHES J.C. & LUND J.W.G. (1962) The rate of growth of *Asterionella formosa* Hass. in relation to its ecology. *Arch. Mikrobiol.*, **42**, 117–29.

233 HURLBERT S.H., ZEDLER J. & FAIRBANKS D. (1971) Ecosystem alteration by mosquito fish (*Gambusia affinis*) predation. *Science*, **175**, 639–41.

234 HUTCHINSON G.E. (1957) *A treatise on limnology. I. Geography, physics and chemistry.* 1015 pp. Wiley, New York.

235 HUTCHINSON G.E. (1961) The paradox of the plankton. *Amer. Nat.*, **95**, 137–46.

236 HUTCHINSON G.E. (1967) *A treatise on Limnology, Vol. II. Introduction to lake biology and the limnoplankton.* 1115 pp. Wiley, New York.

237 HUTCHINSON G.E. (1969) Eutrophication, past and present, in *Eutrophication: causes, consequences, correctives,* pp. 17–26. Acad. Sci., Washington D.C.

238 HUTCHINSON G.E. (1973) Eutrophication. *Amer. Sci.*, **61**, 269–79.

239 HUTCHINSON G.E. (1975) *A treatise on limnology. III. Limnological botany.* 660 pp. Wiley, New York.

240 HUTCHINSON G.E., BONATTI E., COWGILL U.M., GOULDEN C.E., LEVENTHAL E.A., MALLETT M.E., MARGARITORA F., PATRICK R., RACEK A.A., ROBACK S.A., STELLA E., WARD-PERKINS J.B., WELLMAN T.R. (1970) Ianula: an account of the history and development of the Lago di Monterosi, Latium, Italy. *Trans. Am. phil. Soc.*, **60**(4), 1–178.

241 HYNES H.B.N. (1966) *The biology of polluted waters.* 202 pp. Liverpool University Press, Liverpool.

242 HYNES H.B.N. (1970) *The ecology of running waters.* 555 pp. Liverpool University Press, Liverpool.

243 HYNES H.B.N. & KAUSHIK N.K. (1969) The relationship between dissolved nutrient salts and protein production in submerged autumnal leaves. *Verh. int. Verein theor. angew Limnol.*, **17**, 95–103.

244 HYNES H.B.N., KAUSHIK N.K., LOCK M.A., LUSH D.L., STOCKER Z.S.J., WALLACE R.R. & WILLIAMS D.D. (1974) Benthos and allochthonous organic matter in streams. *J. Fish. Res. Bd. Can.*, **31**, 545–63.

245 ILLIES J. (ed.) (1978) *Limnofauna Europaea.* 2nd edn. *A checklist of the animals inhabiting European inland waters with accounts of their distribution and ecology.* 532 pp. Gustav-Fischer Verlag, Stuttgart.

246 IMHOF G. (1973) Aspects of energy flow by different food chains in a reed bed. A review. *Pol. Arch. Hydrobiol.*, **20**, 165–8.

247 IMBRIE J. & NEWELL N (1964) *Approaches to paleoecology.* 432 pp. Wiley, New York.

248 IMEVBORE A.M.A. & ADEGOKE O.S. (eds) (1975) *The ecology of Lake Kainji.* 209 pp. University of Ife Press.

249 INGOLD C.T. (1966) The tetraradiate fungal spore. *Mycologia*, **58**, 43–56.

250 INTERNATIONAL UNION FOR THE CONSERVATION OF NATURE (1962) *Project MAR.* IUCN publications N.S. 3.

251 IVLEV V.S. (1961) *Experimental ecology of the feeding of fishes.* Yale University Press, London.

252 JAAG O. & AMBUHL H. (1964) The effect of the current on the composition of biocoenoses in flowing water streams. *Int. Conf. Wat. Pollut. Res. Lond.* pp. 31–49. Pergamon Press, London.

253 JANNASCH H. (1974) Steady state and the chemostat in ecology. *Limnol. Oceanogr.*, **19**, 716–20.

254 JASSBY A.D. & GOLDMAN C.R. (1974) Loss rates from a lake phytoplankton community. *Limnol. Oceanogr.*, **19**, 618–27.

255 JAWED M. (1969) Body nitrogen and nitrogenous excretion in *Neomysis rayii* Murdoch and *Euphausia pacifica* Hansen. *Limnol. Oceanogr.*, **14**, 748–54.

256 JENKIN P.M. (1942) Seasonal changes in the temperature of Windermere (English Lake District). *J. anim. Ecol.*, **11**, 248–69.

257 JOHNSON M.G. & BRINKHURST R.O. (1971) Associations and species diversity in benthic macro invertebrates of Bay of Quinte and Lake Ontario. *J. Fish Res. Bd. Can.* **28**, 1683–97.

258 JOHNSON M.G. & BRINKHURST R.O. (1971) Production of benthic macroinvertebrates of Bay of Quinte and Lake Ontario. *J. Fish Res. Bd. Can.*, **28**, 1699–1714.

259 JOHNSON M.G. & BRINKHURST R.O. (1971) Benthic community metabolism in Bay of Quinte and Lake Ontario. *J. Fish Res. Bd. Can.*, **28**, 1715–25.

260 JONASSON P.M. (1972) Ecology and production of the profundal benthos in relation to phytoplankton in Lake Esrom. *Oikos*, Suppl. **14**, 1–148.

261 JONASSON P.M. (1977) Lake Esrom research, 1867–1977. *Folia Limnol. Scand.*, **17**, 67–90.

262 JONASSON P.M. (1978) Zoobenthos of lakes. *Verh. int. Verein theor. angew. Limnol.*, **20**, 13–37.

263 JONES J.C. (1978) The feeding behaviour of mosquitos. *Sci. Amer.*, **238**, 138–48.

264 JONES J.G. (1972) Studies of freshwater bacteria: Association with algae and alkaline phosphatase activity. *J. Ecol.*, **60**, 59–75.

265 JONES J.R.E. (1940) A study of the zinc-polluted river Ystwyth in north Cardiganshire, Wales. *Ann. appl. Biol.*, **27**, 368–78.

266 JONES J.R.E. (1964) *Fish and river pollution.* Butterworth, London.

267 JONES J.W. (1961) *The Salmon.* Collins, London.

268 KAUSHIK N.K. & HYNES H.B.N. (1968) Experimental study on the role of autumn shed leaves in aquatic environments. *J. Ecol.*, **56**, 229–43.

269 KAUSHIK N.K. & HYNES H.B.N. (1971) The fate of dead leaves that fall into streams. *Arch. Hydrobiol.*, **68**, 465–515.

270 KESSEL J.F. VAN (1976) Influence of denitrification in aquatic sediments on the nitrogen content of natural waters. *Agric. Res. Reports Wageningen*, **858**, 1–52.

271 KIRK R.G. (1967) The fishes of Lake Chilwa. *Soc. Malawi J.*, **20**, 1–14.

272 KLEIN L. (1959) *River pollution. Vol. 1 Chemical Analysis, Vol. 2. Causes and effects.* Butterworth, London.

273 KRENKEL P.A. & PARKER F.L. (1969) *Biological aspects of thermal pollution.* 407 pp. Vandebilt University Press, Nashville.

274 KROES H.W. (1971) Growth interactions between *Chlamydomonas globosa* Snow and *Chlorococcum ellipsoideum* Deason and Bold under different experimental conditions with special attention to the role of pH. *Limnol. Oceanogr.*, **16**, 869–79.

275 KROKHIN E.M. (1975) Transport of nutrients by salmon migrating from the sea into lakes. In *Coupling of land and water systems*, pp. 153–156, (ed.) Hasler A.D. Springer-Verlag, Berlin.

276 KUZNETSOV S.I. (1977) Trends in the development of ecological microbiology. *Adv. aquat. microbiol.*, **1**, 1–48.

277 LAGLER K.F., BARDACH J.E. & MILLER R.R. (1962) *Ichthyology.* Wiley, New York.

278 LAMBERT J.M., JENNINGS J.N., SMITH C.T., GREEN C. & HUTCHINSON J.N.

(1960) The making of the Broads: a reconsideration of their origin in the light of new evidence. *Roy. Geogr. Soc. Memoir.* **3**, 1–242.

279 LARSEN D.P. & MERCIER H.T. (1976) Phosphorus retention capacity in lakes. *J. Fish. Res. Bd. Can.*, **33**, 1742–50.

280 LAWSON G.W. (1970) Lessons of the Volta—a new man-made lake in tropical Africa. *Biol. Cons.*, **2**, 90–6.

281 LEAH R.T., MOSS B. & FORREST D.E. (1978) Experiments with large enclosures in a fertile, shallow, brackish lake, Hickling Broad, Norfolk, United Kingdom. *Int. Rev. ges. Hydrobiol.*, **63**, 291–310.

282 LEAH R.T., MOSS B. & FORREST D.E. (1980). The role of predation in causing major changes in the limnology of a hyper-eutrophic lake. *Int. Rev. ges. Hydrobiol.*

283 LEAN D.R.S. (1973) Phosphorus dynamics in lake water. *Science (N.Y.)*, **179**, 678–80.

284 LE CREN E.D. (1964) The interactions between freshwater fisheries and nature conservation. *Proc. of the MAR conference.* IUCN publications, 431–7.

285 LE CREN E.D. (1969) Estimates of fish populations and production in small streams in England. In *Symposium on salmon and trout in streams*, pp. 269–80 (ed.) Northcote T. University of British Columbia.

286 LE CREN E.D. (1972) Fish production in freshwaters. *Symp. zool. Soc. Lond.*, **29**, 115–33.

287 LEEDALE G.F. (1967) *Euglenoid flagellates.* 242 pp. Prentice Hall. Englewood Cliffs (N.J.).

288 LEFEVRE M. (1964) Extracellular products of algae. In *Algae and Man*, pp. 337–67. (ed.) Jackson D.F. Plenum, New York.

289 LEHMAN J.T. (1976) Ecological and nutritional studies on *Dinobryon* (Ehrenb.) Seasonal periodicity and the phosphate toxicity problem. *Limnol. Oceanogr.*, **21**, 646–58.

290 LEHMAN J.T., BORKIN D.B. & LIKENS G.E. (1975) The assumptions and rationales of a computer model of phytoplankton population dynamics. *Limnol. Oceanogr.*, **20**, 343–64.

291 LEWIN R.A. (1976) Prochlorophyta as a proposed new division of algae. *Nature (Lond.)*, **261**, 697–8.

292 LEITH H. & WHITTAKER R.L. (eds.) (1975) *Primary productivity of the biosphere.* 339 pp. Ecological Studies V. 14. Springer-Verlag. N.Y.

293 LIKENS G.E. (ed.) (1972) *Nutrients and Eutrophication.* Special Symposia I, American Society of Limnology and Oceanography. 328pp. Lawrence,Kansas.

294 LIKENS G.E. & BORMANN F.H. (1974) Linkages between terrestrial and aquatic ecosystems. *Bioscience*, **24**, 447–56.

295 LIKENS G.E., BORMANN F.H., PIERCE R.S., EATON J.S. & JOHNSON N.M. (1977) *Biogeochemistry of a forested ecosystem.* 146 pp. Springer-Verlag, New York.

296 LIVINGSTONE D. & CAMBRAY R.S. (1978) Confirmation of ^{137}Cs dating by algal stratigraphy in Rostherne Mere. *Nature, Lond.*, **276**, 259–61.

297 LIVINGSTONE D.A. (1975) Late quaternary climatic change in Africa. *Ann. Rev. Ecol. Systematics*, **6**, 249–80.

298 LOVELESS C.M. (1959) A study of the vegetation in the Florida Everglades. *Ecology*, **40**, 1–9.

299 LOWE-MCCONNELL R.H. (ed.) (1966) Man-made lakes. *Symp. Inst. Biol. London*, **15**. Academic Press, London & New York.

300 LOWE-MCCONNELL R.H. (1975) *Fish communities in tropical freshwaters.* 337 pp. Longman, London.

301 LOWE-MCCONNELL R.H. (1977) *Ecology of fishes in tropical waters.* 64 pp. Arnold, London.

302 LUND J.W.G. (1949) Studies on *Asterionella* I. The origin and nature of the cells producing seasonal maxima. *J. Ecol.*, **37**, 389–419.

303 LUND J.W.G. (1950) Studies on *Asterionella formosa* Hass II. Nutrient depletion and the spring maximum. *J. Ecol.*, **38**, 1–14, 15–35.

304 LUND J.W.G. (1954) The seasonal cycle of the plankton diatom *Melosira italica* (Ehr.) Kütz subsp. *subarctica* O. Müll. *J. Ecol.*, **42**, 151–79.

305 LUND J.W.G. 1961. The algae of the Malham Tarn district. *Field Studies* **1**(3), 85–120.

306 LUND J.W.G. (1964) Primary production and periodicity of phytoplankton. *Ver. int. Verein. theor. angew. Limnol.*, **15**, 37–56.

307 LUND J.W.G. (1971) The seasonal periodicity of three planktonic desmids in Lake Windermere. *Mitt. int. Verein. theor. angew. Limnol.*, **19**, 3–25.

308 MACAN T.T. (1959) *A guide to freshwater invertebrate animals.* 118 pp. Longman, London.

309 MACAN T.T. (1963) *Freshwater Ecology.* 338 pp. Longman, London.

310 MACAN T.T. (1970) *Biological Studies of the English lakes*. 260 pp. Elsevier, Amsterdam.

311 MACAN T.T. (1973) *Ponds and Lakes*. 148 pp. Allen & Unwin, London.

312 MACAN T.T. (1976) A twenty-one-year study of the water-bugs in a moorland fishpond. *J. anim. Ecol.*, **45**, 913–22.

313 MACAN T.T. (1977) The influence of predation on the composition of freshwater animal communities. *Biol. Rev.*, **52**, 45–70.

314 MACAN T.T. & KITCHING A. (1972) Some experiments with artificial substrata. *Verh. int. Verein theor. angew. Limnol.*, **18**, 213–20.

315 MACKAY R.J. & KALFF J. (1973) Ecology of two related species of caddis fly larvae in the organic substrates of a woodland stream. *Ecology*, **54**, 499–511.

316 MACKERETH F.J.H. (1957) Chemical analysis in ecology illustrated from Lake District tarns and lakes. 1. Chemical analysis. *Proc. Linn. Soc. London*, **167**, 159–64.

317 MACKERETH F.J.H. (1958) A portable core sampler for lake deposits. *Limnol. Oceanogr.*, **3**, 181–91.

318 MACKERETH F.J.H. (1965) Chemical investigation of lake sediments and their interpretation. *Proc. roy. Soc. B.*, **161**, 295–309.

319 MACKERETH F.J.H. (1966) Some chemical observations on post-glacial lake sediments. *Phil. Trans. R. Soc. (B.)*, **250**, 165–213.

320 MACKERETH F.J.H. (1971) On the variation in direction of the horizontal component of remanent magnetisation in lake sediments. *Earth and Planetary Sci. Letters*, **12**, 332–8.

321 MACKERETH F.J.H., HERON J. & TALLING J.F. (1978) Water analysis and some revised methods for limnologists. *Sci. Pub. Freshwat. Biol. Ass.*, **36**, 120 pp.

322 MAITLAND P.S. (1972) *Key to British freshwater fishes. Sci. Pub. Freshwat. Biol. Ass.*, **27**, 139 pp.

323 MAITLAND P.S. (1974) The conservation of freshwater fishes in the British isles. *Biol. Cons.*, **6**, 7–14.

324 MAITLAND P.S. (1977) Freshwater fish in Scotland in the 18th, 19th, and 20th centuries. *Biol. Conserv.*, **12**, 265–78.

325 MANN K.H. (1956) A study of the oxygen consumption of five species of leech. *J. exp. Biol.*, **33**, 615–26.

326 MANN K.H. (1965) Energy transformations by a population of fish in the River Thames. *J. anim. Ecol.*, **34**, 253–75.

327 MANN R.H.K. (1971) The populations, growth and production of fish in four small streams in southern England. *J. anim. Ecol.*, **40**, 155–90.

328 MANNY B.A. (1972) Seasonal changes in dissolved organic nitrogen in six Michigan lakes. *Verh. int. Verein. theor. argew. Limnol.*, **18**, 147–56.

329 MANNY B.A., MILLER M.C. & WETZEL R.G. (1971) Ultraviolet combustion of dissolved organic compounds in lake waters. *Limnol. Oceanogr.*, **16**, 71–85.

330 MANNY B.A., WETZEL R.G. & JOHNSON W.C. (1975) Annual contribution of carbon, nitrogen, and phosphorus by migrant Canada geese to a hardwater lake. *Verh. int. Verein theor. angew. Limnol.*, **19**, 949–51.

331 MARSHALL J.S. (ed.) (1975) *Proceedings of the second federal conference on the Great Lakes*. 523 pp. Argonne National Laboratory.

332 MARSHALL T.R. (1977) Morphological, physiological and ethological differences between walleye (*Stizostedion vitreum vitreum* and pikeperch (*S. lucioperca*). *J. Fish. Res. Bd. Can.*, **34**, 1515–23.

333 MASON C.F. & BRYANT R.J. (1975) Changes in the ecology of the Norfolk Broads. *Freshwat. Biol.*, **5**, 257–70.

334 MASON C.F. & BRYANT R.J. (1975) Periphyton production and grazing by chironomids in Alderfen Broad, Norfolk. *Freshwat. Biol.*, **5**, 271–7.

335 MASON C.F. & BRYANT R.J. (1975) Production, nutrient content and decomposition of *Phragmites communis* Trin. & *Typha angustifolia* L. *J. Ecol.*, **63**, 71–95.

336 MASON H.L. (1957) *A flora of the marshes of California*. 878 pp. University of California Press, Berkeley & Los Angeles.

337 MATHEWS C.P. & WESTLAKE D.F. (1969) Estimation of production by populations of higher plants subject to high mortality. *Oikos*, **20**, 156–60.

338 MCLACHLAN A.J. (1974) Development of some lake ecosystems in tropical Africa, with special reference to the invertebrates. *Biol. Rev.*, **49**, 365–97.

339 MCLACHLAN A.J. (1974) Recovery of the mud substrate and its associated fauna following a dry phase in a tropical lake. *Limnol. Oceanogr.*, **19**, 74–83.

340 MCLACHLAN A.J. (1978) Interactions between freshwater animals & micro-organisms. *Ann. Appl. Biol.*, **89**, 162–5.

341 MCLACHLAN A.J. & DICKINSON C.H. (1977) Micro-organisms as a factor in

the distribution of *Chironomus lugubris* Zetterstedt in a bog lake. *Arch. Hydrobiol.*, **80**, 133–46. MSS.

342 McLachlan A.J., Pearce L.J. & Smith J.A. (1979) Feeding interactions and cycling of peat in a bog lake. MSS.

343 McMahon A.F.M. (1946) *Fishlore. British Freshwater Fishes.* Pelican, Harmondsworth.

344 Mellanby H. (1963) *Animal life in freshwater.* 308 pp. Chapman & Hall, London.

345 Merritt R.W. & Cummins K.W. (1978) *An introduction to the aquatic insects of North America.* 441 pp. Kendall/Hunt. Dubuque (Iowa).

346 Milbrink G. (1977) On the limnology of two alkaline lakes (Nakuru and Naivasha) in the East Rift Valley system in Kenya. *Int. Rev. ges. Hydrobiol.*, **62**, 1–17.

347 Mills D.H. (1971) *Salmon & trout: a resource, its ecology, conservation and management.* Oliver & Boyd, Edinburgh.

348 Mitchell D.S. (1972) The Kariba weed: *Salvinia molesta. Brit. Fern Gaz.*, **10**, 251–2.

349 Monod J. (1942) *Recherches sur la croissance des cultures bacteriennes.* Hermann, Paris.

350 Moore P.D. & Bellamy D.J. (1974) *Peatlands.* 221 pp. Elek, London.

351 Moore P.D. & Webb J.A. (1978) *An illustrated guide to pollen analysis.* 133 pp. Hodder & Stoughton, London & Sevenoaks.

352 Morgan N.C. (1972) Problems of the conservation of freshwater ecosystems. *Symp. zool. Soc. Lond.*, **29**, 135–54.

353 Moriarty D.J.W. (1973) The physiology of digestion of blue-green algae in the cichlid fish *Tilapia nilotica. J. Zool., Lond.*, **171**, 25–39.

354 Moriarty D.J.W. (1977) Improved method of using muramic acid to estimate biomass of bacteria in sediments. *Oecologia (Berl.)*, **26**, 317–23.

355 Moriarty D.J.W., Darlington J.P.E.C., Dunn I.G., Moriarty C.M. & Tevlin M.P. (1973) Feeding and grazing in Lake George, Uganda. *Proc. roy. Soc. Lond. B.*, **184**, 227–346.

356 Moriarty D.J.W. & Moriarty C.M. (1973) The assimilation of carbon from phytoplankton by two herbivorous fishes: *Tilapia nilotica & Haplochromis nigripinnis. J. Zool., Lond.*, **171**, 41–55.

357 Morris I. (1967) *An introduction to the algae.* 189 pp. Hutchinson. London.

358 Mortimer C.H. (1941–42) Exchange of dissolved substances between water and mud in lakes. I–IV *J. Ecol.*, **29**, 280–329; **30**, 147–201.

359 Mortimer C.H. (1956) An explorer of lakes, in *E. A. Birge, a memoir*, pp. 165–206. G. C. Sellery. Univ. of Wisconsin Press, Madison.

360 Moss B. (1969) Vertical heterogeneity in the water column of Abbot's pond. II The influence of physical and chemical conditions on the spatial and temporal distribution of the phytoplankton and of a community of epipelic algae. *J. Ecol.*, **57**, 397–414.

361 Moss B. (1969) Limitation of algal growth in some Central African waters. *Limnol. Oceanogr.*, **14**, 591–601.

362 Moss B. (1972) Studies on Gull Lake, Michigan. I Seasonal and depth distribution of phytoplankton. *Freshwat. Biol.*, **2**, 289–307.

363 Moss B. (1973) Diversity in freshwater phytoplankton. *Am. Midl. Nat.*, **90**, 341–55.

364 Moss B. (1973) The influence of environmental factors on the distribution of freshwater algae: an experimental study. IV Growth of test species in natural lake waters & conclusion. *J. Ecol.*, **61**, 193–211.

365 Moss B. (1976) The effects of fertilization and fish on community structure and biomass of aquatic macrophytes and epiphytic algal populations: an ecosystem experiment. *J. Ecol.*, **64**, 313–42.

366 Moss B. (1977) Conservation problems in the Norfolk Broads and rivers of East Anglia, England—phytoplankton, boats, and the causes of turbidity. *Biol. Conserv.*, **12**, 95–114.

367 Moss B. (1977) The state of the Norfolk Broads. *The Ecologist*, **7**, 324–6.

368 Moss B. (1977) Adaptations of epipelic and episammic freshwater algae. *Oecologia (Berl.)*, **27**, 103–8.

369 Moss B. (1979) Algal and other fossil evidence for major changes in Strumpshaw Broad, Norfolk, England in the last two centuries. *Br. phyc. J.* In press.

370 Moss B. (1980) Further studies on the palaeolimnology, and changes in the phosphorus budget of Barton Broad, Norfolk. *Freshwat. Biol.* **10**.

371 Moss B. & Moss J. (1969) Aspects of the limnology of an endorheic African lake (L. Chilwa, Malawi). *Ecology*, **50**, 109–18.

372 Moss B., Wetzel R.G. & Lauff G.H. (1980) Annual productivity and phytoplankton changes between 1969 and 1974 in Gull Lake, Michigan. *Freshwat. Biol.*, **10**.

373 MUIRHEAD-THOMPSON R. (1971) *Pesticides and freshwater fauna*. 248 pp. Academic Press, New York.

374 MÜLLER K. (1974) Stream drift as a chronobiological phenomenon in running water ecosystems. *A. Rev. Ecol. Syst.*, **5**, 309–23.

374a MULLER R. (1975) *Worms and disease*. Heinemann, London.

375 MULLIGAN H.F. & BARANOWSKI A. (1969) Nitrogen and phosphorus greenhouse studies on vascular aquatic plants and phytoplankton. *Verh. int Ver. theor. angew. Limnol.*, **17**, 802–10.

376 MUNK W.H. & RILEY G.A. (1952) Absorption of nutrients by aquatic plants. *J. Mar. Res.*, **11**, 215–40.

377 MUNRO A.L.S. & BROCK R.S. (1968) Distinction between bacterial and algal utilization of soluble substances in the sea. *J. gen. Microbiol.*, **51**, 35–42.

378 NATIONAL ACADEMY OF SCIENCES (1969) *Eutrophication: causes, consequences, correctives*. 661 pp. N.A.S., Washington.

379 NATIONAL ENVIRONMENT RESEARCH COUNCIL (1972) *Research in Freshwater Biology*. Natural Environment Research Council Publications Series B. 3.

380 NATURE CONSERVANCY (1965) *Report on Broadland*. H.M.S.O., London.

381 NELSON J.S. (1976) *Fishes of the World*. 416 pp. Wiley, New York.

382 NICHOLS D.S. & KEENEY D.R. (1976) Nitrogen nutrition of *Myriophyllum spicatum*: uptake and translocation of ^{15}N by shoots and roots. *Freshwat. Biol.*, **6**, 145–54.

383 OBENG L.E. (ed.) (1969) *Man-made lakes, the Accra Symposium*. Accra, Ghana University Press.

384 ODUM H.T. (1956) Primary production in flowing waters. *Limnol. Oceanogr.*, **1**, 102–17.

385 OGLESBY R.T. (1977) Relationships of fish yield to lake phytoplankton, standing crop, production, and morphoedaphic factors. *J. Fish. Res. Bd. Can.*, **34**, 2271–9.

386 OGLESBY R.T., CARISON C.A. & MCCANN J.A. (eds.) (1972) *River Ecology and Man*. 465 pp. Academic Press, New York.

387 OHWADA K. & TAGA N. (1973) Seasonal cycles of vitamin B_{12}, thiamine and biotin in Lake Sagami. Patterns of their distribution and ecological significance. *Int. Rev. ges. Hydrobiol.*, **58**, 851–71.

388 OLDFIELD F., APPLEBY P.G. & BATARBEE R.W. (1978) Alternative ^{210}Pb dating: results from the New Guinea Highlands and Lough Erne. *Nature, Lond.*, **271**, 339–42.

389 OMERNIK J. (1976) The influence of land use on stream nutrient levels—U.S. Environmental Protection Agency Report EPA—600/3—76—014: 1–105.

390 OSBORNE P.L. (1978) *Relationships between the phytoplankton and nutrients in the River Ant and Barton, Sutton and Stalham Broads, Norfolk*. Ph.D. Thesis, Univ. of East Anglia, Norwich.

391 OSBORNE P.L. & MOSS B. (1977) Palaeolimnology and trends in the phosphorus and iron budgets of an old man-made lake, Barton Broad, Norfolk. *Freshwat. Biol.*, **7**, 213–34.

392 OSBORNE P.A. & PHILLIPS G.L. (1978) Evidence for nutrient release from the sediments of two shallow and productive lakes. *Verh. int. Ver. theor. angew Limnol.*, **20**, 654–8.

393 PARKER F.L. & KRENKEL P.A. (eds.) (1969) *Engineering aspects of thermal pollution*. 351 pp. Vandebilt University Press, Nashville.

394 PARKER J.I. & EDGINGTON D.N. (1976) Concentration of diatom frustules in Lake Michigan sediment cores. *Limnol. Oceanogr.*, **21**, 887–93.

395 PATRICK R. (1969) Some effects of temperature on freshwater algae. In *Biological aspects of thermal pollution*. pp. 161–85, (eds.) Krenkel P.A. & Parker F.L. Vandebilt University Press, Nashville.

396 PEARSALL W.H. (1921) The development of vegetation in the English lakes, considered in relation to the general evolution of glacial lakes and rock basins. *Proc. roy. Soc. (B.)*, **92**, 259–84.

397 PENNINGTON W. (1969) *The history of British vegetation*. English Universities Press, London.

398 PENNINGTON W., CAMBRAY R.S., EAKINS J.D. & HARKNESS D.D. (1976) Radionuclide dating of the recent sediments of Blelham Tarn. *Freshwat. Biol.*, **6**, 317–33.

399 PENNINGTON W., CAMBRAY R.S. & FISHER E.M. (1973) Observations on lake sediment using fallout Cs-137 as a tracer, *Nature, Lond.* **242**, 324–6.

400 PENNINGTON W. & LISHMAN J.P. (1971) Iodine in lake sediments in Northern England & Scotland. *Biol. Rev.*, **46**, 279–313.

401 PETERS R. & LEAN D. (1973) The characterization of soluble phosphorus released by limnic zooplankton. *Limnol. Oceanogr.*, **18**, 270–9.

401a PETERS R.H. & RIGLER F.H. (1973) Phosphorus release by *Daphnia*. *Limnol. Oceanogr.*, **18**, 821-39

401b PETERS W. & GILLES H.M. (1977) *A colour atlas of tropical medicine and parasitology*. Wolfe Medical Publications, London.

402 PETIT D. (1974) ^{210}Pb and stable lead isotopes in lake sediments. *Earth and Planet. Sci. Letters.*, **23**, 407.

403 PETR T. (1969) Fish population changes in Volta Lake over the period January 1965–September 1966. In *Man-made lakes, the Accra Symposium*, (ed.) Obeng L.E. Ghana University Press, Accra.

404 PETR T. (1978) Tropical man-made lakes —their ecological impact. *Arch. Hydrobiol.*, **81**, 368–85.

405 PHILLIPS G.L. (1977) The mineral levels in three Norfolk Broads differing in trophic status and an annual mineral content budget for one of them. *J. Ecol.*, **65**, 447–74.

406 PHILLIPS G.L., EMINSON D.F. & MOSS B. (1978) A mechanism to account for macrophyte decline in progressively eutrophicated freshwaters. *Aquat. Bot.*, **4**, 103–26.

407 PICKETT-HEAPS J. (1975) *Green algae*. Sinauer Ass., Hartford (Conn.).

407a PIGOTT C.D. & PIGOTT M.E. (1963) Late-glacial and post-glacial deposits at Malham, Yorkshire. *New Phytol.*, **62**, 317–34.

408 PIGOTT M.E. & PIGOTT C.D. (1959) Stratigraphy and pollen analysis of Malham tarn and Tarn Moss. *Fld. Stud.*, **1** (1) 17 pp.

409 PORTER K.G. (1973) Selective grazing and differential digestion of algae and zooplankton. *Nature (Lond.)*, **244**, 179–80.

410 PORTER K.G. (1976) Enhancement of algal growth and productivity by grazing zooplankton. *Science, N.Y.*, **192**, 1332–4.

411 PORTER K.G. (1977) The plant-animal interface in freshwater ecosystems. *Am. Scient.*, **65**, 159–69.

412 PORTER K.S. (ed.) (1975) *Nitrogen and phosphorus: food production, waste and the environment*. 372 pp. Ann Arbor Science Publishers, Ann Arbor.

413 PREPAS E. & RIGLER F.H. (1978) The enigma of *Daphnia* death rates. *Limnol. Oceanogr.*, **23**, 970–88.

414 PRESCOTT G.W. (1962) *Algae of the Western Great Lakes Area*. 977 pp. W.C. Brown, Dubuque (Iowa).

415 PROCTOR V.W. (1957) Studies of algal antibiosis using *Haematococcus* and *Chlamydomonas*. *Limnol. Oceanogr.*, **2**, 125–39.

416 PROWSE G.A. (1959) Relationship between epiphytic algal species and their macrophyte hosts. *Nature, Lond.*, **183**, 1204–5.

417 QUIGLEY M. (1977) *Invertebrates of streams and rivers*. 84 pp. Edward Arnold, London.

418 RAVEN J.A. (1970) Exogenous inorganic carbon sources in plant photosynthesis. *Biol. Rev.*, **45**, 167–22.

419 RAVEN J.A. & GLIDEWELL S.M. (1975) Photosynthesis, respiration and growth in the shade alga *Hydrodictyon africanum*. *Photosynthetica*, **9**, 361–71.

420 RAWSON D.S. (1955) Morphometry as a dominant factor in the productivity of large lakes. *Verh. int. Verein theor. angew. Limnol.*, **12**, 164–75.

421 REDFERN M. (1975) Revised field key to the invertebrate fauna of stony hill streams. *Fld Stud.*, **4**, 105–15.

422 REYNOLDS C.S. & WALSBY A.E. (1975) Water-blooms. *Biol. Rev.*, **50**, 437–81.

423 REYNOLDSON T.B. (1966) The distribution and abundance of lake-dwelling triclads—towards a hypothesis. *Adv. Ecol. Res.*, **3**, 1–71.

424 REYNOLDSON T.B. (1967) *A key to the British species of Freshwater Triclads*. *Sci. Pub. Freshwat. Biol. Ass.*, **23**, 32 pp.

425 REYNOLDSON T.B. & BELLAMY L.S. (1970) The establishment of interspecific competition in field populations, with an example of competition in action between *Polycelis nigra* (Mull) and *P. tenuis* (Ijima) (Turbellaria, Tricladida) *Proc. Adv. Study Inst. Dynamics Numbers Population (Oosterbeek)*, 282–97.

426 RHEINHEIMER G. (1974) *Aquatic microbiology*. 184 pp. Wiley, New York.

427 RICH P.H., WETZEL R.G. & THUY N.V. (1971) Distribution, production and role of aquatic macrophytes in a southern Michigan marl lake. *Freshwat. Biol.*, **1**, 3–21.

428 RICHARDSON J.L. & RICHARDSON A.E. (1972) History of an African rift lake and its climatic implications. *Ecol. Monogr.*, **42**, 499–534.

429 RIDLEY J.E.A. (1964) Thermal stratification and thermocline control in storage reservoirs. *Proc. Soc. Wat. Treat. Exam.*, **13**, 275.

430 RIDLEY J.E.A., COOLEY P. & STEEL J.A. (1966) Control of thermal stratification in Thames valley reservoirs. *Proc. Soc. Wat. Treat. Exam.*, **15**, 225–44.

431 RIGLER F.H. (1964) The phosphorus fractions and the turnover time of inorganic phosphorus in different types of lakes. *Limnol. Oceanogr.*, **9**, 511–18.

432 RIGLER F.H. (1968) Further observations inconsistent with the hypothesis that the

molybdenum blue method measures orthophosphate in lake water. *Limnol. Oceanogr.*, **13**, 7–13.

433 RIGLER F.H. (1973) A dynamic view of the phosphorus cycle in lakes. In *Environmental Phosphorus Handbook*, pp. 539–72, (eds.) Griffith E.J. *et al.* Wiley, New York.

434 ROBERTSON W.B. Jr. (1955) *A survey of the effects of fire in Everglades National Park*. 169 pp. (Cyclostyled), U.S. National Park Service. Homestead, Florida.

435 ROBERTSON W.B. (1959) *Everglades—the Park Story*. 95 pp. University of Miami Press, Coral Gables.

436 ROUND F.E. (1961) The diatoms of a core from Esthwaite Water. *New Phytol.*, **60**, 43–59.

437 ROUND F.E. (1964) The ecology of benthic algae. In *Algae and Man.*, Jackson, D.F. (ed.), pp. 138–84. Plenum, New York.

438 ROUND F.E. (1973) *The biology of algae.* Edward Arnold, London.

439 RUSSELL E.S. (1931) Some theoretical considerations on the 'overfishing' problem. *J. Cons. Int. Explor. Mer.*, **6**, 3–20.

440 RYDER R.A., KERR S.R., LOFTUS K.H. & REGIER H.A. (1974) The morpho-edaphic index, a fish yield estimator—review and evaluation. *J. Fish. Res. Bd. Can.*, **31**, 663–88.

441 RZÓSKA J. (1974) The Upper Nile swamps, a tropical wetland study. *Freshwat. Biol.*, **4**, 1–30.

442 RZÓSKA J. (1976) A controversy reviewed. *Nature, Lond.*, **261**, 444–5.

443 RZÓSKA J. (1976) *The Nile—Biology of an ancient river.* 417 pp. Junk, The Hague.

444 SAND-JENSEN K. (1978) Metabolic adaptation and vertical zonation of *Littorella uniflora* (L.) Aschers. and *Isoetes lacustris* L. *Aquat. Biol.*, **4**, 1–10.

445 SANGER J.E. & GORHAM E. (1970) The diversity of pigments in lake sediments and its ecological significance. *Limnol. Oceanogr.*, **15**, 59–69.

446 SATTLER W. (1963) Über den Körperbau und Ethologie der Larve und Puppe von *Macronema* Pict. (Hydropsychidae), ein als Larve sich von 'Mikro-Drift' ernährendes Trichopter aus dem Amazongebiet. *Arch. Hydrobiol.*, **59**, 26–60.

447 SAUNDERS G.W. (1972a) Potential heterotrophy in a natural population of *Oscillatoria aghardii* var. *isothrix*. Skuja. *Limnol. Oceanogr.*, **17**, 704–11.

448 SAUNDERS G.W. (1972b) The transformation of artificial detritus in lake water.

Mem. 1st. Ital. Idrobiol., **29** *Suppl.* 261–88.

449 SCHINDLER D.W. (1974) Eutrophication and recovery in experimental lakes: implications for lake management. *Science, N.Y.*, **184**, 897–8.

450 SCHINDLER D.W. (1977) Evolution of phosphorus limitation in lakes. *Science, N.Y.*, **195**, 260–2.

451 SCHINDLER D.W. (1978) Factors regulating phytoplankton production and standing crop in the world's freshwaters. *Limnol. Oceanogr.*, **23**, 478–86.

452 SCHINDLER D.W., ARMSTRONG F.A.J., HOLMGREN S.K. & BRUNSKILL G.J. (1971) Eutrophication of lake 227, Experimental lakes area, Northwestern Ontario, by addition of phosphate and nitrate. *J. Fish. Res. Bd. Can.*, **28**, 1763–82.

453 SCHINDLER D.W., BRUNSKILL G.J., EMERSON S., BROECKER W.S. & PENY T.H. (1972) Atmospheric carbon dioxide; its role in maintaining phytoplankton standing crops. *Science*, 177, 1192–4.

454 SCHINDLER D.W. & FEE E.J. (1974) Experimental lakes area: whole-lake experiments in eutrophication. *J. Fish. Res. Bd. Can.*, **31**, 937–53.

455 SCHINDLER D.W., WELCH H.E., KALFF J., BRUNSKILL G.J. & KRITSCH N. (1974) Physical and chemical limnology of Char Lake, Cornwallis Island 75° N lat.) *J. Fish. Res. Bd. Can.*, **31**, 585–607.

456 SCHINDLER J.E. (1971) Food quality and zooplankton nutrition. *J. Anim. Ecol.*, **40**, 589–95.

457 SCOTT W.B. & CROSSMAN E.J. (1973) Freshwater Fishes of Canada. *Bull. Fish Res. Bd. Can.*, **184**, 966 pp.

458 SCOWFIELD D.J. & HARDING J.P. (1966) *A key to the British species of freshwater Cladocera*. 3rd edn. *Sci. Pub. Freshwat. Biol. Ass.*, **5**, 1–55.

459 SCULTHORPE C.D. (1967) *The biology of aquatic vascular plants.* 610 pp. Edward Arnold, London.

460 SERVICE M.W. (1976) *Mosquito Ecology. Field Sampling methods.* 583 pp. Applied Science Publishers, Barking.

461 SHANNON C.E. (1948) A mathematical theory of communication. *Bell Syst. Tech. J.*, **27**, 379–423, 623–56.

462 SHAPIRO J. (1957) Chemical and biological studies on the yellow organic acids of lake water. *Limnol. Oceanogr.*, **2**, 161–79.

463 SHELFORD V.E. (1963) *The ecology of North America.* 610 pp. University of Illinois Press, Urbana.

464 SINKER C.A. (1962) The North Shropshire Meres and Mosses: a background for ecologists. *Fld. Stud.*, **1**(4), 101–7.

465 SMITH S.I. (1972) Factors of ecologic succession in oligotrophic fish communities of the Laurentian Great Lakes. *J. Fish. Res. Bd. Can.*, **29**, 717–30.

466 SMITH S.I. (1962) The future of salmonid communities in the Laurentian Great Lakes. *J. Fish. Res. Bd. Can.*, **29**, 951-7

467 SMYLY W.J.P. (1955) Comparison of the Entomostraca of two artificial moorland ponds near Windermere. *Verh. int. Verein theor. angew. Limnol.*, **12**, 421–4.

468 SOROKIN Y.I. & KADOTA H. (1972) *Microbial production and decomposition in freshwaters.* IBP Handbook No. 23. 112 pp. Blackwell Scientific Publications, Oxford.

469 SPENCE D.H.N. (1964) The macrophytic vegetation of lochs, swamps and associated fens. In *The Vegetation of Scotland*, pp. 306–425. (ed.) Burnett J.H. Oliver & Boyd, Edinburgh.

470 SPENCE D.H.N. (1976) Light and plant response in fresh water. In *Light as an ecological factor* II., (eds.) Evans G.C., Bainbridge R. & Rackham O. Blackwell Scientific Publications, Oxford.

471 SPENCE D.H.N. & CHRYSTAL J. (1970) Photosynthesis and zonation of freshwater macrophytes I. Depth distribution and shade tolerance. *New Phytol.*, **69**, 205–15.

472 SPENCE D.H.N. & CHRYSTAL J. (1970) Photosynthesis and zonation of freshwater macrophytes. II. Adaptability of species of deep and shallow waters. *New Phytol.*, **69**, 217–27.

473 SPENCE D.H.N., MILBURN T.R., NALAWULA-SENYIMBA M. & ROBERTS E. (1971) Fruit biology and germination of two tropical *Potamogeton* species. *New Phytol.*, **70**, 197–212.

474 STEEL J.A. (1972) The application of fundamental limnological research in water supply system design and management. *Symp. zool. Soc. Lond.*, **29**, 41–67.

475 STEELE J.H. (1976) Comparative studies of beaches. *Phil. Trans. R. Soc. (B.)*, **274**, 401–15.

476 STEEMAN NIELSEN E. (1952) The use of radioactive carbon (C^{14}) for measuring organic production in the sea. *J. Cons. perm. int. Explor. Mer.*, **18**, 117–40.

477 STEWART K. & MATTOX K. (1975) Comparative cytology, evolution and classification of the green algae with some consideration of the origin of other organisms with chlorophylls a & b. *Bot. Rev.*, **41**, 104–35.

478 STEWART W.D.P. (ed.) (1974) *Algal physiology and biochemistry.* 989 pp., Blackwell Scientific Publications, Oxford.

479 STEWART W.D.P. & DAFT M.J. (1977) Microbial pathogens of cyanophycean blooms. *Adv. Aquat. Microbiol.*, **1**, 177–218.

480 STEWART W.D.P., FITZGERALD G.P. & BURRIS R.H. (1967) *In situ* studies on N$_2$-fixation using the acetylene reduction technique. *Proc. nat. Acad. Sci.*, **58**, 2071–8.

481 STEWART W.D.P., HAYSTEAD A. & PEARSON H.W. (1969) Nitrogenase activity in heterocysts of filamentous blue-green algae. *Nature (Lond.)*, **224**, 226–8.

482 STEWART W.D.P. & LEX M. (1970) Nitrogenase activity in blue-green alga, *Plectonema boryanum* strain 594, *Arch. Mikrobiol.*, **73**, 250–60.

483 STOCKNER J.G. & BENSON W.W. (1967) The succession of diatom assemblages in the recent sediments of Lake Washington. *Limnol. Oceanogr.*, **12**, 513–32.

484 STRICKLAND J.D.H. & PARSONS T.R. (1968) *A practical handbook of sea-water analysis. Bull. Fish. Res. Bd. Can.*, **167**, 311 pp.

485 STUMM W. & MORGAN J.J. (1970) *Aquatic chemistry; an introduction emphasising chemical equilibria in natural waters.* 583 pp. Wiley, New York.

486 SUBERKROPP K. & KLUG M.J. (1976) Fungi and bacteria associated with leaves during processing in a woodland stream. *Ecology*, **57**, 707–19.

487 TALLING J.F. (1957) The phytoplankton population as a compound photosynthetic system. *New Phytol.*, **56**, 133–49.

488 TALLING J.F. (1957) The longitudinal succession of water characteristics in the White Nile. *Hydrobiologia*, **11**, 73–89.

489 TALLING J.F. (1957c) Diurnal changes of stratification and photosynthesis in some tropical African waters. *Proc. roy. Soc. (B.)*, **147**, 57–83.

490 TALLING J.F. (1966) The annual cycle and stratification and phytoplankton growth in Lake Victoria (E. Africa). *Int. Rev. ges. Hydrobiol.*, **51**, 545–621.

491 TALLING J.F. (1969) The incidence of vertical mixing and some biological and chemical consequences in tropical African lakes. *Verh. int. Verein theor. angew. Limnol.*, **17**, 998–1012.

492 TALLING J.F. (1976) The depletion of carbon dioxide from lake water by phytoplankton. *J. Ecol.*, **64**, 79–121.

493 TALLING J.F. & TALLING I.B. (1965) The chemical composition of African lake waters. *Int. Revue ges. Hydrobiol.*, **50**, 421–63.

494 TAUB F.B. & DOLLAR A.M. (1968) The nutritional inadequacy of *Chlorella* and *Chlamydomonas* as food for *Daphnia pulex*. *Limnol. Oceanogr.*, **13**, 607–17.

494a TAYLOR N. (1975) Medical aspects of nitrate in drinking water. *Wat. Treat. Exam.*, **24**, 196–200.

495 THESIGER W. (1964) *The Marsh Arabs*. Longman, London. (Also Penguin Books, 1967.)

496 THOMPSON R. (1973) Palaeolimnology and palaeomagnetism. *Nature*, **242**, 182–4.

497 THORPE J.E. (1974) Trout and perch populations at Loch Leven, Kinross. *Proc. roy. Soc. Edinb.* (*B.*), **74**, 295–313.

498 TILMAN D. (1977) Resource competition between planktonic algae: an experimental and theoretical approach. *Ecology*, **58**, 338–48.

499 TIPPETT R. (1964) An investigation into the nature of the layering of deep water sediments in two Eastern Ontario lakes. *Can. J. Bot.*, **42**, 1693–709.

500 TIPPETT R. (1970) Artificial surfaces as a method of studying populations of benthic micro-algae in freshwaters. *Br. phyc. J.*, **5**, 187–99.

501 TITMAN D. (1976) Ecological competition between algae: Experimental confirmation of resource-based competition theory. *Science*, **192**, 463–65.

502 TOMS R.G. (1975) Management of river water quality. *River Ecology*, pp. 538–64, (ed.) Whitton B.A. Blackwell Scientific Publications, Oxford.

503 TOWNSEND C.R. & HILDREW A.G. (1976) Field experiments on the drifting, colonization and continuous redistribution of stream benthos. *J. anim. Ecol.*, **45**, 759–72.

504 TUTIN T.G. (1940) The Percy Sladen Trust Expedition to Lake Titicaca in 1937 under the leadership of Mr. H. Cary Gilson. M.A., X. The macrophytic vegetation of the lake. *Trans. Linn. Soc. Lond.*, 3rd Ser., **1**, 161–89.

505 UNITED STATES ENVIRONMENTAL PROTECTION AGENCY (USEPA) (1973) *Measures for the restoration and enhancement of quality of freshwater lakes*. 238 pp. USEPA, Washington DC.

506 VALLENTYNE J.R. (1969) Sedimentary organic matter and palaeolimnology. *Mitt. int. Ver. theor. angew. Limnol.* **17**, 104–10.

507 VALLENTYNE J.R. (1974) *The algal bowl lakes and man*. 186 pp. Misc. Publ. Dept. of the Environment, Fisheries and Marine Service, Ottawa.

508 VARLEY M.E. (1967) *British freshwater fishes*. 148 pp. Fishing News (Books) Ltd, London.

509 VINER A.B. (1975) The supply of minerals to tropical rivers and lakes (Uganda). In *Coupling of land and water systems*, Haster A.D. (ed.), pp. 227–62. Springer-Verlag, New York.

510 VINER A.B. & SMITH I.R. (1973) Geographical, historical and physical aspects of Lake George. *Proc. roy. Soc. B.*, **184**, 235–70.

511 VOLLENWEIDER R.A. (ed.) (1969) *A manual on methods for measuring primary production in aquatic environments*. 213 pp. Blackwell Scientific Publications, Oxford.

512 VOLLENWEIDER R.A. (1970) *Scientific fundamentals of the eutrophication of lakes and flowing waters, with partic.lar reference to nitrogen and phosphorus as factors in eutrophication*. 159 pp. O.E.C.D., Paris.

513 VOLLENWEIDER R.A. (1975) Input–output models with special reference to the phosphorus loading concept in limnology. *Schweiz Z. Hydrol.*, **37**, 53–84.

514 VOLLENWEIDER R.A. & DILLON P.J. (1974) The application of the phosphorus loading concept to eutrophication research. 43 pp. National Research Council of Canada Publication No. 13690.

515 WALSBY A.E. (1965) Biochemical studies on the extracellular polypeptides of *Anabaena cylindrica* Lemm. *Br. phycol. Bull.*, **2**, 514–15.

516 WALSBY A.E. (1978) The properties and buoyancy-providing role of gas vacuoles in *Trichodesmium* Ehrenberg. *Br. phyc. J.*, **13**, 103–16.

517 WALSHE B.M. (1950) The function of haemoglobin in *Chironomus plumosus* under natural conditions. *J. exp. Biol.*, **27**, 73–95.

518 WAUTIER J. & PATTEE E. (1955) Expérience physiologique et expérience ecologique. L'influence du substrat sur la consommation d'oxygène chez les larves d'Ephéméroptères. *Bull. mens. Soc. linn. Lyon.*, **24**, 178–83.

519 WEIR J.S. (1972) Diversity and abundance of aquatic insects reduced by introduction of the fish *Clarias gariepinus* to pools in Central Africa. *Biol. Cons.*, **4**, 169–75.

520 WELCH H.E. (1968) Relationships between assimilation efficiencies and
growth efficiencies for aquatic consumers. *Ecology*, **49**, 755–9.

521 WERNER D. (1977) *The biology of diatoms*. 498 pp. Blackwell Scientific
Publications, Oxford.

522 WERNER E.E. & HALL D.J. (1976) Niche
shifts in sunfishes: Experimental evidence and significance. *Science*, *N.Y.*,
191, 404–6.

523 WERNER E.E. & HALL D.J. (1977)
Competition and habitat shift in two
sunfishes (Centrarchidae). *Ecology*, **58**,
869–76.

524 WERNER E.E., HALL D.J., LAUGHLIN
D.R., WAGNER D.T., WILSMANN L.A. &
FUNK F.C. (1977) Habitat partitioning
in a freshwater fish community. *J. Fish
Res. Bd. Can.*, **34**, 360–70.

525 WEST W. & WEST G.S. (1904–1912) *A
monograph on the British Desmidiaceae*,
vols. 1–4. Ray Society, London.

526 WESTLAKE D.F. (1963) Comparisons of
plant productivity. *Biol. Rev.*, **38**, 385–
425.

527 WETZEL R.G. (1964) A comparative
study of the primary productivity of
higher aquatic plants, periphyton and
phytoplankton in a large shallow lake.
Int. Rev. Ges. Hydrobiol., **49**, 1–61.

528 WETZEL R.G. (1970) Recent and post-
glacial production rates of a marl lake.
Limnol. Oceanogr., **15**, 491–503.

529 WETZEL R.G. (1971) The role of carbon
in hard water marl lakes. In *Nutrients
and eutrophication*, pp. 84–97, in (ed.)
Likens G.E. *Am. Soc. for Limnol. &
Oceanogr.*, Symposium, **1**.

530 WETZEL R.G. (1975) *Limnology*. 743 pp.
Saunders, Philadelphia.

531 WETZEL R.G. & HOUGH R.A. (1973)
Productivity and role of aquatic macro-
phytes in lakes. An assessment. *Pol.
Archs. Hydrobiol.*, **20**, 9–19.

532 WETZEL R.G. & MANNY B.A. (1972)
Secretion of dissolved organic carbon
and nitrogen by aquatic macrophytes.
Verh. int. Verein theor. angew. Limnol.,
18, 162–70.

533 WETZEL R.G. & MANNY B.A. (1979)
Postglacial rates of sedimentation, nutrient and fossil pigment deposition in a
hardwater marl lake of Michigan. *Mitt.
int. Ver. theor. angew. Limnol.* (in press).

534 WETZEL R.G., RICH P.H., MILLE & M.C.
& ALLEN H.L. (1972) Metabolism of
dissolved and particulate detrital carbon
in a temperate hard-water lake. *Mem.
Ist. Ital. Idrobiol.*, **29**, Suppl. 185–243.

535 WHITTAKER R.H. (1969) New concepts
of kingdoms of organisms. *Science*, **163**,
150–60.

536 WHITTON B.A. (1975) *River ecology.*
725 pp. Blackwell Scientific Publications,
Oxford.

537 WHITTON B.A. & SAY P.J. (1975) Heavy
metals. In *River Ecology*, pp. 286–311,
(ed.) Whitton B.A. Blackwell Scientific
Publications, Oxford.

538 WILLOUGHBY L.G. & SUTCLIFFE D.W.
(1976) Experiments on feeding and
growth of the amphipod *Gammarus
pulex* (L.) related to its distribution in
the River Duddon. *Freshwat. Biol.*, **6**,
577–86.

539 WINBERG G.G. (1968) *Methods for the
estimation of production of aquatic
animals* (Transl. A. Duncan, 1971). *Adv.
Ecol. Res.* Academic Press, London.

540 WINBERG G.G., BABITSKY V.A., GAV-
RILOV S.I., GLADKY G.V., ZAKHAREN-
KOV I.S., KOVALEVSKAYA R.Z., MIK-
HEEVA T.M., NEVYADOMSKAYA P.S.,
OSTAPENYA A.P., PETROVICH P.G.,
POTAENKO J.S. & YAKUSHKO O.F. (1970)
Biological productivity of different types
of lakes. In *Productivity problems of
freshwaters*, pp. 383–404, (eds.) Kajak
Z. & Hillbricht-Ilkowska A., PVW,
Warsaw.

541 WINTER T.C. & WRIGHT H.E. Jr. (1977)
Paleohydrologic phenomena recorded by
lake sediments. *EOS*, *Trans. Am.
Geophys. Un.*, **58**, 188–96.

542 WIUM-ANDERSEN S. & ANDERSEN J.M.
(1972) The influence of vegetation on the
redox profile of the sediment of Grane
Langsø, a Danish Lobelia lake. *Limnol.
Oceanogr.*, **17**, 948–52.

543 WORTHINGTON E.B. (1964) Conservation
of water and fisheries in 1970. *Salmon
and Trout Assoc. Lond. Conf.*, **1**, 1–7.

544 WORTHINGTON S. & WORTHINGTON E.B.
(1933) *Inland waters of Africa*. MacMillan, London.

545 WRIGHT R.T. (1975) Studies on glycolic
acid metabolism by freshwater bacteria.
Limnol. Oceanogr., **20**, 626–33.

546 WYATT J.T. & SILVEY J.K.G. (1969)
Nitrogen fixation by Gloeocapsa.
Science, *N.Y.*, **165**, 908–9.

547 YOUNG J.O., MORRIS I.G. & REYNOLD-
SON T.B. (1964) A serological study of
Asellus in the diet of lake-dwelling
triclads. *Arch. Hydrobiol.*, **60**, 366–
73.

548 ZARET T.M. (1969) Predation-balanced
polymorphism of *Ceriodaphnia cornuta*
Sars. *Limnol. Oceanogr.*, **14**, 301–3.

INDEX TO
WATER BODIES

INDEX TO
GENERA AND SPECIES

GENERAL INDEX